AF412875

DIFFUSIVE GRADIENTS IN THIN-FILMS FOR ENVIRONMENTAL MEASUREMENTS

The diffusive gradients in thin-films (DGT) technique is a means of measuring the concentration and speciation of elements in natural waters. Edited by one of the pioneers of the technique, this unique volume provides a complete and authoritative guide to the theory and applications of DGT.

The book includes explanations of the fundamental principles of DGT, accessible to readers with a modest background in chemistry, as well as more advanced chapters that provide a thorough treatment of the physical and chemical dynamics of this technique and evaluate how well it mimics the biological uptake process. Chapters on natural waters, soils and sediments illustrate the applications of DGT, and detailed instructions are included on how to use DGT in practice.

Combining the fundamentals of DGT with more advanced principles, this is an indispensable text for students, researchers and professional scientists interested in the chemistry of natural waters, soils and sediments.

WILLIAM DAVISON is Emeritus Professor of Environmental Chemistry at Lancaster Environment Centre, Lancaster University, UK, where he has previously served as Head of Department. His studies, complemented by research in environmental analytical chemistry, focus on the biogeochemistry of carbon, iron, manganese, sulphur and trace metals in natural waters. Professor Davison, with his colleague Hao Zhang, developed the techniques of diffusive equilibration in thin-films (DET) and diffusive gradients in thin-films (DGT) and pioneered their initial applications in waters, soils and sediments. He has contributed to over 250 publications and the applications of his work have been recognized by the Pollution Abatement Technology Awards and the RSC Sustainable Water Award.

DIFFUSIVE GRADIENTS IN THIN-FILMS FOR ENVIRONMENTAL MEASUREMENTS

Edited by

WILLIAM DAVISON
Lancaster University

CAMBRIDGE
UNIVERSITY PRESS

University Printing House, Cambridge CB2 8BS, United Kingdom

Cambridge University Press is part of the University of Cambridge.

It furthers the University's mission by disseminating knowledge in the pursuit of
education, learning and research at the highest international levels of excellence.

www.cambridge.org
Information on this title: www.cambridge.org/9781107130760

© Cambridge University Press 2016

This publication is in copyright. Subject to statutory exception
and to the provisions of relevant collective licensing agreements,
no reproduction of any part may take place without the written
permission of Cambridge University Press.

First published 2016

Printed in the United Kingdom by TJ International Ltd. Padstow Cornwall

A catalog record for this publication is available from the British Library

Library of Congress Cataloging-in-Publication data
Davison, William, 1948– editor.
Diffusive gradients in thin-films for environmental measurements / edited by
William Davison.
Cambridge : Cambridge University Press, [2016]
Cambridge environmental chemistry series   Includes index.
LCCN 2016017704   ISBN 9781107130760
LCSH: Environmental geochemistry.   Sediments (Geology)
LCC QE516.4.D54   2016   DDC 551.9028/7 – dc23
LC record available at https://lccn.loc.gov/2016017704

ISBN 978-1-107-13076-0 Hardback

Cambridge University Press has no responsibility for the persistence or accuracy
of URLs for external or third-party internet websites referred to in this publication,
and does not guarantee that any content on such websites is, or will remain,
accurate or appropriate.

Contents

The colour plates are to be found between pages 114 and 115.

Contributors

Maja Arsic
Environmental Futures Research Institute and Griffith School of Environment, Griffith University, Queensland, Australia

William W. Bennett
Environmental Futures Research Institute and Griffith School of Environment, Griffith University, Queensland, Australia

William Davison
Lancaster Environment Centre, Lancaster University, Lancaster, UK

Fien Degryse
School of Agriculture, Food & Wine, The University of Adelaide, Glen Osmond, Australia

Josep Galceran
Departament de Química, Universitat de Lleida and Agrotecnio, Lleida, Spain

Yue Gao
Analytical, Environmental and Geochemistry, Chemistry Department, Vrije Universiteit Brussel, Brussels, Belgium

Johan Ingri
Civil, Environmental and Natural Resources Engineering, Luleå University of Technology, Luleå, Sweden

Dianne F. Jolley
School of Chemistry, University of Wollongong, Wollongong, Australia

Niklas J. Lehto
Department of Soil and Physical Sciences, Faculty of Agriculture and Life Sciences, Lincoln University, Lincoln, New Zealand

Sean Mason
School of Agriculture, Food & Wine, University of Adelaide, Glen Osmond, Australia

Heléne Österlund
Civil, Environmental and Natural Resources Engineering, Luleå University of Technology, Luleå, Sweden

Jared G. Panther
Environmental Futures Research Institute and Griffith School of Environment, Griffith University, Queensland, Australia

Jaume Puy
Departament de Química, Universitat de Lleida and Agrotecnio, Lleida, Spain

Carlos Rey-Castro
Departament de Química, Universitat de Lleida and Agrotecnio, Lleida, Spain

Jakob Santner
Department of Forest and Soil Sciences, Institute of Soil Research, University of Natural Resources and Life Sciences, Vienna, Austria

Department of Crop Sciences, Division of Agronomy, University of Natural Resources and Life Sciences, Vienna, Austria

Erik Smolders
Katholieke Universiteit Leuven, Heverlee, Belgium

Peter R. Teasdale
Environmental Futures Research Institute and Griffith School of Environment, Griffith University, Queensland, Australia

David T. Welsh
Environmental Futures Research Institute and Griffith School of Environment, Griffith University, Queensland, Australia

Anders Widerlund
Civil, Environmental and Natural Resources Engineering, Luleå University of Technology, Luleå, Sweden

Paul N. Williams
Institute for Global Food Security, Queen's University Belfast, Belfast, UK

Hao Zhang
Lancaster Environment Centre, Lancaster University, Lancaster, UK

Preface

As scientists from a range of disciplines begin to use DGT, they inevitably ask, "How do I learn about it? What should I read?" Until now the answer has been a range of publications, as none alone have provided a comprehensive overview or the practical details essential to getting started. Hopefully this book will do just that. Hao Zhang and I first started to write such a book about seven years ago, but we soon put it on hold when we realized firstly that there seemed to be more questions than answers and secondly that the increasing breadth of the applications of DGT was stretching our expertise. So when Laura Sigg on behalf of Cambridge University Press approached us shortly after the 2013 DGT Conference to write this book we were initially reluctant, while recognizing the need. Eventually we realized that the answer to our problems might be an edited volume. We still wanted a "text book", so we designed the chapters accordingly and selected key authors who we knew would bring expertise and thoroughness to their allocated chapter topics. Hao chose to relinquish her editorial role due to her extensive commitments, but has contributed to valuable chapters. Sometimes it is wise to wait and this is certainly the case for this book. There have been many developments of the method and new applications in the past seven years. Most of the contents of three chapters, dealing with understanding the DGT measurement in the presence of dynamic complexes, appreciating its ability to mimic some types of biological uptake processes, and two-dimensional imagery of analytes in soils and sediments, have only emerged in the past few years. This book is certainly timely with respect to understanding the DGT measurement, but that does not mean that progress has plateaued. Far from it, as the capability for measuring an increasingly wider range of analytes is being matched by new, high-performance binding layers, I have little doubt that development of this truly versatile technique will continue at pace. Hopefully this book will still be useful in providing the foundations for the theory and practice of DGT.

I would personally like to thank the contributing authors. They have made good use of their specialist expertise while adhering well to the brief, to produce chapters that are authoritative and yet accessible to non-specialists. In its short history many people have shared in the development of DGT. I am especially grateful to my many collaborators and to those students, research staff and visitors who have spent time in the laboratory of Hao and I and shared their thoughts and contributions on DGT.

Symbols and Abbreviations

Units

C	coulombs
g	gram
L	litre
M	moles L^{-1}
s	second
d	day
V	volt
m	milli (10^{-3})
μ	micro (10^{-6})
n	nano (10^{-9})
p	pico (10^{-12})
f	femto (10^{-15})

Symbols		example unit
A	area of DGT sampling interface (used in DGT calculation)	cm^2
A_p	geometric area of the DGT device window	cm^2
A_E	effective area of the binding layer that accumulates analyte	cm^2
b	soil buffer power	
c	concentration of solute	nM, ng L^{-1}
c_e	concentration of analyte in the eluent	nM, ng L^{-1}
c_E	effective concentration in soil or sediment	nM, ng L^{-1}
c^{soln}, c^*	concentration of solute in the deployment or bulk solution	nM, ng L^{-1}
c^g	steady-state concentration at the outer surface of the diffusive gel	nM, ng L^{-1}
c^f	steady-state concentration at the outer surface of the filter	nM, ng L^{-1}
c^i	instantaneous concentration at DGT–medium interface	nM, ng L^{-1}
c^{ls}	concentration of labile analyte in the solid phase[1]	nM, ng L^{-1}
c^s	sorbed concentration	nM, ng L^{-1}
$c_{T,L}$	total concentration of all ligand species	nM, ng L^{-1}

Symbols		example unit
c_{lab}	concentration of all labile species[1]	nM, ng L^{-1}
c_{DGT}	concentration from DGT simple equation	nM, ng L^{-1}
c^{bio}	internal concentration in biota	nM, ng L^{-1}
c^{bio}_{min}	minimum internal concentration in biota	mg, kg^{-1}
c^{bio}_{tox}	toxic threshold concentration in biota	mg, kg^{-1}
c_{diff}	mean interfacial concentration for diffusional supply only	nM, ng L^{-1}
c^{gel}	concentration of analyte in gel	nM, ng L^{-1}
c^{dyn}_{max}	dynamic maximum concentration	nM, ng L^{-1}
c_M	free metal ion concentration in bulk solution	nM, ng L^{-1}
c_{M0}	free metal ion concentration at the root surface	nM, ng L^{-1}
D	diffusion coefficient of analyte	cm^2 s^{-1}
D_i	diffusion coefficient of species i	cm^2 s^{-1}
D^{mdl}	diffusion coefficient of analyte in material diffusion layer	cm^2 s^{-1}
D^g	diffusion coefficient of analyte in the diffusive gel layer	cm^2 s^{-1}
D^f	diffusion coefficient of analyte in the membrane filter	cm^2 s^{-1}
D^w	diffusion coefficient of analyte in water (e.g. in DBL)	cm^2 s^{-1}
D^s	diffusion coefficient in soil or sediment	cm^2 s^{-1}
F	Faraday constant	C mol^{-1}
f_e	elution factor	
f_u	uptake factor	
g^p_{kin}	kinetic term in units of distance	cm
h	number of ligands	
i	interface	
I	ionic strength	mol L^{-1}
J	flux	fmoles s^{-1} m^{-2}
J_{diff}	area-based diffusive flux	fmoles s^{-1} m^{-2}
J_{free}	diffusion flux if only the free ion contributed	fmoles s^{-1} m^{-2}
J_{labile}	flux if all metal species are labile	fmoles s^{-1} m^{-2}
J_{max}	maximum flux (Michaelis–Menten)	fmoles s^{-1} m^{-2}
J_{tot}	diffusion flux if all metal behaves as the free ion	fmoles s^{-1} m^{-2}
J_{upt}	uptake flux	fmoles s^{-1} m^{-2}
K	stability constant	M^{-1}
K'	normalized concentration of complex ($= c^*_{ML}/c^*_M = K/c^*_L$)	
K_d	partition coefficient	cm^3 g^{-1}

---

[1] There is a difficulty in using concentrations of labile species, because different species have different labilities, which will vary depending on the measurement used. However, sometimes it is convenient to represent a concentration of labile species. When this is done, the imprecise nature of such a symbol is recognised in the text.

Symbols		example unit
K_{dl}	labile partition coefficient	$cm^3\ g^{-1}$
K_i	binding constant for competing ion X_i	M^{-1}
K_m	Michaelis–Menten constant	$mol\ L^{-1}$
K_M	binding constant for metal	M^{-1}
K^{os}	outer sphere stability constant	M^{-1}
$K_{Xi\ BL}$	biotic ligand binding constant of ion X_i	M^{-1}
k_1	sorption rate constant	s^{-1}
k_{-1}	desorption rate constant	s^{-1}
k_a	association rate constant	$M^{-1}\ s^{-1}$
k_d	dissociation rate constant	s^{-1}
$k_{a,R}$	rate constant for association to resin	$M^{-1}\ s^{-1}$
$k_{d,R}$	rate constant for dissociation from resin	s^{-1}
k_L	association rate with surface ligand	$cm^3\ mol^{-1}\ s^{-1}$
k_{-w}	rate constant for removal of a water molecule	s^{-1}
L	ligand	
L_T	density of surface ligands	$mol\ cm^{-2}$
l	length, distance	cm
M	metal	
M	mass of analyte on the binding layer	pg
M_f	final mass	pg
M_i	initial mass	pg
M_p	mass of particles	g
M_{wt}	molecular mass	$g\ mol^{-1}$
m	disequilibrium parameter	cm
n	number of moles of analyte in the binding layer	pmol
P_c	particle concentration (weight in volume of soil/sediment)	$g\ cm^{-3}$
P_{diff}	diffusional permeability	$cm\ s^{-1}$
P_m	membrane permeability	$cm\ s^{-1}$
Q	concentration quotient	M^{-1}
Q_i	internal concentration	μM
$Q_{i(min)}$	minimum internal concentration	μM
$Q_{i(tox)}$	toxic threshold of internal concentration	μM
q_i	normalized concentration of species i (e.g. $q_{ML} = c_{ML}/c_{ML}^*$)	
R	resin	
R	correlation coefficient or gas constant	
R_c	c_{DGT}/c^{soln} in soils and sediments	
r_o	radius	cm
s	slope	
T	temperature	^{o}C
T^K	absolute temperature	^{o}K
t_c	response time to perturbation	s
t	time	s
t_e	equilibration time	s

Symbols		example unit
t_{ss}	time to achieve steady state	s
u_g	specific growth rate	s^{-1}, d^{-1}
U	flow velocity	$cm\ s^{-1}$
V^{bl}	volume of binding layer	mL
V_e	volume of eluent	mL
V_s	volume of soil	cm^3
V_{MWHC}	soil maximum water holding capacity	% (W/W)
v_0	rate of water advection	$cm\ d^{-1}$
W	weight	g
$W_{1/2}$	width of peak at half height	cm
x	perpendicular distance from plastic base	cm
Xi	activity of ion i	M
X_i	ion i	
y	intercept	
z_C	valence of cations	
z_{SE}	valence of supporting electrolyte	
z_i	number of charge of species i	
α	root absorption power	$cm\ s^{-1}$
γ_i	activity coefficient of species i	
δ	geochemical delta value of isotope	
δ	general diffusion layer thickness	cm
δ^g	thickness of the diffusive gel layer	cm
δ^f	thickness of the membrane filter	cm
δ^{mdl}	thickness of material diffusion layer ($= \delta^g + \delta^f$)	cm
Δg	thickness of material diffusion layer ($= \delta^{mdl}$)	cm
δ^r	thickness of binding layer	cm
δ^{dbl}	thickness of the diffusive boundary layer	cm
ε	normalized complex diffusion coefficient (D_{ML}/D_M)	
η	viscosity of water	Pa s
θ	volumetric moisture content	
θ_T	tortuosity	
λ_M	metal penetration parameter	cm
λ_{ML}	complex penetration parameter	cm
μ	reaction layer thickness	cm
υ	kinematic viscosity	$cm^2\ s^{-1}$
Π	Donnan partition factor	
ξ	lability degree	
ρ^g	charge density of gel	$C\ cm^{-3}$
ρ^b	soil bulk density	$g\ cm^{-3}$
ρ^s	density of solid phase material	$g\ cm^{-3}$
φ	porosity	
ψ	Donnan potential	V

Abbreviations

AAS	atomic absorption spectroscopy
APA	polyacrylamide gel cross-linked with an agarose derivative
APS	ammonium persulphate
AVS	acid volatile sulphur
BLM	biotic ligand model
CID	computer imaging densitometry
DBL	diffusive boundary layer
DET	diffusive equilibration in thin-films
DGT	diffusive gradients in thin-films
DOC	dissolved organic carbon
DOM	dissolved organic matter
$EC50_{free}$	50% effect concentration (free ion based)
$EC50_{free}{}^*$	50% effect (free) concentration in absence of competing ions
FIAM	free ion activity model
FIP	first ionisation potential
FA	fulvic acid
GF-AAS	graphite furnace atomic absorption spectroscopy
HA	humic acid
HPW	high purity water
HS	humic substances
HSI	hyperspectral imaging
IAP	ion activity product
IC	ion chromatography
ICP-MS	inductively coupled plasma mass spectrometry
LA	laser ablation
LOD	limit of detection
MDL	material diffusion layer (diffusive gel and membrane filter)
MUF	4-methylumbelliferyl
OM	organic matter
PIXE	proton induced x-ray emissions
SD	standard deviation
SEM	simultaneously extracted metals
SI	saturation index
SPR-IDA	suspended particulate reagent iminodiacetic acid
TEMED	tetramethylethylenediamine
TWA	time weighted average concentration
UPW	ultra pure water
UV	ultraviolet
WHAM	Windermere humic acid model

1

Introduction to DGT

WILLIAM DAVISON AND HAO ZHANG

1.1 Origins

DGT has its origins in sediment geochemistry. From 1970 to 1990, there were notable advances in the chemistry of sediments as scientists tried to unravel the processes controlling early diagenesis. The seminal works of Aller [1] and other workers on anoxic sediments provided a framework that led to refined models of oxidation of organic material by a series of electron acceptors [2]. Much of the understanding relied on measurements of solutes in porewaters, obtained by squeezing or centrifuging a slice of sediment typically about 1 cm deep and from 3 to 10 cm diameter. An alternative approach of inserting into sediments assemblies of dialysis cells, referred to by Hesslein [3] et al. as peepers, was developed. Solutes were allowed to equilibrate between the porewaters and individual compartments of solution in Perspex blocks, separated from the sediment by a filter membrane. These compartments were quite large, with typically 1 cm $\times$ 10 cm windows, resulting in equilibration times of two to three weeks. Inspired by mm-scale measurements of authigenetic solid phases on plastic substrates [4], hydrogels were used for dialysis instead of peepers [5]. In this new technique, which became known as DET (diffusive equilibrium in thin-films) [6], solutes were equilibrated with the non-bound water of the hydrogels. Like the most commonly encountered hydrogels, contact lenses, the DET gels can be easily handled. They can be sliced at the mm scale and the solutes can be back-equilibrated for analysis [7]. Consequently measurements can be made at a much higher spatial resolution than for peepers, and their small scale ($\sim$1-mm thick gel) allows equilibration within days [8].

The idea for the technique of DGT (diffusive equilibration in thin-films), which includes a binding layer to accumulate solutes (Figure 1.1), stemmed from the need to improve the sensitivity of DET, especially for trace metals present in porewaters at very low concentrations. However, the first step in its development was as a solution device for measuring trace metals in seawater [9]. The parallels between DGT and voltammetry were clear at the outset and an analogous theory was developed. It is now recognised that, from a physical chemistry perspective, DGT and voltammetry can both be classed as dynamic techniques that initiate and respond to a flux of solute to the device [10]. Whereas DET relies on establishing equilibrium between solutes in the solution and in the device, in DGT solutes

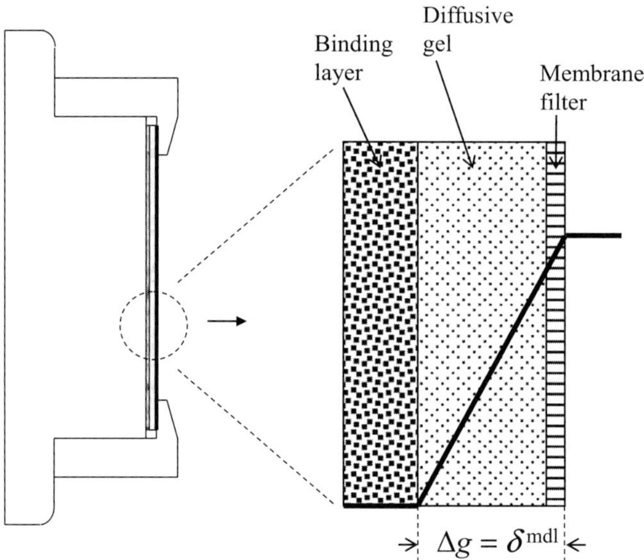

Figure 1.1 Schematic diagram of a DGT piston assembly, with an exploded view of the binding and diffusion layers, showing the concentration gradient of the analyte.

continuously diffuse across a well-defined diffusion layer and are progressively accumulated. The ideas behind DGT were patented worldwide with a filing date in 1993, and the trademark DGT® is registered in many countries and should be displayed when DGT is used commercially. However, this does not prevent the research use of DGT, and the simple initials DGT will be used throughout this book.

In principle, there is little difference between DGT and a range of passive samplers that have been developed for monitoring components in air [11] and organic substances in water [12]. However, the theoretical approach to the interpretation of the DGT measurement has taken a different path, and DGT has been used much more extensively than other devices for the measurement of predominantly inorganic solutes in natural waters. The detailed interpretation of the measurement in terms of equilibrium speciation and dissociation kinetics of complexes has been possible because of the firm theoretical developments. Moreover, the use of DGT has been extended to soils and sediments where it has provided unique information. These applications in particular demonstrate that the description 'passive sampler' is inappropriate. As DGT continually removes solute from its deployment medium, it perturbs the system. The accumulated solute that is eventually measured reflects this perturbation. It is this feature that allows it to provide information on speciation in solution and solid–solution interactions in soils and sediments.

This chapter sets out a simplified view of the basic principles of DGT that have been at the heart of most of its applications for measuring solutes in solution. The later chapters provide more thorough treatments of the principles and practise of the measurement in waters, sediments and soils and consider in detail the underlying assumptions.

1.2 Basic Principles

There are really only two essential parts of DGT: a layer which selectively binds the solute or solutes of interest, known as the binding layer, and a layer which permits diffusion of the solutes prior to binding, known as the diffusion layer. Details of the required properties of these layers and of the types that have been used are provided in Chapters 3 and 4. The original work, aimed at measuring trace metals, used a polyacrylamide hydrogel for both layers [9]. Hydrogels, as their name implies, are largely composed of water. The particular type used was 95% water, allowing virtually unimpeded diffusion of simple cations. Chelex resin binds trace metals more strongly than major cations (Ca^{2+}, Mg^{2+}, Na^+, K^+). It was incorporated into a hydrogel during the casting process to form a dense layer of resin beads at one surface of the resin sheet. These binding and diffusion layers were held together in a plastic holder, with only the diffusion layer exposed to solution through a circular 'exposure window'. As the first deployments in estuarine conditions showed that the diffusive gel layer was prone to clogging by particles, it was overlain by a protective membrane filter, which had similar diffusion properties to the gel. Figure 1.1 shows a cross-section through a commonly used, commercially available version of the device for measurements in solution. The cylindrical plastic assembly, which is injection moulded in ABS (acrylonitrile butadiene styrene) plastic, is sometimes referred to as a piston holder. The top cap with the exposure window fits tightly over the base, which supports the sandwich of binding layer, diffusion layer and membrane filter.

The simple and robust DGT device is readily deployed in situ in natural waters or in solutions in the laboratory. After deployment, the piston holder is washed and pulled apart to retrieve the binding layer, which, for measurement of metals, is immersed in a small volume (typically 1 mL) of usually 1 M nitric acid. The mass of metal, M, on the binding layer, of volume V^{bl}, is calculated from the measured concentration of metal, c_e, in the acid eluent, of volume V_e, using equation 1.1. The fraction of bound metal released, f_e, known as the elution factor, can be established for controlled conditions:

$$M = \frac{c_e(V^{bl} + V_e)}{f_e} \tag{1.1}$$

As the DGT device is deployed for a known time, t, and the area that is exposed to solution is determined by the area of the window in the device cap, A_p, it is simple to calculate the flux of solute, J (mass per unit area per time), passing through the device (equation 1.2).

$$J = \frac{M}{A_p t} \tag{1.2}$$

To appreciate how to convert the measured mass and associated flux to the concentration in solution during deployment, we consider Figure 1.1. It shows the steady-state concentration gradient of the solute being measured that is established through the device when it is deployed in solution. Fick's first law of diffusion [13] states that the flux, J, through such a system is simply the diffusion coefficient of the solute in the diffusion layer, D^{mdl}, times the concentration gradient, dc/dx, where c is the concentration of solute and x represents distance (equation 1.3). Material diffusion layer (MDL) is used in the superscript to make the distinction between diffusion in the materials of the device and diffusion in solution, as explained in more detail in Chapter 2.

$$J = D^{\text{mdl}} \frac{\text{d}c}{\text{d}x} \tag{1.3}$$

Usually the solute of interest binds so strongly to the binding layer that its concentration at the interface between the diffusion and binding layers is negligibly small. To a first approximation, convection ensures that the concentration of solute in solution, c^{soln}, is effectively constant. Consequently for the region corresponding to the thickness of the diffusion layer, the gradient is simply $c^{\text{soln}}/\Delta g$, allowing equation 1.3 to be expressed in known quantities (equation 1.4).

$$J = \frac{D^{\text{mdl}} c^{\text{soln}}}{\Delta g} \tag{1.4}$$

Although Δg has usually been used to designate the total thickness of the MDL, comprising gel and filter membrane, for consistency in nomenclature, and to distinguish it from individual layers where diffusion occurs (see Chapter 2), it will usually be designated as δ^{mdl} in the rest of the book. If it is assumed that the time required to reach steady state is negligible, an expression for the concentration in solution can be obtained by equating 1.2 and 1.4 and rearranging (equation 1.5). The term c_{DGT} is used in equation 1.5 because this DGT-measured concentration is an interpreted quantity which, depending on the conditions, may not equate to c^{soln}. For example, in more complex media not all species of a solute might diffuse and bind.

$$c_{\text{DGT}} = \frac{M \Delta g}{D^{\text{mdl}} A_{\text{p}} t} \tag{1.5}$$

In principle, however, equation 1.5, which has become known as the DGT equation, allows the concentration in the deployment medium to be calculated by simply measuring the amount of solute that binds to the gel. The diffusion layer thickness and area are obtained from the geometry of the device, while time can be accurately measured. The diffusion coefficients in the MDL of a wide range of solutes of interest are known for a range of temperatures (Chapter 3 and Appendix). For simple solutions, the value of D^{mdl} can be regarded as a calibration factor, as it can be obtained by deploying DGT devices in known solutions. A sample calculation is shown in Table 1.1, which illustrates that concentrations should be expressed per millilitre for conformity with distance units of centimetre.

Equation 1.5 has been shown to work well for deployments in stirred or flowing solutions when using the standard device equipped with a 0.8-mm thick gel and 0.14-mm filter membrane [14] (see Chapter 2.2). It predicts that the measured mass increases linearly with time, which has been verified by numerous workers [6]. The slope of this plot, $c^{\text{soln}} D^{\text{mdl}} A_{\text{p}}/\Delta g$, provides an estimate of c^{soln} if it is unknown or of D^{mdl} where deployments are made in known solutions. The conditions that give rise to deviations from linearity are treated in Chapters 4 and 5. According to equation 1.5 the measured mass is inversely

Table 1.1 *Example values and units used in calculating the DGT-measured concentration,* c_{DGT}, *where the concentration in the diluted eluent measured analytically is 0.9 mg L^{-1}.* *The deployment time and concentration in the eluent for a particular measurement are shown with typical units (unit 1) and the units necessary for the calculation (unit 2).*

Set values		Experiment	Unit 1	Unit 2	Calcn	eqn	value
D^{mdl} (cm^2/s)	5.29E–06	Time	24 h	86400 s	M	1.1	12.7 µg
A_p (cm^2)	3.14	c (analysis)	0.9 mg L^{-1}	0.9 µg mL^{-1}	c_{DGT}	1.5	0.83 µg mL^{-1}
Δg (cm)	0.094	Dilution	10×				
f_e	0.8	c_e	9.0 mg L^{-1}	9 µg mL^{-1}			
V_e (mL)	1						
V^{bl} (mL)	0.126						

proportional to Δg. Again this has often been verified, but we will see later (Chapter 2); there are more appropriate equations that should be used when diffusive gel thicknesses differ from 0.8 mm or the water is not fast flowing.

1.3 Measurements in Natural Waters

The robust design of DGT devices makes them easy to deploy in situ. Moreover for simple solutions, where the diffusion coefficient of the solute being measured is known, no calibration other than normal quality control checks is necessary. When the environmental concentration changes with time, as might occur in a river, the DGT device provides the time weighted average (TWA) concentration for the deployment time. These characteristics, allied to the low cost of DGT devices, have contributed to their use in monitoring programmes [15, 16]. The fairly long diffusive path, approaching 1 mm in a standard device, ensures that the DGT measurement is almost insensitive to the rate of solution flow above a threshold flow of 2 cm s^{-1}. A further attractive feature for monitoring purposes is the ease of analysis. If the binding layer for a standard device is eluted into 1 mL of 1M HNO$_3$, which is then diluted by a factor of 10, the matrix is ideal for analysis by atomic absorption spectroscopy (AAS) and inductively coupled plasma (ICP) (both optical and mass spectrometry (MS)). For a twenty-four-hour deployment of the standard device fitted with a 0.8-mm thick diffusive gel and 0.14-mm thick filter, the concentration measured by the analytical technique in the diluted eluent will be similar to the concentration in the deployment solution, assuming a diffusion coefficient of 5×10^{-6} cm^2 s^{-1} and an elution factor of 0.8, typical for transition metals, as illustrated in Table 1.1. Unless deployment times extend to several days or weeks, the analytical advantage of the DGT measurement lies in the well-defined matrix as much as the enhancement of the concentration.

Although DGT was originally used for measuring trace metals, it has proved to be very versatile. Garmo et al. [17] showed that the original binding layer containing Chelex resin can be used to measure fifty-five elements. Alternative binding layers have been used for sulphide [18], Cs [19], Hg [20], Tc [21] and oxyanions (e.g. phosphate [22] and arsenate [23]). Chapter 4 considers in detail the range of possible analytes, which have recently been extended to hydrophilic organic compounds [24].

From the outset, it was recognised that DGT can be used as a speciation tool. At a simple level, it can be used to provide a direct measure of solutes that are both mobile and labile [10]. The term mobile refers to the fact that they must be capable of diffusing at a reasonable rate through the diffusion layer. The term labile is used to denote species which can interconvert, within the timescale of their diffusional transport, to a form that can bind. Gradually the theory has advanced, as detailed in Chapter 5, to allow more sophisticated interpretation in terms of the distribution of species in solution and the rates at which they interconvert [25].

1.4 Applications to Soils and Sediments

DGT has been used extensively to perform measurements in sediments [26] and in soils that are usually fully hydrated [27]. Solutes are supplied from the porewaters, but as there is no convective supply to sustain their concentration, they are depleted adjacent to the device. Interpretation in terms of the concentrations in the porewaters prior to the DGT perturbation is then not so simple. The solute accumulated by DGT is determined by both the concentration of labile and mobile solutes in the porewaters and the rate of supply of solute from the solid phase, as detailed in Chapter 7. The positive aspect of this dependency is that, with a systematic set of measurements and appropriate modelling, information can be obtained about the dynamics of solute interaction with the solid phase [28]. The advantage of using DGT to do this is that it is readily deployed in situ in sediments and in soils that are hydrated to their maximum water-holding capacity.

While the DGT measurement can be treated analytically to derive its component contributions, including kinetic and speciation effects, it can also be viewed holistically. In this case, the interpretation is simply that c_{DGT} reflects directly the response of the soil or sediment system to the perturbation imposed by DGT of removing solute in a controlled way. This approach has been used most successfully by considering that DGT mimics the way that plant roots perturb solutes in a soil system, leading to DGT being used as a tool to predict uptake of solutes by plants and to understand how the dynamics of uptake affects the soil [29]. The underlying processes and assumptions that determine relationships between uptake by DGT and by biota are discussed in Chapter 9.

Most work in soils has involved homogenisation of the sample prior to deployment. The intense redox gradients in sediments and their sensitivity to oxygen prevent this approach. The focus has been in performing DGT measurements at high spatial resolution to inform understanding of early diagenesis. The complexities of the different contributions to the accumulated DGT mass, the steep concentration gradients and the inherent heterogeneity of solute distributions have hindered progress. Measurements at high spatial resolution (sub-mm scale) have revealed apparently stochastic distributions of localised high concentrations of solutes, termed microniches. Ironically, while the technique of DGT grew out of sediment work, advancing understanding of the processes occurring in this medium has been most difficult. However, with more systematic experiments [30], the use of rapid colorimetric techniques of two-dimensional imagery [31], and the development of three-dimensional models of sediments to simulate the localised DGT signals [32] and the distribution of solutes in microniches [33], progress is now being made, as demonstrated in Chapter 8. Two-dimensional imagery of solute concentrations has been extended to soils where it is being used to advance understanding of processes at plant roots [34, 35].

1.5 This Book

Even with this brief introduction, it is clear that DGT is a tool that can be used at various levels of application and interpretation For simple monitoring, where the aim is to obtain the DGT measurement of labile and mobile solutes, the theory represented by equations 1.1–1.5 will suffice in many cases. The following chapters examine in detail the assumptions associated with these basic forms of the equations and provide more complete treatments. The generic performance characteristics of DGT are examined followed by the consideration of specific factors associated with individual solutes. Further chapters explore the interactions of DGT with its deployment media and provide the principles for understanding the dynamic exchange between species in solution and the dynamics of solute interactions between solution and the solid phases in soils and sediments. Key examples of the application of these principles are provided, with particular emphasis on the information that can be obtained at various spatial scales. The role of DGT in mimicry of biological uptake is considered as an example of holistic measurement. Important in all these developments is sound experimental practice, and so the last chapter is devoted to the practicalities of using DGT.

References

1. R. C. Aller, Quantifying solute distributions in the bioturbated zone of marine sediments by defining an average microenvironment, *Geochim. Cosmochim. Acta* 44 (1980), 1955–1962.
2. B. P. Boudreau, *Diagenetic models and their implementation* (Berlin: Springer, 1997), 414pp.
3. R. H. Hesslein, An in situ sampler for close interval pore water studies, *Limnol. Oceanog.* 21 (1976), 912–915.
4. N. Belzile, R. R. DeVitre and A. Tessier, In situ collection of diagenetic iron and manganese oxyhydroxides from natural sediments, *Nature* 340 (1989), 376–377.
5. W. Davison, G. W. Grime, J. A. W. Morgan and K. Clarke, Distribution of dissolved iron in sediment pore waters at submillimeter resolution, *Nature* 352 (1991), 352, 323–325.
6. W. Davison, G. Fones, M. Harper, P. Teasdale and H. Zhang, Dialysis, DET and DGT: In situ diffusional techniques for studying water, sediments and soils, In *In situ monitoring of aquatic systems: Chemical analysis and speciation*, ed. J. Buffle and G. Horvai (Chichester: Wiley, 2000), pp. 495–569.

7. M. D. Krom, P. Davison, H. Zhang and W. Davison, High-resolution pore-water sampling with a gel sampler, *Limnol. Oceanogr.* 39 (1994), 1967–1972.
8. M. P. Harper, W. Davison and W. Tych, Temporal, spatial, and resolution constraints for in situ sampling devices using diffusional equilibration: Dialysis and DET, *Environ. Sci. Technol.* 31 (1997), 3110–3119.
9. W. Davison and H. Zhang, In-situ speciation measurements of trace components in natural-waters using thin-film gels, *Nature* 367 (1994), 546–548.
10. H. P. van Leeuwen, R. M. Town, J. Buffle et al., Dynamic speciation analysis and bioavailability of metals in aquatic systems, *Environ. Sci. Technol.* 39 (2005), 3545–3556.
11. S. J. Hayward, T. Gouin and F. Wania, Comparison of four active and passive sampling techniques for pesticides in air, *Environ. Sci. Technol.* 44 (2010), 3410–3416.
12. R. Greenwood, G. Mills and B. Vrana (eds), *Passive sampling techniques in environmental modelling* (Oxford: Elsevier, 2007).
13. A. Fick, On liquid diffusion, *Phil. Mag. J. Sci.* 10 (1855), 30–39.
14. K. W. Warnken, H. Zhang and W. Davison, Accuracy of the diffusive gradient in thin-films technique: Diffusion boundary layer and effective sampling area considerations, *Anal. Chem.* 78 (2006), 3780–3787.
15. J. Warnken, R. J. K. Dunn and P. R. Teasdale, Investigation of recreational boats as a source of copper at anchorage sites using time-integrated diffusive gradients in thin films and sediment measurements, *Mar. Poll. Bull.* 49 (2004), 833–843.
16. K. W. Warnken, A. J. Lawlor, S. Lofts et al., In situ speciation measurements of trace metals in headwater streams, *Environ. Sci. Technol.* 43 (2009), 7230–7236.
17. O. A. Garmo, O. Roysett, E. Steinnes and T. P. Flaten, Performance study of diffusive gradients in thin films for 55 elements, *Anal. Chem.* 75 (2003), 3573–3580.
18. P. R. Teasdale, S. Hayward and W. Davison, In situ, high-resolution measurement of dissolved sulfide using diffusive gradients in thin films with computer-imaging densitometry, *Anal. Chem.* 71 (1999), 2186–2191.
19. C. Murdock, M. Kelly, L. Y. Chang, W. Davison and H. Zhang, DGT as an in situ tool for measuring radiocesium in natural waters, *Environ. Sci. Technol.* 35 (2001), 4530–4535.
20. C. Fernández-Gómez, B. Dimock, H. Hintelmann and S. Díez, Development of the DGT technique for Hg measurement in water: Comparison of three different types of samplers in laboratory assays, *Chemosphere* 85 (2011), 1452–1457.
21. A. French, H. Zhang, J. M. Pates, S. E. Bryan and R. C. Wilson, Development and performance of the diffusive gradients in thin-films technique for the measurement of technetium-99 in seawater, *Anal. Chem.* 77 (2005), 135–139.
22. H. Zhang, W. Davison, R. Gadi and T. Kobayashi, In situ measurement of dissolved phosphorus in natural waters using DGT, *Anal. Chim. Acta.* 370 (1998), 29–38.
23. W. W. Bennett, P. R. Teasdale, J. G. Panther, D. T. Welsh and D. F. Jolley, Speciation of dissolved inorganic arsenic by diffusive gradients in thin-films: Selective binding of As-III by 3-mercaptopropyl-functionalized silica gel, *Anal. Chem.* 83 (2011), 8293–8299.
24. C-E Chen, H. Zhang and K. C. Jones, A novel passive water sampler for in situ sampling of antibiotics, *J. Environ. Monit.* 14 (2012), 1523–1530.
25. K. W. Warnken, W. Davison and H. Zhang, Interpretation of in situ speciation measurements of inorganic and organically complexed trace metals in freshwater by DGT, *Environ. Sci. Technol.* 42 (2008), 6903–6909.

26. H. Zhang, W. Davison, S. Miller and W. Tych, In situ high resolution measurements of fluxes of Ni, Cu, Fe and Mn and concentrations of Zn and Cd in porewaters by DGT, *Geochim. Cosmochim. Acta* 59 (1995), 4181–4192.

27. H. Zhang, W. Davison, B. Knight and S. P. McGrath, In situ measurements of solution concentrations and fluxes of trace metals in soils using DGT, *Environ. Sci. Technol.* 32 (1998), 704–710.

28. H. Ernstberger, H. Zhang, A. Tye, S. Young and W. Davison, Desorption kinetics of Cd, Zn, and Ni measured in soils by DGT, *Environ. Sci. Technol.* 39 (2005), 1591–1597.

29. H. Zhang, F. J. Zhao, B. Sun, W. Davison and S. P. McGrath, A new method to measure effective soil solution concentration predicts copper availability to plants, *Environ. Sci. Technol.* 35 (2001), 2602–2607.

30. A. Widerlund and W. Davison, Size and density distribution of sulfide-producing microniches in lake sediments. *Environ. Sci. Technol.* 41 (2007), 8044–8049.

31. W. W. Bennett, P. R. Teasdale, J. G. Panther et al., Investigation of arsenic speciation in sediments with DGT and DET: A mesocosm evaluation of oxic-anoxic transitions, *Environ. Sci. Technol.* 46 (2012), 3981–3989.

32. L. Sochaczewski, W. Davison, H. Zhang and W. Tych, Understanding small-scale features in DGT measurements in sediments, *Environ. Chem.* 6 (2009), 477–485.

33. L. Sochaczewski, A. Stockdale, W. Davison, W. Tych and H. Zhang, A three-dimensional reactive transport model for sediments, incorporating microniches, *Environ. Chem.* 5 (2008), 218–225.

34. J. Santner, H. Zhang, D. Leitner et al., High-resolution chemical imaging of labile phosphorus in the rhizosphere of Brassica napus L. cultivars, *Environ. Exp. Bot.* 77 (2012), 219–226.

35. P. N. Williams, J. Santner, M Larsen et al., Localized flux maxima of arsenic, lead, and iron around root apices in flooded lowland rice, *Environ. Sci. Technol.* 48 (2014), 8498–8506.

2

Principles of Measurements in Simple Solutions

WILLIAM DAVISON AND HAO ZHANG

2.1 Introduction

This chapter sets out the principles of DGT that are common to all measurements, irrespective of the determinand, including physical processes concerned with geometry and transport. Specific chemical processes, such as individual binding mechanisms and the effect of solution pH, are treated in Chapters 4 and 5. The solute that can be measured is assumed to be present in a simple solution, so that possible speciation effects need not be considered. As the measurement of metals dominated the early development of DGT, the theoretical development and testing tended to use metals as an example, but the principles are applicable to any analyte.

The theory has been progressively developed and refined with time, with the version presented here relying chiefly on a key paper by Warnken et al. [1] and a recent assessment of our understanding of the basic principles of DGT [2].

2.2 Assumptions

In using the simple DGT equation (equation 1.5) to calculate concentration, several assumptions are implicitly made:

1. The mass accumulated in the initial transient period before steady state is achieved is negligible compared to the total accumulated mass.
2. The transport of analyte through the device can be described by planar diffusion.
3. The measured quantities of accumulated mass, diffusion layer thickness, diffusion coefficient of analyte in the diffusion layer, the area through which analyte diffuses and the deployment time are accurately known.
4. Interactions, such as charge effects and specific binding of the analyte and solution constituents with the diffusive gel and the membrane filter, are negligible.
5. The analyte interacts rapidly with the binding layer so that it does not penetrate beyond its surface.

To examine the validity of these assumptions requires a detailed understanding of the relatively simple processes that operate when DGT is deployed in a solution, as addressed

10

in this chapter. Interactions of solution components with the gel and membrane will be dealt with in detail in Chapter 3, while Chapter 5 considers in more detail than here the possible penetration of analyte into the binding layer.

Even if the above assumptions are met, there remains the question of the meaning of c_{DGT} and its relationship to the concentration in solution. DGT is an operational measurement and consequently c_{DGT} is operationally defined by equation 1.5. As shown in Chapter 5, "c_{DGT} can be regarded as the effective free analyte concentration in a virtual system with only analyte present (no complexes), that would result in the measured accumulation". In a very simple system, c_{DGT} might equate to c^{soln}, but often this is not the case. Interpretation of the DGT measurement so that c_{DGT} more closely approaches c^{soln} requires further equations that describe the system more accurately.

2.2.1 Diffusive Boundary Layer

It was recognised when DGT was first proposed that the assumption of a diffusion gradient only being established within the diffusive gel and membrane filter, referred to collectively as the material diffusion layer, MDL, is a simplification [3]. At any solid surface in a flowing solution there will be a layer close to the surface where there is effectively no flow. As transport of solutes in this layer is restricted to diffusion, it is known as the diffusive boundary layer (DBL) [4]. Although the definition of the thickness of the DBL is arbitrary, because the concentration of an analyte in the layer approaches asymptotically the value in the bulk solution, it is common to refer to a DBL thickness, δ^{dbl}. With increasing rate of solution flow the layer becomes thinner, but even when the deployment solution for DGT devices is vigorously stirred, a thin residual DBL will remain. Therefore, to ensure optimal accuracy, it should be taken into account when calculating the concentration measured by DGT.

Figure 2.1 shows a more complete representation of the analyte concentration gradients associated with a DGT device at steady state than the simplified version of Figure 1.1. It includes the concentration gradient through the DBL and allows for different values of analyte diffusion coefficients through the individual layers of the DGT device, namely the filter membrane, D^f, of thickness δ^f, the diffusive gel layer, D^g, of thickness δ^g and the DBL, of thickness δ^{dbl}, where the diffusion coefficient is that in water, D^w. In many situations D^f and D^g are so similar that they can be set equal, but D^w is usually 10–30% higher than D^f or D^g. Diffusion coefficients for various analytes are given in Chapter 3, and the Appendix of this book provides parameters that allow simple calculation of D^g for commonly measured ions at various temperatures. At steady state, the flux, J, in and out of each layer must be constant and equal to the flux through the adjoining layer, as given by equation 2.1, where c^f, c^g and c^r represent the concentrations of the analyte at the surface of the filter, gel and binding layers, respectively, and c^{soln} is the concentration in bulk solution.

$$J = \frac{D^g(c^g - c^r)}{\delta^g} = \frac{D^f(c^f - c^g)}{\delta^f} = \frac{D^w(c^{soln} - c^f)}{\delta^{dbl}} \tag{2.1}$$

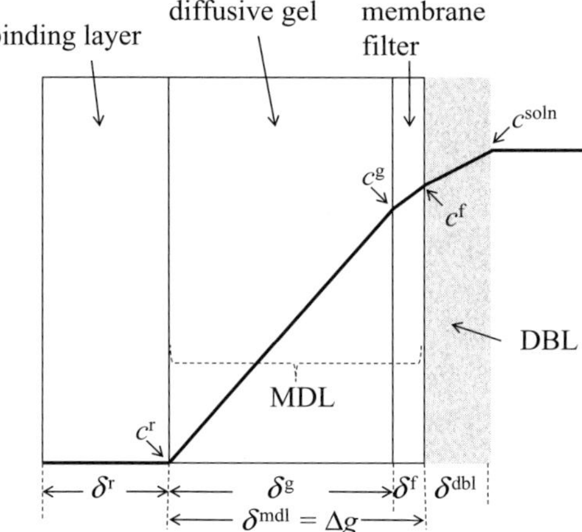

Figure 2.1   Schematic representation of the steady state concentration gradient of the analyte through the binding and diffusion layers within a DGT device, indicating symbols and terminology used in the text.

As the flux is proportional to the diffusion coefficient times the concentration gradient (Fick's Law), the concentration gradients differ in each layer, according to the inverse of D. If the analyte reacts rapidly and strongly with the binding layer, c^r can be set to zero, provided the binding sites do not approach saturation.

As the concentration in solution is simply the sum of the concentration difference in each layer, equation 2.2 can be deduced from the three parts of equation 2.1.

$$c^{soln} = J \left(\frac{\delta^g}{D^g} + \frac{\delta^f}{D^f} + \frac{\delta^{dbl}}{D^w} \right) \tag{2.2}$$

Substituting for the flux being equal to mass per unit area per unit time (equation 1.2) gives equation 2.3, where t is the deployment time. The effective area, A_E, which is explained below, is assumed to apply to each of the three layers:

$$c_{DGTe} = \frac{M}{A_E t} \left(\frac{\delta^g}{D^g} + \frac{\delta^f}{D^f} + \frac{\delta^{dbl}}{D^w} \right) \tag{2.3}$$

When the concentration is derived using this extended equation, from the measured accumulated mass in the binding layer, M, we can call it c_{DGTe}. Like c_{DGT}, calculated using equation 1.5, it is defined operationally. The value of c_{DGTe} is likely to be closer to c^{soln}, but it may not equal it because, for example, of complexation effects.

To use equation 2.3 to calculate c_{DGTe} requires knowledge of the DBL thickness and the diffusion coefficients of the solute in water, gel and filter. Usually a simpler approach is adopted. For the most commonly used diffusion layer combination, of a polyacrylamide gel cross-linked with an agarose derivative and a 0.45-μm polysulfone filter membrane, the diffusion coefficients of simple metal cations for each layer are indistinguishable. This allows a single value of diffusion coefficient, D^{mdl}, to be used for the total thickness of the diffusion layer within the DGT device, which we refer to as the material diffusion layer, MDL; hence equation 2.4.

$$c_{\text{DGTe}} = \frac{M}{A_{\text{E}}t}\left(\frac{\delta^{\text{mdl}}}{D^{\text{mdl}}} + \frac{\delta^{\text{dbl}}}{D^{\text{w}}}\right) \tag{2.4}$$

This equation is often used with the MDL thickness, $\delta^{\text{mdl}} = \delta^{\text{g}} + \delta^{\text{f}}$ shown as Δg. Equation 2.4 is the more complete version of the regularly used simplified DGT equation (equations 1.5 and 2.20), where the DBL was neglected, with δ^{dbl} being typically 0.2–0.5 mm (see Table 2.1) and δ^{mdl} typically 0.94 mm. This fuller equation should generally be preferred for deployments in solution, but we will see later that there are exceptions where the simplified equation is appropriate for measurements in solution. When DGT is deployed in sediments and soils equation 2.4 is inappropriate, as there is no DBL. The approaches in these cases are considered in Chapters 7 and 8.

2.2.2 Effective Area

The derivations of equations 2.1–2.4 assume that the diffusion of analyte through DGT occurs in one dimension perpendicular to the surface of the device, with no lateral diffusion (Figure 2.2). However, in the most commonly used DGT devices supplied by DGT Research Ltd, the areas of the membrane filter and diffusive and binding gels are equal (typically 4.91 cm^2) and larger than the area of the exposure window, A_{p} (typically 3.14 or 2.54 cm^2). Moreover, the diffusive boundary layer will extend to all surfaces of the device and will not be restricted to the exposure window. Consequently the diffusive transport of the solute is not constrained to a parallel path, as represented schematically by Figure 2.2. It is well known that curvature of the diffusional path, as occurs at the edge of the window, will enhance mass transfer in comparison to the planar flux to the same exposure area [5]. The spreading of the diffusion pathway at the rim of the window also allows analyte to accumulate over an area of binding gel greater than the geometric area of the exposure window. The resulting enhanced flux can be accommodated in the DGT equations, including equations 2.1–2.4, by assuming that there is an effective area, A_{E}, which is larger than the physical area of the exposure window, A_{p}. The value of A_{E} will depend on the design of the DGT device.

Warnken et al. [1] determined values of A_{E} by performing a series of careful DGT measurements in solutions with known concentrations of analyte. They found that the effective area did not appear to depend on the MDL thickness (0.14–1.34 mm). For a standard solution device supplied by DGT Research Ltd, with a 2-cm diameter window ($A_{\text{p}} = 3.14$ cm^2), a value of $A_{\text{E}} = 3.8$ cm^2 was obtained. For devices often used for deployments directly in soils, with a window diameter of 1.8 cm ($A_{\text{p}} = 2.54$ cm^2), values of $A_{\text{E}} = 3.08$ cm^2 [1] and 3.00 cm^2 [6] were independently found when they were used in solution. By modelling the pathway of diffusion through a DGT device and its boundary

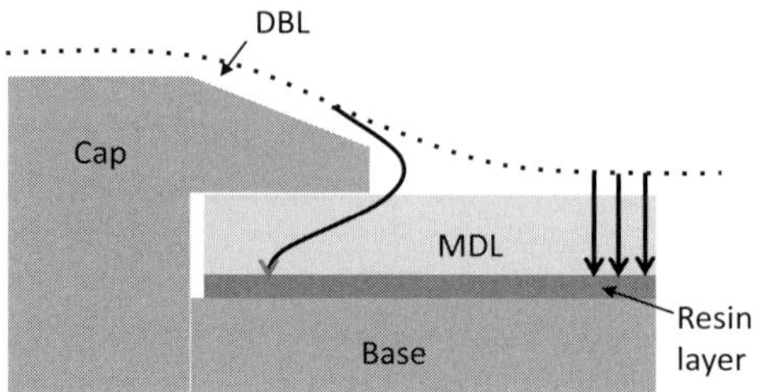

Figure 2.2   Two-dimensional representation of diffusion pathways into part of a DGT device. The surmised extent of the diffusive boundary layer (DBL) is shown by a dotted line. Away from the edge, planar diffusion applies, as represented by the parallel arrows. Close to the edge, diffusion can also occur laterally, both in the MDL and the DBL, as indicated schematically by the curved arrow.

layer, Garmo [7] confirmed the value of $A_E = 3.8$ cm^2 for standard solution devices. For these two devices it is realistic to set $A_E = 1.21A_p$.

Figure 2.2 suggests that the diffusion pathway may be linear through the central region of the exposure window. Warnken et al. [1] confirmed this by showing that the accumulated mass in a 1.89-cm diameter disc in the centre of the resin gel could be calculated accurately from the known concentration in solution using this physical area. When a DGT device was constructed so that it used diffusive and binding gel layers with the same diameter as the exposure window, the accumulated mass was predicted reasonably well using A_p, demonstrating that lateral diffusion within the gels contributes substantially to the larger effective area [8].

Santner et al. [9] have correctly pointed out that the pragmatic approach adopted by Warnken et al. [1] of assuming that A_E does not depend on diffusion layer thickness is an oversimplication. Their numerical modelling of the effects of lateral diffusion in the MDL on the accumulated mass showed that A_E is indeed dependent on δ^{mdl} and to a lesser extent on other factors, such as D^{mdl} and δ^{dbl}. For very thin MDLs, A_E approaches A_p. A consequence of using a constant value of A_E is likely to be slight underestimation of δ^{dbl}. The modelled predictions of Santner et al. [9] did not exactly fit the data of Warnken et al. [1], perhaps because they could not account for possible lateral diffusion occurring within the DBL. More precise DGT measurements over a wide range of δ^{mdl} with well-controlled stirring are required to produce a pragmatic adjustment of A_E to δ^{mdl} or to parameterise the equation of Garmo et al. [10] (see below).

2.2.3 Basic Validation of DGT

Although most users will wish to use DGT to calculate concentration, it is accumulated mass that is measured directly. Equation 2.4 can be rearranged to provide a relationship for

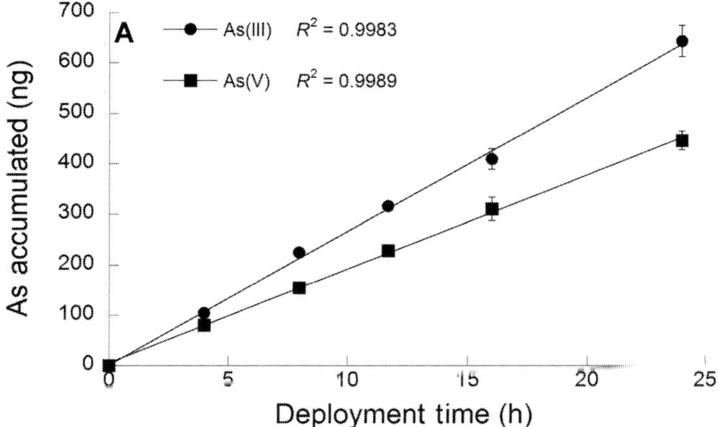

Figure 2.3   Mass of As(III) and As(V) accumulated by the TiO_2 (Metsorb) binding layers of DGT devices deployed for various times in a well-stirred solution containing 20 µg L^{-1} of each As species in 0.01M $NaNO_3$ solution at pH 7. Different slopes reflect different diffusion coefficients of the two species. Adapted with permission from Ref. [25]. Copyright (2010) American Chemical Society.

the reciprocal of the measured mass (equation 2.5) or the measured mass directly (equation 2.6).

$$\frac{1}{M} = \frac{1}{c_{\text{DGTe}} A_{\text{E}} t} \left(\frac{\delta^{\text{mbl}}}{D^{\text{mdl}}} + \frac{\delta^{\text{dbl}}}{D^{\text{w}}} \right) \tag{2.5}$$

$$M = \frac{c_{\text{DGTe}} A_{\text{E}} t}{\left(\dfrac{\delta^{\text{mdl}}}{D^{\text{mdl}}} + \dfrac{\delta^{\text{dbl}}}{D^{\text{w}}} \right)} = \frac{A_{\text{E}} D^{\text{mdl}} D^{\text{w}} c_{\text{DGTe}} t}{D^{\text{mdl}} \delta^{\text{dbl}} + D^{\text{w}} \delta^{\text{mdl}}} \tag{2.6}$$

The more complete forms, where D^{f} and D^{g} differ, are simply equations 2.7 and 2.8:

$$\frac{1}{M} = \frac{1}{c_{\text{DGTe}} A_{\text{E}} t} \left(\frac{\delta^{\text{g}}}{D^{\text{g}}} + \frac{\delta^{\text{f}}}{D^{\text{f}}} + \frac{\delta^{\text{dbl}}}{D^{\text{w}}} \right) \tag{2.7}$$

$$M = \frac{c_{\text{DGTe}} A_{\text{E}} t}{\left(\dfrac{\delta^{\text{g}}}{D^{\text{g}}} + \dfrac{\delta^{\text{f}}}{D^{\text{f}}} + \dfrac{\delta^{\text{dbl}}}{D^{\text{w}}} \right)} = \frac{A_{\text{E}} D^{\text{g}} D^{\text{f}} D^{\text{w}} c_{\text{DGTe}} t}{D^{\text{g}} D^{\text{f}} \delta^{\text{dbl}} + D^{\text{g}} D^{\text{w}} \delta^{\text{f}} + D^{\text{f}} D^{\text{w}} \delta^{\text{g}}} \tag{2.8}$$

In evaluating the DGT technique, a range of deployment times has often been used. In simple solutions, the accumulated mass increases linearly with time, as predicted by equation 2.6 (Figure 2.3). Causes of non-linearity, including competitive binding and capacity-related effects, are discussed in Chapter 5. Measurements have also been made using several devices deployed for the same time, but with different thicknesses of gel layer. Linear plots of M against $1/\delta^{\text{mbl}}$ have often been obtained, which initially was taken as validation of equation 1.5 [11]. However, as equation 2.6 indicates, data can only be fitted linearly by a line

Table 2.1 *Values of the DBL thickness, δ^{dbl}, measured in the laboratory and in situ using DGT devices with several MDL thicknesses. Specified for each analyte are: the deployment time and conditions, the type of binding layer and the exposure window diameter, the equation used in the calculation and whether the effective area, A_E, or physical area, A_p was used.*

δ^{dbl} (mm)	Analyte	Deployment	Time	DGT	Eqn	Ref
0.11–0.18	Cd	On plastic mounts in 3 L vigorously stirred solution	1 d	Chelex 18 mm	2.5, A_E	[13]
0.20	Ln	On plastic mounts in 2.5 L vigorously stirred solution	1 d	Chelex 18 mm	2.13	[6]
0.19–0.22	Eu	On plastic mounts in 2.5 L vigorously stirred solution	24 or 76 h	Chelex 18 mm	2.13	[10]
0.23±0.03[a] 1.5±0.13[b]	Cd	On plastic mounts in 2.5 L solution stirred at 100–800[a], 0 rpm[b]	50 h	Chelex 20 mm	2.5, A_E	[1]
0.23±0.03	Cd	Bespoke, miniature DGT deployed in a stirred solution	8 h	Chelex 5 mm	2.15, A_E	[8]
0.25±0.05[c] 0.25±0.04[d] 0.44±0.10[e]	Cd, Co, Cu, Mn, Ni, Pb, Zn	In 5 L containers at three stirring rates: 0 rpm[e]; 50 rpm[c]; 400 rpm[d]	1 d	Chelex 20 mm	2.13 (known c_{DGT})	[14]
0.28±0.03	Co, Ni, Cu, Cd, Pb	On plastic mounts in well-stirred solutions	1 or 2 d	Chelex 18 mm	2.5, A_E	[15]
0.30–0.38	Eu	On plastic mounts in 2.5 L vigorously stirred solution	24 or 76 h	Chelex 18 mm	2.14	[10]
0.45±0.29[f] 0.52±0.31[g] 0.65±0.21[h]	Cd, Co, Cu, Mn, Ni, Pb, Zn	In 5 L containers at three stirring rates: 0 rpm[f]; 50 rpm[g]; 400 rpm[h]	1 d	Chelex 20 mm	2.14	[14]
0. 3	U	In situ; fast flowing stream, on plastic sheet on a line	4 d	Metsorb 20 mm	2.15, A_E	[16]
0.14± 0.03[j] 1.15± 0.10[k]	Cd	In situ; river; on a plastic sheet in a net; high[j] and lower[k] flow	14 d	Chelex 20 mm	2.13	[17]

0.23	antibiotic	In situ; inflow to wastewater treatment works	10 d	XAD18 20 mm	2.15, A_p	[18]
0.26±0.01	Cd	In situ; fast flowing river on a flat plate parallel to flow	3 d	Chelex 18 mm	2.5, A_E	[19]
0.29	Cd	In situ; fast flowing river on a flat plate parallel to flow	3 d	Chelex 18 mm	2.5, A_E	[20]
0.31 ± 0.22	Ca, Mg	In situ; lake water column, tied to nylon line	13–14 d	Chelex 20 mm	2.5 A_p	[21]
0.35±0.19	U	In situ; high salinity estuary site on plastic mounts on a buoyed line	5 d	Metsorb 20 mm	2.15 A_E	[22]
0.37–1.41	U	In situ; edge of variable flow stream mounted on a plastic sheet	7 d	Metsorb 20 mm	2.15 A_E	[23]
0.39	P	In situ; mounted on cords 20 cm below surface of a small pond	8 h	Ferrihyd. 20 mm	2.18	[12]
0.46±0.06	U	In situ; freshwater stream; plastic mounts on a line; water plants	5 d	Metsorb 20 mm	2.15 A_E	[22]
0.47±0.09[g] 0.65±0.21[h]	Cu	In situ; coastal sites; mounted on a line[g] and on a flat plate[h]	2–30 d	Chelex 20 mm	2.18	[24]
0.48±0.36[i] 0.58±0.19[j] 0.64±0.92[k]	Co	In situ; channel of treated wastewater. DGTs placed at three flow rates: 0.07[i]; 1[j]; 3 cm s^{-1}[k]	15 d	Chelex 20 mm	2.14	[14]
0.62–0.86	U	In situ; edge of fast flowing stream mounted on a plastic sheet	7 d	Metsorb 20 mm	2.15 A_E	[23]
0.67 ± 0.07	As	In situ; coastal marina, 30 cm below surface, on plastic sheet on a line	4 d	Metsorb 20 mm	2.15 A_E	[25]
0.68±0.02	P	In situ; estuarine marina, on plastic sheet on a line	4 d	Metsorb 20 mm	2.5 A_E	[26]
0.78±0.12	P	In situ; freshwater stream, on plastic sheet on a line	4 d	Metsorb 20 mm	2.5 A_E	[26]
0.81 ± 0.13	As	In situ; freshwater stream, 30 cm below surface, on plastic sheet	4 d	Metsorb 20 mm	2.15 A_E	[25]

Note: Superscript letters shown in columns 1 and 3 link the particular conditions to each measurement set within a given reference.

passing through the origin when the proportional contribution of δ^{dbl} to the total diffusion layer thickness is small compared to the precision errors of the experiment. When there is an appreciable DBL, plots of mass versus $1/\delta^{mbl}$ are not linear, as shown in Figure 2.4a [1]. The deviation from linearity in this figure is particularly evident for the lowest value of δ^{mbl}, which was obtained using only a 0.14-mm thick filter membrane and no diffusive gel. The proportional effect of the DBL (equation 2.6) is then large and conversely it is usually small for large (>1 mm) values of δ^{mbl}. Rather than plot M against $1/\delta^{mbl}$ to test the validity of DGT, it is better to plot $1/M$ against δ^{mbl}, as explained in the next section.

2.3 Dealing Practically with the DBL

2.3.1 Estimation of DBL Thickness

When data are available for different gel layer thicknesses, plotting $1/M$ versus δ^{mbl} enables estimation of both an unknown concentration and δ^{dbl} [12, 13]. As equation 2.5 shows, the slope is related to concentration (equation 2.9), while the intercept (equation 2.10) is related to concentration and δ^{dbl}.

$$s = \frac{1}{c_{DGTe}\,A_E\,D^{mdl}t} \tag{2.9}$$

$$y = \frac{\delta^{dbl}}{c_{DGTe}\,A_E\,D^{w}t} \tag{2.10}$$

The thickness of the DBL and c_{DGTe} can be calculated from the measured values of s and y (equations 2.11 and 2.12).

$$\delta^{dbl} = \frac{y\,D^{w}}{s\,D^{mdl}} \tag{2.11}$$

$$c_{DGTe} = \frac{1}{s\,D^{mdl}\,A_E t} \tag{2.12}$$

This data treatment is very useful because, while δ^{g} and δ^{f} can be measured directly using optical microscopy, obtaining a value for δ^{dbl} is more difficult. A further challenge is that the convective regime is unlikely to be constant in natural waters, causing δ^{dbl} to vary during the course of the deployment. Simply deploying in situ a set of devices, which have a range of diffusion layer thicknesses allows estimation of a mean value of δ^{dbl} that is effective while DGT is accumulating analyte. Each device needs to be exposed to a similar flow regime to ensure that they all have a very similar effective DBL and the analyte measured must be fully labile to ensure that kinetic effects do not affect the measurement of δ^{dbl}, as explained in detail in Chapter 5.

Good straight lines for plots of $1/M$ versus δ^{mdl} have been obtained from deployments of multiple DGT devices in the laboratory and field, as illustrated by Figure 2.4b and referenced in Table 2.1. As the thickness of the DBL depends on the rate of flow of solution across the device, the value of δ^{dbl} is affected by the experimental setup in the laboratory

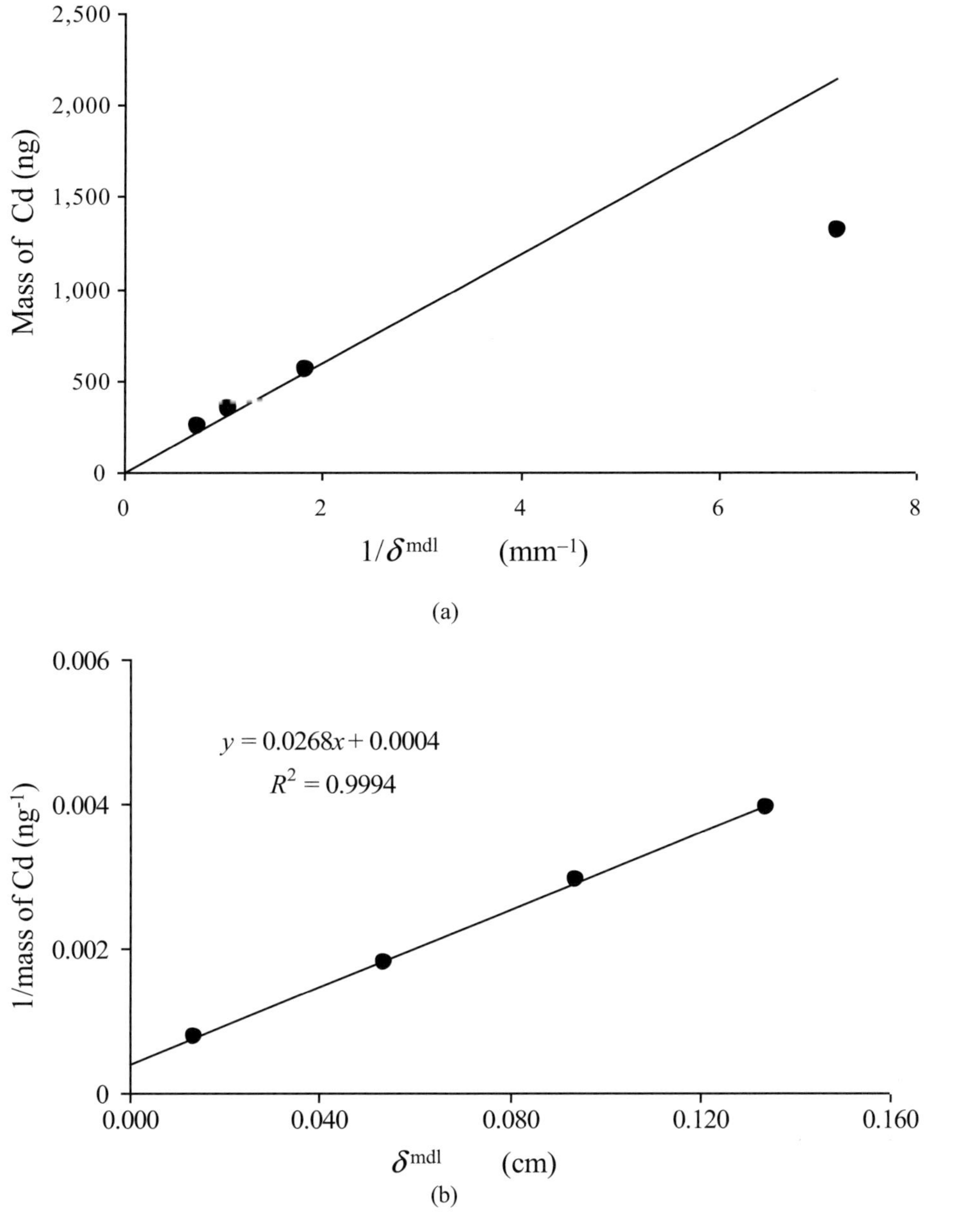

Figure 2.4 Example plots for the same dataset of (a) accumulated mass versus the reciprocal of the thickness of the material diffusion layer, δ^{mdl}, showing non-linearity, and (b) reciprocal of mass versus δ^{mdl}. Adapted with permission from Ref. [1]. Copyright (2010) American Chemical Society.

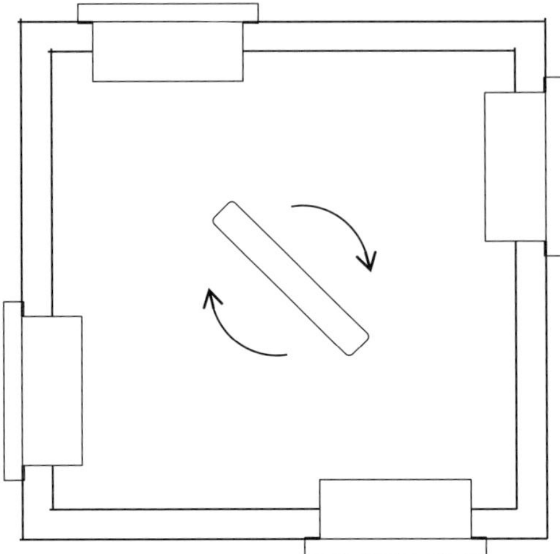

Figure 2.5   Plan view showing DGT devices mounted on a plastic support that can be placed inside a container stirred with a magnetic follower. Adapted with permission from Ref. [1]. Copyright (2010) American Chemical Society.

and the method used for in situ deployments. Laboratory measurements are frequently undertaken by mounting DGT devices in simple plastic housings, which are placed at the boundary of a round or square container where the solution is stirred with a magnetic follower (Figure 2.5). The DGT devices should be freely exposed to the solution flow. Electronically controlled magnetic stirrers, which allow precise selection of rotation speeds, enable reproducible selection of the convective regime. There should be adequate volume of solution to prevent the amount of metal accumulated by DGT significantly depleting its concentration during the deployment. Typically, if four DGT devices ($D^{mdl} = 5 \times 10^{-6}$ cm^2 s^{-1}) are deployed for 24 h, in 2 L of solution, the concentration of analyte will be lowered by 6% by the end of the deployment time. The precise method of in situ deployment of DGT devices will also affect δ^{dbl}. When devices are exposed to full flow by dangling them from a line they are likely to experience a thinner DBL than when they are mounted on a supporting plate, which might reduce convection at the DGT surface. Table 2.1 lists example values of δ^{dbl} which have been determined using DGT in both laboratory and in situ conditions. In the latter case, data are only included if there is reason to believe that any kinetic limitation of the supply of analyte is negligible (see Chapter 5).

Several approaches have been adopted in calculating δ^{dbl} from DGT measurements. Garmo et al. [10] noted that it is equation 2.6 which provides the direct expression for the measured quantity, M. Values of c_{DGTe} and δ^{dbl} can be obtained by using a least squares fitting routine for equation 2.6 directly. In undertaking a transformation of equation 2.6 to an inverse function that provides a straight line (equation 2.5), the error distribution of the dependent variable ($1/M$) undergoes a nonlinear transformation. Consequently, where there are errors in the data, values of c_{DGTe} and δ^{dbl} obtained using equation 2.6 will differ slightly from those obtained using equation 2.5, with the direct fit (equation 2.6) values being most accurate. If there are no errors in the data, both equations will give exactly the same values. While Garmo et al. [10] observed consistently lower values for both c_{DGTe} and δ^{dbl} when using the direct fit approach for their laboratory measurement of Eu, Uher et al. [14] found no significant difference for laboratory measurements of a range of metals. These latter authors found they could estimate δ^{dbl} most accurately if c_{DGTe} was treated as a known quantity rather than obtaining it by fitting. They found that the scatter of their in situ measurements with different diffusion layer thicknesses was sufficient to prevent direct M fits (equation 2.6), whereas sensible inverse plots, which provided estimates of both c_{DGTe} and δ^{dbl}, were obtained.

Garmo et al. [10] actually used a variant of equation 2.6, namely equation 2.13, which considered that diffusion through the DBL/membrane boundary occurs through the physical area, A_p, whereas diffusion through the diffusive gel/binding gel boundary occurs through the effective area, A_E.

$$M = \frac{A_E A_P D^{mdl} D^w c_{DGTe1} t}{A_E D^{mdl} \delta^{dbl} + A_P D^w \delta^{mdl}} \tag{2.13}$$

In recognition of the different calculation, c_{DGTe1} is used. Other workers have followed this approach [14, 17] and the use of these two area terms has been extended to the inverse relationship in the form of equation 2.14 [10].

$$\frac{1}{M} = \frac{1}{c_{DGTe1} t}\left(\frac{\delta^{mdl}}{A_E D^{mdl}} + \frac{\delta^{dbl}}{A_P D^w}\right) \tag{2.14}$$

Conceptually this approach is appealing. However, the value of A_E commonly used was obtained experimentally and is a single effective area for the DGT experiment, as defined by equation 2.5, which was used to obtain A_E. Therefore, if using previously determined values of A_E, for consistency equations 2.5 and 2.6 should be preferred. Note that when equations 2.13 and 2.14 are used to determine δ^{dbl}, the value will be smaller by the fraction ($A_p/A_E = 0.83$) than the values obtained using equations 2.5 and 2.6.

The idea of using DGT devices with a range of gel layer thicknesses was introduced by Zhang et al. [11], who assumed that the diffusion coefficient in water, D^w, could be taken to be the same as the value in the diffusive layer, D^{mdl}, and that the physical area of the exposure window, A_p, could be used throughout the device. The relevant equations 2.15 and 2.16 are then much simpler.

$$\frac{1}{M} = \frac{\delta^{mdl} + \delta^{dbl}}{c_{DGTe2} A_P D^{mdl} t} \tag{2.15}$$

$$M = \frac{c_{DGTe2} A_P D^{dl} t}{\delta^{mdl} + \delta^{dbl}} \tag{2.16}$$

As neglecting the effective area provides a different operational definition, c_{DGTe2} has been used. Later applications of equation 2.15 have usually used A_E. As δ^{dbl} is simply the ratio of intercept, y, to slope, s, making this simplification will tend to underestimate δ^{dbl}, depending on the ratio of $D^{\mathrm{mdl}}/D^{\mathrm{w}}$ (see equation 2.11).

When only two diffusive layer thicknesses, δ_1^{mdl} and δ_2^{mdl}, are used, with accumulated masses of M_1 and M_2, respectively, c_{DGTe2} and δ^{dbl} can be calculated directly using equations 2.17 and 2.18.

$$c_{\mathrm{DGTe2}} = \frac{M_1 M_2 \left(\delta_1^{\mathrm{mdl}} - \delta_2^{\mathrm{mdl}}\right)}{D^{\mathrm{mdl}} A_{\mathrm{P}} t \left(M_1 - M_2\right)} \tag{2.17}$$

$$\delta^{\mathrm{dbl}} = \frac{M_1 \left(\delta_1^{\mathrm{mdl}} - \delta_2^{\mathrm{mdl}}\right) - \delta_2^{\mathrm{mdl}} \left(M_1 - M_2\right)}{\left(M_1 - M_2\right)} \tag{2.18}$$

The physical area was used in the original version of equation 2.17, but ideally A_{E} should be used [11]. Although in principle this approach is sound, in practise the estimates of δ^{dbl} tend to be imprecise [24], because reliance is placed on the difference between two measurements and the errors for the terms in equation 2.18 are additive. Ideally data should be available for a wide range of diffusion layer thicknesses to define the slope of the plot of $1/M$ versus δ^{mdl}. The intercept is most accurately estimated if there are high precision data for very thin diffusion layers [1].

2.3.2 Effect of Solution Flow

If the diffusion layer thickness did not vary with the deployment conditions it would be simple to interpret DGT measurements accurately in terms of c_{DGTe} by assigning a value to δ^{dbl} in one of the equations provided above. However, δ^{dbl} depends on the flow rate and Table 2.1 shows a wide range of values of δ^{dbl} have been obtained from both in situ and laboratory deployments of devices with a range of MDL thicknesses. Different workers have used different methods of calculation, which are indicated in Table 2.1 by the equation they have chosen to use. The calculation method affects the value of δ^{dbl} obtained, but this contribution to the observed variations is likely to be small compared to differences in the flow rate of the solution and the method of mounting DGT. The local flow rate at the DGT surface is what matters and this is likely to be substantially moderated by mounting the device in large structures, such as large flat sheets of plastic or deploying near obstructions, including water weeds [22].

Various studies have examined the effect of flow rate on δ^{dbl} and DGT measurements. With the laboratory setup shown in Figure 2.5, δ^{dbl} was virtually constant, at 0.23 mm, when stir rates were above 100 rpm [1]. For stir rates below 100 rpm, δ^{dbl} increased markedly with decreasing rate and attained a value of 1.5 ± 0.13 mm in quiescent solutions. When DGT has been deployed in similar well-stirred conditions, values of δ^{dbl} have generally been in the range 0.2–0.4 mm (see first eight rows of Table 2.1).

There are well-established theories of fluid mechanics that allow calculation of the thickness of the DBL and the much larger hydrodynamic boundary layer for certain surfaces and flow conditions [27]. To apply them to DGT, as undertaken in [10, 11], requires some severe assumptions. The surface of the DGT device is assumed to be a flat plane, with solution flowing parallel to it with a velocity of U (cm s^{-1}) in the bulk solution. Equation 2.19 was used by Garmo et al. [10] to estimate δ^{dbl}.

$$\delta^{\text{dbl}} \approx 5\left(\frac{D^{\text{w}}}{v}\right)^{1/3}\left(\frac{vx}{U}\right)^{1/2} \tag{2.19}$$

The distance from the leading edge is x (cm), and v (cm^2 s^{-1}) is the kinematic viscosity of the fluid. The expression predicts that δ^{dbl} is very dependent on flow for $U < 1$ cm s^{-1}, but not very dependent for $U > 2$ cm s^{-1}. Although, the precise numerical values should be regarded as estimates, because the DGT device differs considerably from the theoretically required planar surface, being a disc which is not flat, DGT measurements undertaken in a flume with known flow rates [28] provided some support for this predicted flow dependence. Once flow exceeded 2 cm s^{-1}, c_{DGT} was constant to within 10%. Perhaps surprisingly, a further finding of this work was that c_{DGT} was not very sensitive to the orientation of the device in the flow. Contrary to these laboratory and modelling findings, estimates of δ^{dbl} obtained from in situ deployments of DGT in a variable width channel receiving treated wastewater were not significantly different (0.48–0.64 mm) at flow rates of 0.07, 1 and 2 cm s^{-1} (Table 2.1) [14], possibly reflecting the poor accuracy of the measurements. As the last eight rows of Table 2.1 illustrate, several studies have reported in situ δ^{dbl} values in the range 0.4–0.8 mm. Some of these have been for high flow (>2 cm s^{-1}) conditions, which has led to speculation about the possible role of particulates, biological activity and organic matter on estimates of δ^{dbl} [23], but it also illustrates the need to ensure a clear exposure of DGT to the water flow. Although lower values of δ^{dbl}, approaching those obtained in well-stirred solution in the laboratory, have generally corresponded to high flow conditions during in situ deployments, this has not always been the case. An exception is deployment in lakes from buoys where the wave action is likely to continuously oscillate the devices.

If biofilm formation increased the thickness of the diffusion layer, the biofilm thickness would be accommodated in δ^{dbl}, so that estimates of c_{DGTe} from deployments of DGTs with several MDL thicknesses would be unaffected, provided the biofilm did not significantly impede diffusion [12]. Although biofilms can affect the DGT measurement (see Chapter 6), this may be due to active interactions with the analyte, rather than simply physically increasing the diffusion layer [29]. In fact biofilms are generally much thinner than the MDL, but the thickness of those on DGT devices has not been measured in situ.

2.3.3 Use of the Simple DGT Equation (without δ^{dbl})

When DGT devices with binding layer diameters greater than that of the exposure window are used with a range of gel layer thicknesses, it is essential that the effective area is used and the thickness of the DBL is taken into account in the calculation. The original DGT work used the simplified equation 2.20 [2].

$$c_{\text{DGT}} = \frac{M}{A_{\text{p}}t}\frac{\delta^{\text{mdl}}}{D^{\text{mdl}}} \tag{2.20}$$

In using A_p rather than A_E, c_{DGT} will overestimate c_{DGTe} and hence c^{soln}, while ignoring the DBL will underestimate c_{DGTe}. These two effects balance one another if the DBL thickness is 0.2 mm and DGT measurements are made using the most commonly used gel layer thickness of 0.8 mm ($\delta^{mdl} = 0.94$ mm) [1]. Table 2.1 shows that for most measurements in well-stirred solutions, δ^{dbl} lies within the range of 0.15–0.25 mm. Therefore using the simple equation 2.20 in these situations to interpret the DGT measurement should provide a value of c_{DGT} equal to c_{DGTe} within $\pm$ 5%, which is generally less than the expected analytical errors. If solutions are not vigorously stirred, δ^{dbl} is likely to be much greater than 0.2 mm, which would introduce greater errors if equation 2.20 is used. Using devices with different values of δ^{mdl} would also add to the errors, especially for thinner diffusion layers. In both these cases it would be preferable to use equation 2.4, even if it means estimating a value of δ^{dbl} by simply observing the flow conditions.

In situ deployments have provided a wide range of δ^{dbl} (Table 2.1). Key factors are the flow rate and the possible moderation of flow at the surface of the device due to the method of mounting or possible accumulation of debris. Values of δ^{dbl} in the range of 0.15–0.3 mm have often been observed in fast flowing streams, in which case equation 2.20 or equation 2.4 with an estimated values of δ^{dbl} could be used. For the exceptionally high reported value of 1.15 mm, DGT devices were within a net, which may possibly have been obscured by debris for part of the deployment. A substantial set of the values of δ^{dbl} obtained in situ are within the range 0.3–0.8 mm. These examples make the case for using DGT with multiple values of δ^{mdl} so that δ^{dbl} can be estimated. Such an approach was advocated in a study of U deployed in a river under a wide range of flow conditions [23]. However, several workers [14, 17] have considered that while there are errors associated with using equation 2.20, they are generally less than the analytical errors, suggesting that this approach is still appropriate for simple monitoring exercises. This will be especially the case if deployment procedures maximise the exposure of the devices to flow.

2.4 Establishing Steady State

In this chapter all the equations that relate the measured accumulated mass to concentrations have been derived assuming that steady state conditions apply. However, several processes occur during a transient period before a steady state is established. We consider what happens when a DGT device is immersed in a simple solution comprising a solute that can bind to the binding layer and a simple salt (e.g. KCl) whose cations and anions do not bind. Initially, depending on the pre-treatment procedures, there are either no solutes in the water that fills the pore spaces of the binding gel, diffusive gel and filter, or they contain the ions of a simple conditioning salt (e.g. $NaNO_3$). Immediately on immersion, solutes diffuse between the solution and the diffusive layer. For the non-binding solutes, the system tends towards a state of chemical equilibrium, which is achieved when the concentration of each solute in each layer is equal to their concentration in the deployment solution. This is in effect a DET (diffusive equilibration in thin-films) experiment [30]. The progress towards this equilibrium state is illustrated in Figure 2.6a by the concentration profiles through the

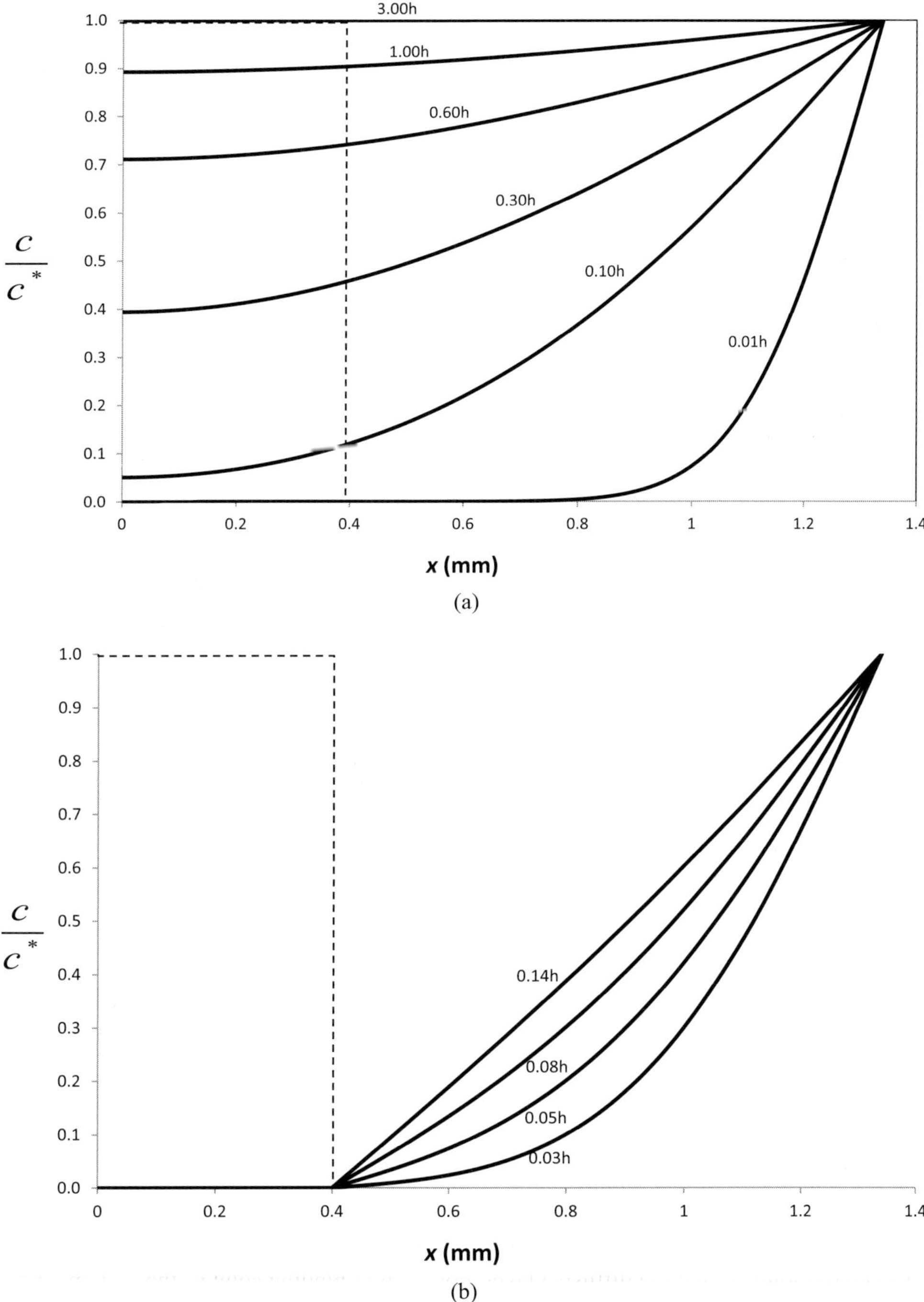

Figure 2.6   Profiles of concentration, normalised with respect to the concentration in solution, c^*, through a DGT device at various times from immersion for (a) ions which do not interact with the binding layer and (b) analytes which are bound instantaneously. The binding layer is 0.4 mm thick and the material diffusion layer is 0.94 mm thick. It was assumed that a diffusion coefficient of 5×10^{-6} cm^2 s^{-1} applied to all layers, that the diffusive boundary layer was negligible, that the analyte or ion was not complexed and that there was no interaction between the analyte or ion and the MDL. (a) represents the approach to equilibrium of a DET experiment with a total diffusion layer thickness of 1.34 mm and (b) represents the approach to steady state in a typical DGT experiment. Figures were kindly supplied by Josep Galceran, Martin Jiménez-Piedrahita and Jaume Puy.

DGT device as a solute initially absent from the DGT layers diffuses in. The diffusion coefficient of 5×10^{-6} cm^2 s^{-1} is assumed to be the same in the binding layer (0.4 mm), diffusive gel (0.8 mm) and membrane filter (0.14 mm).

For the solute that does bind, equilibrium is never achieved because the binding layer continually removes the solute, maintaining a negligible concentration at the interface between the binding layer and the MDL. The solute progressively enters the MDL until a steady state is achieved, where the rate or flux of solute entry is equal to the rate or flux of solute removal (Figure 2.6b). The concentration profile at steady state is linear, with a constant gradient throughout the MDL.

For the most commonly used DGT device, with a 0.8-mm thick diffusive gel and a 0.14-mm thick membrane filter ($\delta^{\mathrm{mdl}} = 0.94$ mm), steady state is effectively reached within 11 minutes. The gradient at the boundary between the binding and diffusion layers is then within 95% of the steady state value. Generally, this time will be inversely proportional to the diffusion coefficient and directly proportional to the square of the MDL thickness, according to Einstein-type relationships for planar diffusion [31]. For other commonly used MDL thicknesses (mm), the times (in parentheses) to establishing a gradient within 95% of the steady state value, assuming $D = 5 \ 10^{-6}$ cm^2 s^{-1}, will be 0.14 (15 s), 0.30 (67 s) 0.54 (216 s), 1.34 (22 min), 2.14 (57 min).

It is apparent from Figure 2.6 that it takes longer for a solute to reach equilibrium than steady state. This is partly because of the longer diffusion pathway (diffusion in the binding layer), but even when the diffusion paths are equal, the time taken for the mass of solute in the gel and filter layers to be within 95% of the equilibrium value is approximately three times the time taken to approach within 95% of steady state. Variations of the well-known Einstein equation can be used to estimate these times for other situations. Harper et al. [30] have shown, by fitting an equation to their numerical modelled outputs, that the time required for the mean concentration in the diffusion layer to be 95% of the solution concentration is given by equation 2.21. Here x represents diffusive distance.

$$t = 1.25x^2/D^{\mathrm{mdl}} \tag{2.21}$$

For the DGT situation, the time to achieve 95% of the gradient at steady state, t_{ss}, can be calculated using equation 2.22, which has again been derived from numerical simulations of the approach to steady state, as illustrated in Figure 2.6.

$$t_{\mathrm{ss}} = x^2/2D^{\mathrm{mdl}} \tag{2.22}$$

In this and the remaining equations in this subsection D^{mdl} is used, but it is the total thickness of the diffusion layer, including the DBL, which should be considered, with the diffusion coefficient being an aggregated value for the whole layer. The solutes that do not bind usually have little effect on the DGT measurement. Their behaviour is included here to provide a more complete picture of what happens when DGT is deployed.

During the transient period, analyte is accumulated, so the potential error associated with the assumption of instantaneous steady state is not simply the ratio of the transient time to the deployment time. By using a dynamic numerical model which accommodated the time to steady state Lehto et al. [32] were able to calculate the realistic accumulated mass for various values of t, δ^{mdl} and D^{mdl} compared to the mass calculated using equation 2.20. The difference between the two calculations provides the error associated with the assumed instantaneous steady state. To a good approximation it can be expressed as a percentage by equation 2.23.

$$\text{error} = \frac{100\left(\delta^{\text{mdl}} + \delta^{\text{dbl}}\right)^2}{6D^{\text{mdl}}t} \tag{2.23}$$

For simple solutes, with diffusion coefficients near to or greater than 5×10^{-6} cm^2 s^{-1}, the error is less than 3% for deployments of 4 h or greater using devices with standard MDL thicknesses (<1 mm), indicating that steady state equations are fully applicable. With thicker MDLs, up to 2.4 mm, deployment times of ≥ 16 h are necessary to ensure negligible ($\sim 4\%$) errors. Large molecules can have much lower diffusion coefficients. For normal thickness gels with $D^{\text{mdl}} = 10^{-6}$ cm^2 s^{-1}, errors are negligible for 24 h deployments, but for thicker gels deployment times of up to 3 days may be required to allow use of the simple equation.

The times to reach steady state considered in this section assume no complexation in solution and no interactions of the solution components with the MDL. Both these factors can substantially extend the time needed to reach steady state, as discussed in Chapters 3 and 5.

2.5 Temperature Effects

Consideration of equation 2.4 indicates that the effect of temperature on the measurement is likely to be mainly associated with the diffusion coefficient. The temperature dependence of diffusion coefficients is linked to the viscosity of water, η, and the absolute temperature, T^{K}, by the Stokes–Einstein equation [33] (equation 2.24), where subscripts indicate parameter values at two different temperatures.

$$\frac{D_1 \eta_1}{T_1^{\text{K}}} = \frac{D_2 \eta_2}{T_2^{\text{K}}} \tag{2.24}$$

The viscosity of water is represented accurately by equation 2.25 [34], where T is temperature in °C.

$$\log \frac{\eta_{25}}{\eta_{\text{T}}} = \frac{1.37023(T - 25) + 0.000836(T - 25)^2}{109 + T} \tag{2.25}$$

The diffusion coefficient measured at 25°C can be corrected to any temperature, T, using equation 2.26.

$$\log D_{\text{T}} = \frac{1.37023\,(T - 25) + 0.000836(T - 25)^2}{109 + T} + \log \frac{D_{25}\,(273 + T)}{298} \tag{2.26}$$

Pragmatically, the temperature dependence of diffusion coefficients derived from equations 2.25 and 2.26 can be represented by simple quadratic equations, as shown in the Appendix for a selection of ions commonly measured by DGT. Early work, which showed that the temperature dependence of the measured flux to the DGT device is well predicted using equation 2.26 within the range 4–35°C [11, 35], has been extended to 60°C [36]. Therefore other effects associated with possible size changes, such as expansion or contraction of the moulding, which could affect the area of the window, and the gel and filter thickness, appear to be negligible.

For laboratory studies the temperature correction is effectively built into the measurement, as the appropriate value of D for the measured temperature is used. However, for in situ deployments, it is not so straightforward, as the temperature may vary substantially during the measurement. For each degree increment in the temperature range 10–20°C, D increases by about 3%. Therefore some knowledge of the temperature range during deployment is required. This is most easily acquired using a commercial temperature probe with data logger. A mean temperature can usually be used if the concentration in solution is unlikely to vary very much. The problem arises when both temperature and analyte concentration can change appreciably, as most accumulation can occur in an unknown short time period which may not be well represented by the mean temperature. If the temperature range does not exceed 5°C it is probably acceptable to use the mean temperature, as the error in c_{DGT} will most probably still be within 10%. Larger fluctuations illustrate the problems of this approach, as illustrated in Chapter 6. Whether it can still be used depends on the system under study and the purpose of the work. For simple monitoring programmes, reduced accuracy may be acceptable, but for some research projects ancilliary spot samples may be required to define more clearly the likely errors.

An alternative approach to correcting for in situ changes in temperature and the presence of a DBL has been used for measurement of hydrophobic organic compounds with passive samplers [37]. Similar organic compounds to the analyte, known as performance reference compounds (PRC), are incorporated into the binding phase. They diffuse out into the solution during deployment and their dissipation is measured. As their release is sensitive to the DBL and changes in temperature, the in situ calibration automatically accommodates these effects. Isotopically labelled versions of the compounds being measured are often used as PRCs. The PRC approach has not been used for the measurement of inorganic substances and it is generally not recommended for the measurement of polar organic compounds using samplers, such as POCIS [37]. It is unlikely that PRCs will prove useful for DGT.

2.6 Summary and Recommendations

This chapter has considered how basic physical effects influence DGT measurements. When calculating c_{DGT}, the effect of temperature is readily accommodated by using an appropriate value for the diffusion coefficient, as provided in Chapter 3 and the Appendix of this book. To ensure accurate application of the steady state DGT equations, deployment times in simple solutions with standard devices should exceed 4 h, but for complex solutions and in situ deployments they should be several days. If standard DGT devices ($\delta^{mdl} =$ 0.94 mm) are deployed in high flow conditions, such as when solutions are vigorously stirred or devices are mounted unshielded in a fast flowing river, the simple equation 2.20 can be used with a good degree of accuracy. For devices with different MDL thicknesses, or where flow rates are lower, it is advisable to determine the DBL thickness by deploying multiple devices with a range of δ^{mdl}. Generally it is then acceptable to use equations 2.4, 2.5 and 2.6 for the calculation.

References

1. K. W. Warnken, H. Zhang and W. Davison, Accuracy of the diffusive gradient in thin-films technique: Diffusion boundary layer and effective sampling area considerations, *Anal. Chem.* 78 (2006), 3780–3787.
2. W. Davison and H. Zhang, Progress in understanding the use of diffusive gradients in thin-films – back to basics, *Environ. Chem.* 9 (2012), 1–13.
3. W. Davison and H. Zhang, In-situ speciation measurements of trace components in natural-waters using thin-film gels, *Nature* 367 (1994), 546–548.
4. J. K. Gundersen and B. B. Jorgensen, Microstructure of diffusive boundary layers and oxygen uptake of the sea floor, *Nature* 345 (1990), 604–607.
5. J. Galceran, J. Puy, J. Salvador, J. Cecilia and H. P. van Leeuwen, Voltammetric lability of metal complexes at spherical microelectrodes with various radii. *Journal of Electroanal. Chem.* 505 (2001), 85–94.
6. O. A. Garmo, N. J. Lehto, H. Zhang et al., Dynamic aspects of DGT as demonstrated by experiments with lanthanide complexes of a multidentate ligand, *Environ Sci Technol.* 40 (2006), 4754–4760.
7. O. A. Garmo, Simulating the effect of lateral diffusion and the diffusive boundary layer on uptake in solution and soil type DGTs. Presented at Conference on DGT and the Environment, Lancaster, UK, July 2013.
8. N. Alexa, H. Zhang and J. Lead, Development of a miniaturised diffusive gradients in thin films (DGT) device, *Anal. Chim. Acta* 655 (2009), 80–85.
9. J. Santner, A. Kreuzeder, A. Schnepf and W. W. Wenzel, Numerical evaluation of lateral diffusion inside diffusive gradients in thin films samplers, *Environ. Sci. Technol.* 49 (2015), 6109–6116.
10. O. A. Garmo, K. R. Naqvi, O. Royset and E. Steinnes, Estimation of diffusive boundary layer thickness in studies involving diffusive gradients in thin films (DGT), *Anal. Bioanal. Chem.* 386 (2006), 2233–2237.
11. H. Zhang and W. Davison, Performance characteristics of the technique of diffusion gradients in thin-films (DGT) for the measurement of trace metals in aqueous solution, *Anal. Chem.* 67 (1995), 3391–3400.
12. H. Zhang, W. Davison, R. Gade and T. Kobayashi, In situ measurement of phosphate in natural waters using DGT, *Anal. Chim. Acta.* 370 (1998), 29–38.
13. J. Levy, H. Zhang, W. Davison and R. Groben, Using diffusive gradients in thin films to probe the kinetics of metal interaction with algal exudates, *Environ. Chem.* 8 (2011), 517–524.

14. E. Uher, M. Tusseau-Vuillemin and C. Gourley-France, DGT measurement in low flow conditions: Diffusive boundary layer and lability consideration, *Environ. Sci. Process. Impacts* 15 (2013), 1351–1358.
15. J. L. Levy, H. Zhang, W. Davison, J. Galceran and J. Puy, Kinetic signatures of metals in the presence of Suwannee River fulvic acid, *Environ. Sci. Technol.* 46 (2012), 3335–3342.
16. C. M. Hutchins, J. G. Panther, P. R. Teasdale et al., Evaluation of a titanium dioxide-based DGT technique for measuring inorganic uranium species in fresh and marine waters, *Talanta* 97 (2012), 550–556.
17. R. Buzier, A. Charriau, D. Carona et al., DGT-labile As, Cd, Cu and Ni monitoring in freshwater: Toward a framework for interpretation of in situ deployment. *Environ. Poll.* 192 (2014), 52–58.
18. C-E. Chen, H. Zhang, G-G. Ying and K. C. Jones, Evidence and recommendations to support the use of a novel passive water sampler to quantify antibiotics in wastewaters, *Environ. Sci. Technol.* 47 (2003), 13587–13593.
19. K. W. Warnken, W. Davison, H. Zhang, J. Galceran and J. Puy, *In situ* measurements of metal complex exchange kinetics in freshwater, *Environ. Sci. Technol.* 41 (2007), 3179–3185.
20. K. W. Warnken, W. Davison and H. Zhang, Interpretation of *in situ* speciation measurements of inorganic and organically complexed trace metals in freshwater by DGT, *Environ. Sci. Technol.* 42 (2008), 6903–6909.
21. M. C. Alfaro-De la Torre, P. Y. Beaulieu and A. Tessier, *In situ* measurement of trace metals in lakewater using the dialysis and DGT techniques, *Anal. Chim. Acta* 418 (2000), 53–68.
22. G. S. C. Turner, G. A. Mills, P. R. Teasdale et al., Evaluation of the DGT technique for measuring inorganic uranium species in natural waters: interferences, deployment time and speciation. *Anal. Chim. Acta* 739 (2012), 37–46.
23. G. S. C. Turner, G. A. Mills, M. J. Bowes et al., Evaluation of DGT as a long term water quality monitoring tool in natural waters; uranium as a case study. *Environ. Sci. Process. Impacts* 16 (2014), 393–403.
24. A. W. Webb and M. J. Keough, Quantification of copper doses to settlement plates in the field using diffusive gradients in thin films, *Sci. Total Environ.* 298 (2002), 207–217.
25. W. W. Bennett, P. R. Teasdale, J. G. Panther, D. T. Welsh and D. F. Jolley, New diffusive gradients in thin film technique for measuring inorganic arsenic and selenium(IV) using a titanium dioxide adsorbent, *Anal. Chem.* 82 (2010), 7401–7407.
26. J. D. Panther, P. R. Teasdale, W. M. Bennett, D. T. Welsh and H. J. Zhao, Titanium dioxide-based DGT technique for in situ measurement of dissolved reactive phosphorus in fresh and marine waters, *Environ. Sci. Technol.* 44 (2010), 9419–9424.
27. H. P. van Leeuwen and J. Galceran, Biointerfaces and mass transfer, In *Physicochemical kinetics and transport at chemical-biological surfaces*, ed. H. P. van Leeuwen and W. Köster, IUPAC Series on Analytical and Physical Chemistry of Environmental Systems (Chichester: Wiley, 2004).
28. J. Gimpel, H. Zhang, W. Hutchinson and W. Davison, Effect of solution composition, flow and deployment time on the measurement of trace metals by the diffusive gradients in thin films technique, *Anal. Chim. Acta* 448 (2001), 93–103.
29. C. Pichette, H. Zhang and S. Sauve, Using diffusive gradients in thin-films for in situ monitoring of dissolved phosphate emissions from freshwater aquaculture, *Aquaculture* 286 (2009), 198–202.

30. M. P. Harper, W. Davison and W. Tych, Temporal, spatial, and resolution constraints for in situ sampling devices using diffusional equilibration: dialysis and DET, *Environ. Sci. and Technol.* 31 (1997), 3110–3119.

31. H. P. van Leeuwen, Dynamic aspects of in situ speciation processes and techniques, In *In situ monitoring of aquatic systems: Chemical analysis and speciation*, ed. J. Buffle, G. Horvai and H. P. van Leeuwen, vol. 6, IUPAC Series on Analytical and Physical Chemistry of Environmental Systems (Chichester: Wiley, 2000), pp. 253–277.

32. N. J. Lehto, W. Davison, H. Zhang and W. Tych, An evaluation of DGT performance using a dynamic numerical model, *Environ. Sci. Technol.* 40 (2006), 6368–6376.

33. J. H. Simpson and H. Y. Carr, Diffusion and nuclear spin relaxation in water, *Phys. Rev.* 111 (1958), 1201–1202.

34. P. W. Atkins and J. de Paula, *Atkins' physical chemistry*, 8th edn. (Oxford: Oxford University Press, 2006).

35. C. Murdock, M. Kelly, I. Y. Chang, W. Davison and H. Zhang, DGT as an in situ tool for measuring radiocesium in natural waters, *Environ. Sci. Technol.* 35 (2001), 4530–4535.

36. Cheng, H. Investigation of diffusion and binding properties for extending applications of the DGT technique, PhD Thesis, University of Lancaster, UK, 2014.

37. C. Harman, I. J. Allan and E. L. M. Vermeirssen, Calibration and use of a polar organic integrative sampler – A critical review, *Environ. Toxicol. Chem.* 31 (2012), 2724–2738.

3

Diffusion Layer Properties

WILLIAM DAVISON AND HAO ZHANG

3.1 Introduction

A key feature of DGT is that solutes are transported through the diffusion layer in a controlled predictable manner. For deployments in solution, diffusional transport operates in both the water layer adjacent to the surface of the device, known as the diffusive boundary layer (DBL), and in the diffusion layer within the device, known as the material diffusion layer (MDL). Ideally, the material of this layer should have an open structure that allows virtually unimpeded diffusion, and chemical interactions between the MDL and solutes should be negligible. Additionally it should be capable of fabrication with a well-controlled thickness and have a spatially uniform composition and structure. From the outset hydrogels have formed the major part of the MDL, while membrane filters have provided a protective surface layer. This chapter considers the properties of the hydrogels and membranes used in DGT, including the binding and diffusion of solutes, and their impact on DGT measurements.

3.2 Gel and Membrane Composition

Hydrogels are intermediate between a solid and liquid [1]. They are three-dimensional polymer networks that are both insoluble in water and highly hydrophilic. Their polar functional groups allow them to absorb water and swell, so that they can eventually contain over 95% water. A stable dimension is reached for a given ionic strength, pH and temperature. The water prevents the polymer network from collapsing and the network prevents the water from flowing away. Water in hydrogels has been classified as bound, which is considered to be strongly associated with the polymer chains by hydrogen bonding or dipole dipole interactions, or free, which has the same properties as normal water. As free water usually dominates, one way of regarding the more open pored gels is as 'parcels' of water, which can be handled. Depending on their precise composition, gels vary in consistency from viscous fluids to fairly rigid solids. Although they range from being fragile to quite strong, they are typically soft, flexible and elastic. Three types of gel have commonly been used for the MDL in DGT, agarose, polyacrylamide cross-linked with an agarose derivative and polyacrylamide cross-linked with bis-acrylamide. In all cases, transport of solutes occurs solely by diffusion. As gel properties are very dependent on the precise details of preparation, reproducing measurements using gels prepared in different laboratories can be challenging.

Agarose is a linear polysaccharide isolated from agar (or agar agar), a jelly-like substance obtained from marine red algae, by removing the agaropectin constituent. Although agarose mainly consists of repeating units of agarobiose (D-galactose and 3,6-anhydro-L-galactopyranose), there are impurities, particularly sulphonate, ester sulphate, ketal pyruvate and carboxyl groups. The purity of the material determines the extent of these groups and the precise properties of the gel. When a hot solution of agarose in water is cooled the molecular strands associate to form double helixes. The resulting three-dimensional network entraps water to form a weakly ionic gel. There is a wide-ranging distribution of pore sizes from 1 to 900 nm, depending on the agarose concentration. Agarose gels used for DGT measurements have been prepared from solutions containing 1.5% agarose. This produces a gel with pore sizes ranging from 1 to 480 nm, with average values of 35–47 nm [2]. Differential scanning calorimetry indicated that there is little bound water in the commonly used agarose gel [2]. The large pore size and dominance of free water allows relatively unimpeded diffusion of simple solutes.

The gel most commonly used for the diffusion layer in DGT, often designated APA, is polyacrylamide cross-linked with a commercial agarose derivative [3]. Acrylamide can be polymerised without a cross-linker, but it would simply consist of linear, unbranched chains. Cross-linker molecules can be incorporated into two chains simultaneously, permanently linking them together. Consequently the polyacrylamide grows into a complex web of interconnecting groups and branches. Cross-linker and acrylamide are added to water to form a solution, which is then polymerised by adding ammonium persulphate (APS) and TEMED (tetramethylethylenediamine). In the initial stages of polymerisation APS first reacts with TEMED to form a TEMED molecule with an unpaired electron. This activated molecule can then combine with acrylamide (or the agarose derivative), with the unpaired electron being transferred to this unit, which can then in turn react further to eventually form, as the chain continues, the polymer. As the cross-linker is a commercial product, its precise structure, including the number of repeat saccharide units is unknown, but the agarose derivative has sufficient chain length to give an open structure. The resulting gel is strong, flexible and elastic. While the structure is not quite so open as agarose, it allows relatively free passage of simple solutes.

Different versions of APA have been used [3]. The most common formulation, referred to here as simply APA, uses a gel solution with 15% monomer and 0.3% cross-linker. It is polymerised at 42–45°C by adding ammonium sulphate and TEMED to the gel solution and set for 1 h at 42–45°C. Experimental details for the preparation of this and other commonly used gels have been published extensively and are provided in Chapter 10. In the early DGT work, undertaken prior to 1998, the proprietary cross-linker appeared to be less concentrated, producing a more open pored gel. The properties of early gels were subsequently emulated by using a lower percentage (0.12%) of the post-1998 cross-linker [3].

The acrylamide gels most commonly used in electrophoresis are cross-linked with bis-acrylamide (*N, N'*-methylene-bis-acrylamide), which has two acrylamide molecules that are linked through their aminocarbonyl groups. The effective pore size of these gels is very dependent on the proportion of reagents, but generally it is smaller than for the APA gel, consistent with bis-acrylamide being a smaller bridging unit. A version of bis-acrylamide has been used regularly in DGT measurements where some in situ discrimination against large molecules, such as humic acid, is required [4]. Due to the restricted passage of larger solutes, this particular composition has been referred to as a restricted gel (RG or RES). It is much more fragile than the APA gel.

Membrane filters are polymers that have been fabricated to have a network of pores that provide an average cut-off for filtration. The material most commonly used for DGT is polyethersulphone. Depending on the manufacturer, the thickness of these filters typically varies from 0.13 to 0.15 mm. A pore size of 0.45 µm is usually chosen, but filters with different pore sizes can be used, provided the diffusion coefficients of the solutes of interest through the filter are known. Cellulose acetate and nitrate membranes have also been successfully used in DGT devices. Polyethersulphone is preferred for in situ deployments in natural waters, as it is believed to be more resistant to microbial attack [5].

3.3 Interactions of Solutes

Ions in solution can in principle interact with gels and membranes in two ways. They can specifically bind to sites such as the amides within polyacrylamide. Like any chemical binding, the affinity can vary greatly between ions. Ions can also interact electrostatically with fixed charge groups in gels and membranes. These non-specific interactions may affect the equilibrium partitioning of all ions and will be sensitive to the ionic strength of the solution. The literature on the binding of measured species and their ligands to the gels and membranes used in DGT has been reviewed and discussed [6]. A simple way to test for binding effects is to immerse a gel in a solution with known concentrations of ions. After equilibration, which may take more than a day, depending on gel size and solution flow [7], the gel is removed and the concentrations of ions in the gel are measured. As this is the basic test procedure for measurements made with the technique of diffusive equilibration in thin-films (DET), a substantial body of data is available, as illustrated by the examples shown in Table 3.1 for the ratio of the concentration measured in the gel, c^{gel}, to the concentration measured in solution, c^{soln}.

Gels and membranes generally have very low concentrations of specific binding sites. The value of 50 nmol per litre estimated by Garmo et al. [11] for the binding of Cd to APA gel corresponds to 20 pmol of sites in each diffusion gel typically used in a standard DGT solution holder. When the gel is equilibrated with a solution containing solute at high concentration, the amount of solute occupying the specific binding sites will be negligible compared to the amount in solution within the gel pores. The ratio c^{gel}/c^{soln} is therefore indistinguishable from 1. Ratios of 1 have been obtained for cations and anions, including

Table 3.1 *Examples of measured ratios of concentrations of analytes in a diffusion gel or filter (f) membrane and solution ($c^{gel,f}/c^{soln}$). Where there are additional data by other workers with similar values and conditions the reference and known differences in conditions are shown in italics. APA: polyacrylamide gels cross-linked with an agarose derivative; AGE: agarose; RES: bis cross-linked polyacrylamide gel; PES: polyethersulfone membrane filter; AFA: aquatic fulvic acid; AHA: aquatic humic acid; SHA: soil-derived humic acid; DOM: dissolved organic matter.*

Analyte	Concn (µM)	Medium	pH	Gel or filter	$c^{gel,f}/c^{soln}$	Ref
Ca^{2+}	10,000	0.53 M NaCl + 0.027 M MgSO$_4$[b]	6–7?	APA	0.97	[8]
Ca^{2+}	0.01–0.5	0.0001 M HCl?	4?	APA	1.0	[9]
Cd^{2+}	0.53	<0.01–10 mM NaNO$_3$, *1 mM NaNO$_3$*	5.6?, 5.5	APA	1.0	[10, 11]
Cd^{2+}	0.089	<0.01–10 mM NaNO$_3$, *1 mM NaNO$_3$*	5.6?, 5.5	APA	1.4–2.0	[10, 11]
Cd^{2+}	0.009	<0.01–10 mM NaNO$_3$, *1 mM NaNO$_3$*	5.6?, 5.5	APA	8.8	[10, 11]
Co^{2+}	0.01	1 mM NaNO$_3$, 0.06 mM Ca(NO$_3$)$_2$, 0.04 mM Mg(NO$_3$)$_2$	5.5	APA	0.98	[11]
Co^{2+}	0.1	1 mM NaNO$_3$, 0.06 mM Ca(NO$_3$)$_2$, 0.04 mM Mg(NO$_3$)$_2$	5.5	APA	1.2	[11]
Cs^{+}	0.8?	0.001–1000 mM NaCl	2–9	APA	1.0	[12]
Cr(VI)	2	?	?	APA	1.01	[13]
Cu^{2+}	0.01	1 mM NaNO$_3$, 0.06 mM Ca(NO$_3$)$_2$, 0.04 mM Mg(NO$_3$)$_2$	5.5	APA	14.2	[11]
Cu^{2+}	0.1	1 mM NaNO$_3$, 0.06 mM Ca(NO$_3$)$_2$, 0.04 mM Mg(NO$_3$)$_2$	5.5	APA	3.9	[11]
Fe^{2+}	23–100	1 mM NaHCO$_3$	6.85	APA	1.0	[14]
Fe^{2+}	45–90, *360*	?	?	APA	1.0	[15, 9]
Fe^{2+}	45–90	?	?	APA[aa]	1.0	[15]
K^{+}	0.005–0.13	0.0001 M HCl?	4?	APA	1.0	[9]
Mg^{2+}	0.01–0.2	0.0001 M HCl?	4?	APA	1.0	[9]
Mn^{2+}	180	?	?	APA	1.0	[9]
Mn^{2+}	9–36	18 mM phosphate	6.8	APA	1.0	[14]
Ni^{2+}	0.01	1 mM NaNO$_3$, 0.06 mM Ca(NO$_3$)$_2$, 0.04 mM Mg(NO$_3$)$_2$	5.5	APA	1.5	[11]
Ni^{2+}	0.1	1 mM NaNO$_3$, 0.06 mM Ca(NO$_3$)$_2$, 0.04 mM Mg(NO$_3$)$_2$	5.5	APA	1.3	[11]
Pb^{2+}	0.01	1 mM NaNO$_3$, 0.06 mM Ca(NO$_3$)$_2$, 0.04 mM Mg(NO$_3$)$_2$	5.5	APA	1.9	[11]
Pb^{2+}	0.1	1 mM NaNO$_3$, 0.06 mM Ca(NO$_3$)$_2$, 0.04 mM Mg(NO$_3$)$_2$	5.5	APA	2.6	[11]
Sr^{2+}	0.8	0.001–1000 mM NaCl	2–9	APA	1.0	[12]
UO_2^{2+}	0.4	10 mM NaNO$_3$	7.4	APA	1.8	[16]
Alk.	1,150	0.53 M NaCl + 0.027 M MgSO$_4$[b]	6–7?	APA	0.97	[8]
Br^{-}	795	0.53 M NaCl + 0.027 M MgSO$_4$[b]	6–7?	APA	1.0	[8]
Cl^{-}	1,340	*I = 4 mM?*	6–7?	APA	1.0	[17]
Cl^{-}	527,000	0.53 M NaCl + 0.027 M MgSO$_4$[b]	6–7?	APA	1.0	[8]
ΣCO_2	2,291	0.53 M NaCl + 0.027 M MgSO$_4$[b]	6–7?	APA	1.0	[8]

(cont.)

Table 3.1 (*cont.*)

Analyte	Concn (µM)	Medium	pH	Gel or filter	$c^{gel,f}/c^{soln}$	Ref
NH_4^+	2,640	$I = 4$ mM?	6–7?	APA[a]	1	[17]
NH_4^+	50	0.53 M NaCl + 0.027 M $MgSO_4$[b]	6–7?	APA	1.0	[8]
NO_3^-	766	$I = 4$ mM?	6–7?	APA	1.0	[17]
NO_3^-	101	0.53 M NaCl + 0.027 M $MgSO_4$[b]	6–7?	APA	1.0	[8]
PO_4^{3-}	490	$I = 4$ mM?	6–7?	APA	0.89	[17]
SO_4^{2-}	495	$I = 4$ mM?	6–7?	APA	1	[17]
SO_4^{2-}	27,000	0.53 M NaCl + 0.027 M $MgSO_4$[b]	6–7?	APA	0.97	[8]
AFA	2.6–50 mgL^{-1}	1 mM $NaNO_3$, 1? mM MES buffer	6.1	APA	1	[18]
AFA	30 mgL^{-1}	1 mM NaCl, 1 mM MES buffer	6	APA	3	[19]
AHA	30 mgL^{-1}	1 mM NaCl, 1 mM MES buffer	6	APA	2.7	[19]
SHA	0.6–10 mgL^{-1}	1 mM $NaNO_3$, 1? mM MES buffer	6.1	APA	47–53	[18]
SHA	0.6–10 mgL^{-1}	10 mM $NaNO_3$, 1? mM MES buffer	6.1	APA	11–17	[18]
SHA	30 mgL^{-1}	1 mM NaCl, 1 mM MES buffer	6	APA	12–17	[19]
SHA	40 mgL^{-1}	10 mM $NaNO_3$	6	APA	2.5	[20]
SHA	40 mgL^{-1}	1 mM $NaNO_3$	6	APA	5	[21]
SHA	40 mgL^{-1}	100 mM $NaNO_3$	6	APA	1.25	[21]
DOM	2–60 mgL^{-1}	Various natural softwaters	3.7–7.4	APA	1–2.6	[19]
Cd^{2+}	89	?	?	AGE	0.99	[22]
Cd^{2+}	0.008–5	1 mM $NaNO_3$, 0.06 mM $Ca(NO_3)_2$, 0.04 mM $Mg(NO_3)_2$	5.5	AGE	2.2	[11]
Co^{2+}	0.01	1 mM $NaNO_3$, 0.06 mM $Ca(NO_3)_2$, 0.04 mM $Mg(NO_3)_2$	5.5	AGE	1.8	[11]
Co^{2+}	0.1	1 mM $NaNO_3$, 0.06 mM $Ca(NO_3)_2$, 0.04 mM $Mg(NO_3)_2$	5.5	AGE	2.1	[11]
Cu^{2+}	157	?	?	AGE	0.98	[22]
Cu^{2+}	0.01	1 mM $NaNO_3$, 0.06 mM $Ca(NO_3)_2$, 0.04 mM $Mg(NO_3)_2$	5.5	AGE	7.3	[11]
Cu^{2+}	0.1	1 mM $NaNO_3$, 0.06 mM $Ca(NO_3)_2$, 0.04 mM $Mg(NO_3)_2$	5.5	AGE	3.8	[11]
Li^+	10,000	10 mM LiF	6	AGE	0.95	[2]
Li^+	1,000	1 mM LiF	6	AGE	1.1	[2]
Li^+	100	0.1 mM LiF	6	AGE	2.2	[2]
Mn^{2+}	182	?	?	AGE	1.03	[22]
Na^+	10,000	10 mM NaCl	6	AGE	1.0	[2]
Na^+	1,000	1 mM NaCl	6	AGE	1.3	[2]
Na^+	100	0.1 mM NaCl	6	AGE	3.5	[2]

Ion	Conc.	Medium	pH	Gel	C^{gel}/C^{soln}	Ref.
Ni^{2+}	0.01	1 mM $NaNO_3$, 0.06 mM $Ca(NO_3)_2$, 0.04 mM $Mg(NO_3)_2$	5.5	AGE	1.8	[11]
Ni^{2+}	0.1	1 mM $NaNO_3$, 0.06 mM $Ca(NO_3)_2$, 0.04 mM $Mg(NO_3)_2$	5.5	AGE	2.2	[11]
Pb^{2+}	0.01	1 mM $NaNO_3$, 0.06 mM $Ca(NO_3)_2$, 0.04 mM $Mg(NO_3)_2$	5.5	AGE	7.4	[11]
Pb^{2+}	0.1	1 mM $NaNO_3$, 0.06 mM $Ca(NO_3)_2$, 0.04 mM $Mg(NO_3)_2$	5.5	AGE	6.0	[11]
Br^-	125	?	?	AGE	0.99	[22]
Cl^-	10,000	10 mM NaCl	6	AGE	0.94	[2]
Cl^-	1,000	1 mM NaCl	6	AGE	0.74	[2]
Cl^-	282	?	?	AGE	0.99	[22]
Cl^-	100	0.1 mM NaCl	6	AGE	0.27	[2]
F^-	10,000	10 mM LiF	6	AGE	0.85	[2]
F^-	1,000	1 mM LiF	6	AGE	0.43	[2]
F^-	100	0.1 mM LiF	6	AGE	0.10	[2]
NO_3^-	161	?	?	AGE	0.96	[22]
SO_4^{2-}	104	?	?	AGE	0.94	[22]
Cd^{2+}	0.008–5	1 mM $NaNO_3$, 0.06 mM $Ca(NO_3)_2$, 0.04 mM $Mg(NO_3)_2$	5.5	RES	1.3	[11]
Co^{2+}	0.01	1 mM $NaNO_3$, 0.06 mM $Ca(NO_3)_2$, 0.04 mM $Mg(NO_3)_2$	5.5	RES	0.98	[11]
Co^{2+}	0.1	1 mM $NaNO_3$, 0.06 mM $Ca(NO_3)_2$, 0.04 mM $Mg(NO_3)_2$	5.5	RES	1.2	[11]
Cu^{2+}	0.01	1 mM $NaNO_3$, 0.06 mM $Ca(NO_3)_2$, 0.04 mM $Mg(NO_3)_2$	5.5	RES	7.7	[11]
Cu^{2+}	0.1	1 mM $NaNO_3$, 0.06 mM $Ca(NO_3)_2$, 0.04 mM $Mg(NO_3)_2$	5.5	RES	4.1	[11]
Ni^{2+}	0.01	1 mM $NaNO_3$, 0.06 mM $Ca(NO_3)_2$, 0.04 mM $Mg(NO_3)_2$	5.5	RES	1.0	[11]
Ni^{2+}	0.1	1 mM $NaNO_3$, 0.06 mM $Ca(NO_3)_2$, 0.04 mM $Mg(NO_3)_2$	5.5	RES	1.2	[11]
Pb^{2+}	0.01	1 mM $NaNO_3$, 0.06 mM $Ca(NO_3)_2$, 0.04 mM $Mg(NO_3)_2$	5.5	RES	3.4	[11]
Pb^{2+}	0.1	1 mM $NaNO_3$, 0.06 mM $Ca(NO_3)_2$, 0.04 mM $Mg(NO_3)_2$	5.5	RES	2.1	[11]
Cd^{2+}	0.008–5	1 mM $NaNO_3$, 0.06 mM $Ca(NO_3)_2$, 0.04 mM $Mg(NO_3)_2$	5.5	PES	1.9	[11]
Co^{2+}	0.01, 0.1	1 mM $NaNO_3$, 0.06 mM $Ca(NO_3)_2$, 0.04 mM $Mg(NO_3)_2$	5.5	PES	2.0	[11]
Cu^{2+}	0.01	1 mM $NaNO_3$, 0.06 mM $Ca(NO_3)_2$, 0.04 mM $Mg(NO_3)_2$	5.5	PES	3.8	[11]
Cu^{2+}	0.1	1 mM $NaNO_3$, 0.06 mM $Ca(NO_3)_2$, 0.04 mM $Mg(NO_3)_2$	5.5	PES	4.7	[11]
Ni^{2+}	0.01	1 mM $NaNO_3$, 0.06 mM $Ca(NO_3)_2$, 0.04 mM $Mg(NO_3)_2$	5.5	PES	1.7	[11]
Ni^{2+}	0.1	1 mM $NaNO_3$, 0.06 mM $Ca(NO_3)_2$, 0.04 mM $Mg(NO_3)_2$	5.5	PES	2.1	[11]
Pb^{2+}	0.01, 0.1	1 mM $NaNO_3$, 0.06 mM $Ca(NO_3)_2$, 0.04 mM $Mg(NO_3)_2$	5.5	PES	2.4	[11]

[a] Dipotassium peroxodisulphate used instead of ammonium persulphate.

[b] C^{gel}/C^{soln} was the same with a medium with half these concentrations.

[aa] Different proportions of reagents to produce a non-swelling gel

Fe^{2+} and Mn^{2+}, at quite high concentrations (>20 μmol L^{-1}). While a ratio of 1 was also obtained for the major cations at fairly low concentrations (μmol L^{-1}): Ca^{2+} (0.01–0.5); Mg^{2+} (0.01–0.2); Mg^{2+} (0.01–0.2); K^+ (0.005–0.13) [9], for trace metals the ratio increases as their concentration in solution is lowered (Table 3.1). This dependence on solution concentration, which is demonstrated graphically for Cd (Figure 3.1), is characteristic of specific binding.

To appreciate the nuances of the binding further, it is necessary to consider the effects of charges within the gels and membranes. Within agarose gel there are sites of agarobiose substituted pyruvic acid, which may dissociate to provide a negative charge [2]. Cations from solution will electrostatically associate with these negative charges, raising c^{gel}/c^{soln} above 1. However, at high ionic strength, the attraction of the locally available major cations for the negatively charged sites minimises the effect of the electrostatic interaction for the analyte ions. This screening effect means that c^{gel}/c^{soln} is dependent on ionic strength, and approaches 1 at high ionic strengths, as shown for a simple salt solution in Figure 3.2.

Electrostatic interactions and specific binding, with associated competition effects, occur simultaneously, so interpreting measured values of c^{gel}/c^{soln} may not be simple. After a systematic study, which considered a range of solution compositions and trace metal ions, Garmo et al. [11] were able to establish the general binding characteristics of the different materials used in DGT, forming the basis for the following statements.

At ionic strengths ≥ 1 mmol L^{-1} there is no measureable, charge-related binding of divalent trace metals to the most commonly used APA gel [23] ($c^{gel}/c^{soln} \sim 1$ at high concentrations of Cd in Figure 3.1a). Warnken et al. [10] showed that when these gels were prepared correctly and washed copiously they have a very slight positive charge, but it is only discernible, as $c^{gel}/c^{soln} < 1$, at ionic strengths less than 1 mM. However, without copious washing, residual polymerisation reactants may be left in the gel, conferring a negative charge, which can result in erroneous values of $c^{gel}/c^{soln} > 1$. A review and discussion of these artifacts is available [6].

The APA gel has specific binding sites, with ions such as Cu^{2+} and Pb^{2+}, which have known strong affinities for organic groups, binding most effectively. If these strong binding cations are present, they outcompete other trace metal cations, including Co^{2+}, Ni^{2+}, Cd^{2+} and Mn^{2+}, which consequently have a ratio of c^{gel}/c^{soln} close to 1. Figure 3.3 demonstrates this for the measurement of these ions in seawater. Amide functional groups are believed to be responsible for the binding, although there is also a possibility that some of these sites are hydrolysed to acrylic acid. The presence of acrylic acid groups is highly dependent on the proportion of TEMED used and the setting temperature, illustrating the importance of tightly controlling gel preparation procedures to ensure reproducible properties.

The enhanced binding of metals in the presence of fulvic acid (Figure 3.3) was attributed by Garmo et al. [11] to fulvic acid binding directly to the gel. Subsequently, using spectrophotometric measurements of gels and their immersion solutions, humic substances were shown to bind directly to the APA gel [18, 19]. Fulvic and humic acid extracted from river water were found to bind very slightly, but there was appreciable binding of humic

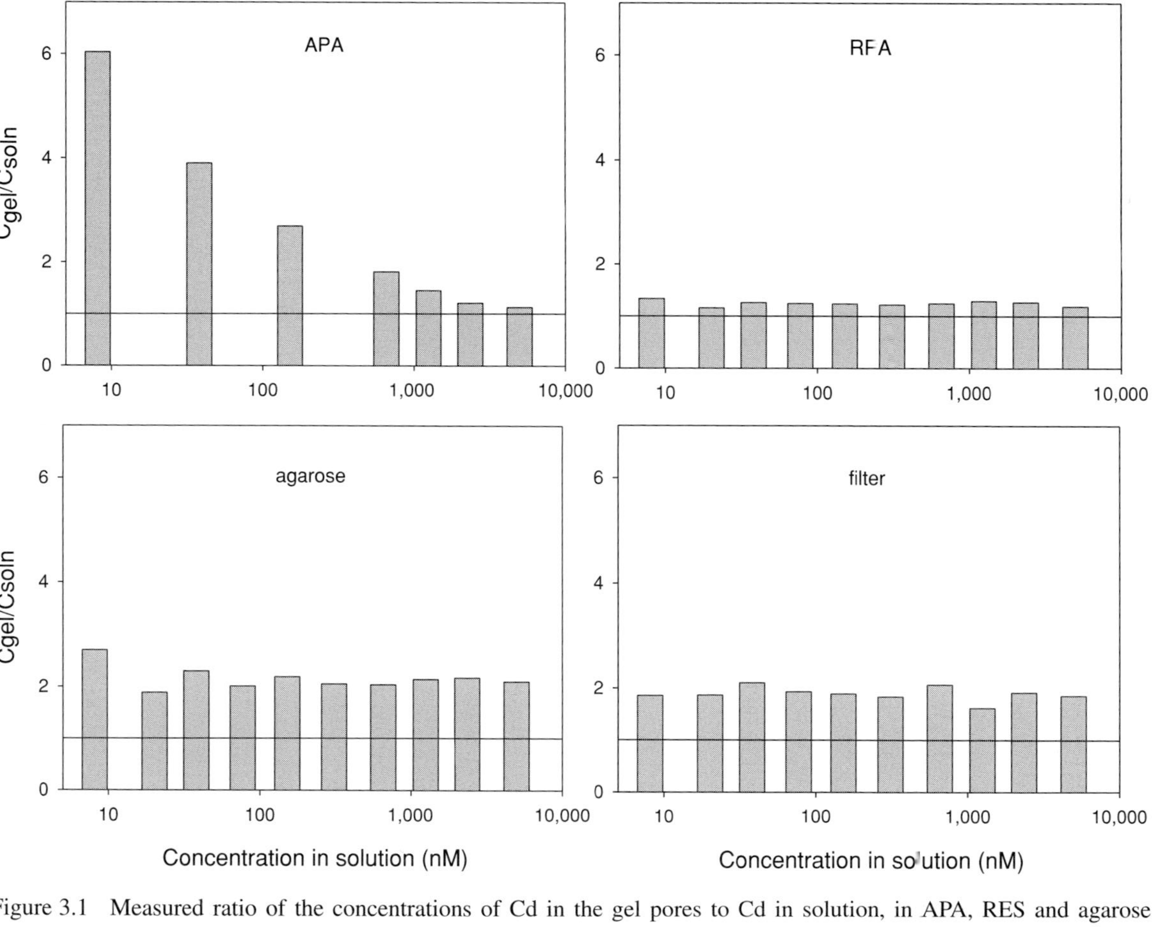

Figure 3.1   Measured ratio of the concentrations of Cd in the gel pores to Cd in solution, in APA, RES and agarose diffusive gels and in a PES membrane filter for a range of concentrations of Cd in a solution of $\_$ mM $NaNO_3$, 0.06 mM $Ca(NO_3)_2$ and 0.04 mM $Mg(NO_3)_2$. Data kindly supplied by O Garmo who previously published them in a different form [11].

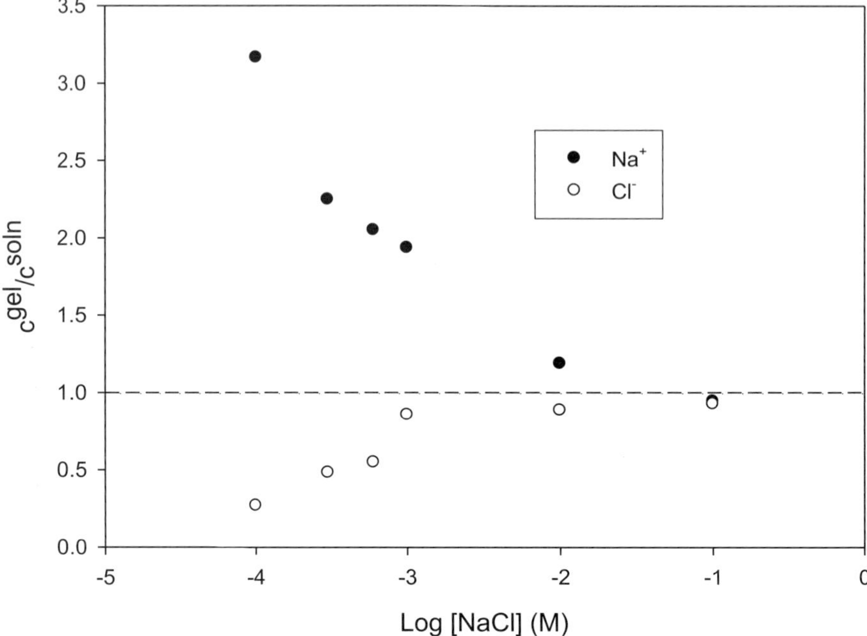

Figure 3.2   The ratio of the concentration of Na or Cl measured in an agarose gel to their concentrations in its immersion solution for various concentrations of NaCl. Data kindly supplied by N. Fatin-Rouge and J. Buffle who previously published part of it in a different form [2].

acid extracted from both forest [18] and peaty soils [19]. A recent study has used confocal laser scanning microscopy to show that binding of humic acid extracted from forest soil is enhanced in a 15-μm surface layer of an APA gel [20, 21]. The ratio of the concentration in the gel to the concentration in solution in this thin surface layer was on average 10 at pH 6 (10 mM NaNO$_3$), whereas it was only 2.5 in the bulk gel. Studies using a micro-electrode in the gel suggested that Cd binds more strongly to humic acid adsorbed to the gel than to humic acid in solution [24].

The charge-related binding of ions to agarose has been systematically investigated [2]. While c^{gel}/c^{soln} for cations increases at low ionic strength, due to the attractive forces (Figure 3.2), for anions it decreases, due to repulsion. The constant positive value of c^{gel}/c^{soln} of ~2 for a wide range of concentrations of Cd shows that there is a high concentration of charged sites (Figure 3.1), and no evidence for specific binding. However, Cu and Pb (Figure 3.3), and Hg [2] bind specifically and electrostatically. Other measurements of c^{gel}/c^{soln} show no evidence of specific binding, with values of 1.0 being reported for Cd (50 μM), Mn (180 μM), chloride (280 μM) and bromide (124 μM), 0.94 for sulphate (100 μM) and 0.96 for nitrate (190 μM) (Table 3.1) [22]. There is no evidence for adsorption of fulvic acid to agarose. In its presence, positive values of c^{gel}/c^{soln} for metals, associated with specific

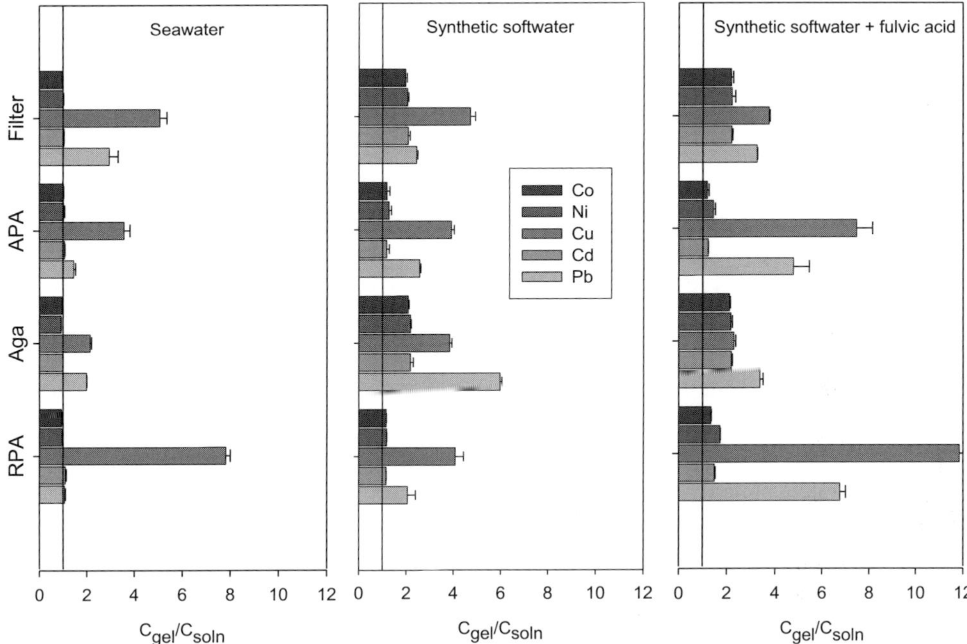

Figure 3.3 Measured ratio of the concentrations of metal in the gel or filter pores to metal in solution, in APA, restricted polyacrylamide (RPA) and agarose diffusive gels and in a PES membrane filter for three solutions: filtered seawater (salinity 20 ppt, pH 7.9), a synthetic softwater (1 mM $NaNO_3$, 60 µM $Ca(NO_3)_2$, and 40 µM $Mg(NO_3)_2$) and the softwater with added fulvic acid (FA) (3 mg L^{-1}). Data kindly supplied by O Garmo who previously published them in a different form [11].

binding, decrease, due to the competitive binding of the metal ions with fulvic acid in solution (Figure 3.3).

The charge density of bis-acrylamide cross-linked gels is known to be much lower than agarose gels and virtually negligible below an ionic strength of 1 mM [25]. This is consistent with small positive values of c^{gel}/c^{soln} of ~ 1.3, which indicate very slight electrostatic binding (in Figure 3.1). The concentration independence of c^{gel}/c^{soln} suggests there is insignificant specific binding of Cd. According to Figure 3.3 there is certainly specific binding of Cu and Pb, with particularly strong binding of Cu in seawater. As for the APA gel, fulvic acid enhances metal binding. Bis-acrylamide cross-linked gels with compositions other than that of the RES gel have been used. For one such gel a value of c^{gel}/c^{soln} greater than 2 was found for 5 mM Cd with virtually no supporting electrolyte [26]. Morford et al. [27] reported only slightly elevated values of c^{gel}/c^{soln} of 1.1 in a bis cross-linked gel in synthetic seawater, for Re, U and Mo at concentrations of 5, 320 and 700 nM, respectively.

Metals are known to interact with most filter membranes [28]. Supor polyethersulphone membranes have a slight negative charge [29]. The effect of charge is evident in the value of c^{gel}/c^{soln} for Cd being fairly constant and similar, at ~ 2, to that of agarose (Figure 3.1d). Figure 3.3 shows the effects of both electrostatic and specific binding. The ratio c^{gel}/c^{soln} is about 2 for all metals in synthetic freshwater, except for Cu, which appears to bind specifically to the membrane, as it does for all types of gel. At the high ionic strength of seawater, where there is negligible electrostatic binding, some specific binding of Pb is revealed. There is no evidence for fulvic acid adsorbing to the filter membrane.

Only recently has DGT been used for the measurement of polar organic compounds [30–32] and so detailed knowledge of the interaction of these analytes with the MBL are lacking. Chen et al. [30] reported negligible adsorption of the antibiotic sulfamethoxazole to agarose gel and the PES membrane, and this was also the case for the APA gel (Chen, pers. comm.). Bisphenols were found to adsorb appreciably to PES, nylon and cellulose ester filters, but not to the agarose gel and polytetrafluoroethylene filters that were chosen for DGT measurements [32].

As yet the binding properties of the gels and membranes used in DGT have been tested for a limited range of analytes and/or a limited range of conditions, even though the binding may depend on pH, concentration of analyte and ionic strength. The most commonly used APA gel has received most attention, but even here information is sporadic. It is dangerous to assume that interactions are minimal without performing systematic laboratory tests. Fortunately those based on measuring c^{gel}/c^{soln} are quite straightforward. The next section will show that binding can influence transport in the MDL. The possible influence of all these effects on DGT measurements is then considered.

3.4 Diffusion

The measured response of a DGT device can be calibrated by deploying it under controlled laboratory conditions in synthetic solutions with known concentrations of analyte, where c_{DGT} (or c_{DGTe}) would be expected to be the same as c^{soln}. The empirically obtained relationship between accumulated mass in the binding layer, M, and concentration in solution, c^{soln}, then provides a calibration curve, in a similar fashion to most analytical procedures. From equation 2.4 the slope is given by equation 3.1.

$$A_E t \left(\frac{D^{mdl} D^{w}}{\delta^{mdl} D^{w} + \delta^{dbl} D^{mdl}} \right) \tag{3.1}$$

As DGT is such a simple device, most parameters required to calculate the slope and therefore the in situ concentration can be readily measured. The deployment time, t, and the thicknesses of the diffusive gel, δ^g, and filter, δ^g, are readily obtained ($\delta^{mdl} = \delta^g + \delta^f$) and will be unaffected by the deployment medium. The thickness of the DBL, δ^{dbl}, and the effective area of the exposure window, A_E, can be estimated (as shown in Chapter 2). Therefore the 'calibration' simply reflects the values of the diffusion coefficients in gel, filter and water. These diffusion coefficients are fixed physical quantities, so if values are known, it is possible to calculate concentrations from in situ DGT deployments without resort to calibration. Furthermore, because the effect of temperature on diffusion coefficients is well established (Chapter 2), the calculation can be performed at any deployment temperature. This ability to interpret DGT measurements without performing a calibration is one of its great strengths for in situ measurements. However, it is advisable with any field campaign to ensure quality control, by performing measurements with some devices from the same batch in standard solutions under controlled laboratory conditions.

3.4.1 Diffusion Coefficients of Ions and Complexes in Water

Several techniques have been used to measure diffusion coefficients of solutes in water and many compilations are available (e.g. [33, 34]). The set of tables in the supporting information of Buffle et al. [35] and the associated discussion are particularly relevant to natural waters and the analytes measured by DGT. The values of diffusion coefficients in water, shown in Table 3.2 and the Appendix of the book, are predominantly taken from this source. For simple inorganic ligands, such as phosphate, it is usually safe to assume that the various protonated forms have the same diffusion coefficient. For simple complexes where inorganic anions replace water in the inner hydration shell of cations, the diffusion coefficient of the hydrated cation can usually be assumed. An exception occurs if the hydration shell is completely replaced by small hydrophobic cations (e.g. $HgCl_4^{2-}$ has a substantially higher diffusion coefficient than $Hg(OH)_4^{2-}$). The diffusion coefficient of a complex increases as the size of the ligand increases, with the value assuming that of the ligand when its molar mass exceeds 300 Da. For molar masses, M_{wt}, in the range 200–10^5 Daltons, diffusion coefficients in $cm^2\ s^{-1}$ can be estimated reasonably well using the semi-empirical equation 3.2 [35].

$$D = 2.84.10^{-5}/M_{wt}^{1/3} \tag{3.2}$$

Ligands derived from natural organic matter, such as fulvic and humic substances, cannot be represented by a single macromolecular formula and have a distribution of sizes. The mean diffusion coefficients of fulvic and humic substances measured by techniques such as flow field flow fractionation and fluorescence correlation spectroscopy are consistent and reproducible and suggest that the diffusional behaviour of the molecules is similar to that of fairly regular spheres. Humic and fulvic substances extracted from a wide range of aquatic origins were shown to have similar diffusion coefficients (within $\pm 20\%$) [53]. The diffusion coefficients were independent of pH above pH 5. Below pH 5 they were about 20% lower, due to aggregation. If humic substances are aggregated their diffusion coefficients can be much lower [21]. The limited data on metal complexes with fulvic and humic substances suggest that they have the same diffusion coefficient as the uncomplexed ligand [35].

3.4.2 Diffusion in Gels and Filters

Values for diffusion coefficients of solutes in gels and filters have been obtained in two main ways. DGT devices have simply been deployed under controlled laboratory conditions in

Table 3.2 *Examples of diffusion coefficients of analytes in a diffusive gel, D^g, measured using either a diffusion cell, DC, or a DGT device. The DGT measurement is of a gel and filter. Unless specified otherwise the filter was PES. In some cases data are shown for measurements of a metal ligand complex using a diffusion cell for the same conditions. Values of the diffusion coefficient in water of the analyte or complex at 25°C, D^w, are shown where available in the critical compilation of Buffle et al. [35] or in Li and Gregory [33] (italics). APA: polyacrylamide gels cross-linked with an agarose derivative; AGE: agarose; RES: bis cross-linked polyacrylamide gel; AFA: aquatic fulvic acid; AHA: aquatic humic acid; SHA: soil-derived humic acid; PHA: peat-derived humic acid.*

Analyte	T°C	Medium	pH	Gel	D^g DC	D^g DGT	Ligand	D^g M–L	D^w	Ref
Al	25	10 mM NaNO$_3$, 0.5 mM NaHCO$_3$	8.3	APA^i		5.83			5.41	[36]
Al	25	10 mM NaNO$_3$	4.0	APA^i	4.14					[36]
As(III)	24	10 mM NaNO$_3$, 25 mM Na acetate	5	APA	6.4	5.95				[37]
As(III)	25	10 mM NaNO$_3$	6.0	APA^i		7.1				[38]
As(III)	25	10 mM NaNO$_3$	6.0	APA^i	7.65					[38]
As(V)	25	10 mM NaNO$_3$, 4 mM Mg(NO$_3$)$_2$	7.04	APA^i		5.59			9.05	[36]
As(V)	25	10 mM NaNO$_3$	4.0	APA^i	5.36					[36]
As(V)	24	10 mM NaNO$_3$, 25 mM Na acetate	5	APA	4.85	4.90				[37]
As(V)	25	10 mM NaNO$_3$	6.8	APA	5.18	5.25				[39]
As(V)	25	10 mM NaNO$_3$	~6.3	APA^d	5.21	5.54				[40]
As(V)	25	10 mM NaNO$_3$	4.1	APA^d		5.48				[40]
As(V)	25	10 mM NaNO$_3$	7.9	APA^d		4.78				[40]
As(V)	25	10 mM NaNO$_3$, 0.5 mM Na acetate	6.06	APA^i		6.78				[41]
As(V)	25	10 mM NaNO$_3$	6.06	APA^i	5.54					[41]
As(V)	25	10 mM NaNO$_3$	6.0	APA^i	6.05	6.10				[38]
Cd^{2+}	25	100 mM NaNO$_3$	5.6?	APA	6.09				7.19	[3]
Cd^{2+}	25	10 mM NaNO$_3$	5.6?	APA	6.45					[3]
Cd^{2+}	25	1 mM NaNO$_3$	5.6?	APA	6.52					[3]
Cd^{2+}	25	0.1 mM NaNO$_3$	5.6?	APA	2.96					[3]
Cd^{2+}	17	100 mM NaNO$_3$?	5.6?	APA		3.99				[42]

Cr(VI)	25	10 mM NaNO$_3$	?	APA[j]		8.18			11.2	[43]
Cr(VI)	25	10 mM NaNO$_3$	?	APA	8.82					[43]
Cs$^+$	25	10 mM NaNO$_3$	5.6?	APA[a]	18.9				20.6	[12]
Cu^{2+}	25	100 mM NaNO$_3$	5.6?	APA	6.42				7.14	[3]
Cu^{2+}	25	10 mM NaNO$_3$	5.6?	APA	6.30					[3]
Cu^{2+}	25	1 mM NaNO$_3$	5.6?	APA	6.45					[3]
Cu^{2+}	25	0.1 mM NaNO$_3$	5.6?	APA	3.29					[3]
Cu^{2+}	17.7	10 mM NaNO$_3$	5.6?	APA	4.95		NTA	4.29		[44]
Hg^{2+}	21.5	10 mM NaNO$_3$	6?	APA[g]		2.84				[45]
CH$_3$Hg$^+$	20	100 mM NaCl	6?	APA[i]		5.1				[46]
CH$_3$Hg$^+$	18	100 mM NaCl	6?	APA[i]		4.7				[46]
Mn^{2+}	25	10 mM NaNO$_3$, 4 mM Mg(NO$_3$)$_2$	7.04	APA[i]		5.43			7.12	[36]
Mn^{2+}	25	10 mM NaNO$_3$	4.0	APA[i]	4.95					[36]
Mo(VI)	25	10 mM NaNO$_3$?	5.6?	APA	6.48				9.91	[47]
Mo(VI)	25	10 mM NaNO$_3$	~6.3	APA[d]	5.96	5.70				[40]
Mo(VI)	25	10 mM NaNO$_3$	4.1	APA[d]		5.14				[40]
Mo(VI)	25	10 mM NaNO$_3$	7.9	APA[d]		3.27				[40]
Mo(VI)	25	10 mM NaNO$_3$, 4 mM Mg(NO$_3$)$_2$	7.04	APA[i]		6.67				[36]
Mo(VI)	25	10 mM NaNO$_3$	4.0	APA[i]	5.58					[36]
Ni^{2+}	25	100 mM NaNO$_3$	5.6?	APA	5.77				7.05	[3]
Ni^{2+}	25	10 mM NaNO$_3$	5.6?	APA	6.18					[3]
Ni^{2+}	25	1 mM NaNO$_3$	5.6?	APA	6.04					[3]
Pb^{2+}	25	100 mM NaNO$_3$	5.6?	APA	8.00				9.45	[3]
Pb^{2+}	25	10 mM NaNO$_3$	5.6?	APA	8.48		DGA	6.28		[3]
Pb^{2+}	25	10 mM NaNO$_3$	5.6?	APA	8.48		NTA	5.95		[3]
Pb^{2+}	25	10 mM NaNO$_3$	5.6?	APA	8.48		AFA	1.7		[3]

(*cont.*)

Table 3.2 (*cont.*)

Analyte	$T°C$	Medium	pH	Gel	D^g DC	D^g DGT	Ligand	D^g M–L	D^w	Ref
Pb^{2+}	25	10 mM $NaNO_3$	5.6?	APA	8.48		AHA	0.66		[3]
Pb^{2+}	25	1 mM $NaNO_3$	5.6?	APA	8.22					[3]
PO_4^{2-}	25	6.25 mM KH_2PO_4	5	APA	6.05				6.1^e	[48]
Se(VI)	25	10 mM $NaNO_3$	6.0	APA^i	7.22	7.1			9.46	[38]
Sb(V)	25	10 mM $NaNO_3$	~6.3	APA^d	5.55	5.25			8.25	[40]
Sb(V)	25	10 mM $NaNO_3$	4.1	APA^d		5.50				[40]
Sb(V)	25	10 mM $NaNO_3$	7.9	APA^d		3.27				[40]
Sb(V)	25	10 mM $NaNO_3$, 4 mM $Mg(NO_3)_2$	7.04	APA^i		6.25				[36]
Sb(V)	25	10 mM $NaNO_3$	4.0	APA^i	5.50					[36]
U(VI)	25	10 mM $NaNO_3$	7.2	APA	5.64	5.64				[16]
V(V)	25	10 mM $NaNO_3$	~6.3	APA^d	6.72	6.69				[40]
V(V)	25	10 mM $NaNO_3$	4.1	APA^d		6.50				[40]
V(V)	25	10 mM $NaNO_3$	7.9	APA^d		6.78				[40]
V(V)	25	10 mM $NaNO_3$	6.0	APA^i	6.7	6.73				[38]
W(VI)	25	10 mM $NaNO_3$	~6.3	APA^d	5.45	5.82			9.23	[40]
W(VI)	25	10 mM $NaNO_3$	4.1	APA^d		5.32				[40]
W(VI)	25	10 mM $NaNO_3$	7.9	APA^d		5.53				[40]
W(VI)	25	10 mM $NaNO_3$, 4 mM $Mg(NO_3)_2$	7.04	APA^i		6.05				[36]
W(VI)	25	10 mM $NaNO_3$	4.0	APA^i	4.28					[36]
La	20	100 mM $NaNO_3$^{b}	2.3	APA	4.33		quin2^b	2.51	6.19	[49]
Ce	20	100 mM $NaNO_3$^{b}	2.3	APA	4.31		quin2^b	2.53		[49]
Pra	20	100 mM $NaNO_3$^{b}	2.3	APA	4.30		quin2^b	2.55		[49]
Nd	20	100 mM $NaNO_3$^{b}	2.3	APA	4.24		quin2^b	2.58		[49]
Sm	20	100 mM $NaNO_3$^{b}	2.3	APA	4.24		quin2^b	2.62		[49]
Eu	20	100 mM $NaNO_3$^{b}	2.3	APA	4.20		quin2^b	2.60		[49]
Gd	20	100 mM $NaNO_3$^{b}	2.3	APA	4.15		quin2^b	2.60		[49]

Species	T (°C)	Medium		Method						Ref
Tb	20	100 mM NaNO$_3$[b]	2.3	APA	4.06		quin2[b]	2.57		[49]
Dy	20	100 mM NaNO$_3$[b]	2.3	APA	4.02		quin2[b]	2.58		[49]
Ho	20	100 mM NaNO$_3$[b]	2.3	APA	3.98		quin2[b]	2.58		[49]
Er	20	100 mM NaNO$_3$[b]	2.3	APA	4.01		quin2[b]	2.63		[49]
Tm	20	100 mM NaNO$_3$[b]	2.3	APA	3.96		quin2[b]	2.64		[49]
Yb	20	100 mM NaNO$_3$[b]	2.3	APA	3.99		quin2[b]	2.73		[49]
Lu	20	100 mM NaNO$_3$[b]	2.3	APA	3.92		quin2[b]	2.66		[49]
SMX	20	10 mM NaCl	5.6?	APA	3.62					[30]
Al	25	10 mM NaNO$_3$, 0.5 mM NaHCO$_3$	8.3	RES[i]		4.01			5.41	[36]
Al	25	10 mM NaNO$_3$,	4.0	RES[i]	2.82					[36]
Cd^{2+}	25	10 mM NaNO$_3$, 4 mM Mg(NO$_3$)$_2$	7.04	RES[i]		4.69			7.19	[36]
Cd^{2+}	25	10 mM NaNO$_3$	4.0	RES[i]	3.87					[36]
Cd^{2+}	25	100 mM NaNO$_3$	5.6?	RES	3.96					[3]
Cd^{2+}	25	10 mM NaNO$_3$	5.6?	RES	4.02					[3]
Cd^{2+}	25	1 mM NaNO$_3$	5.6?	RES	3.87					[3]
Cd^{2+}	25	0.1 mM NaNO$_3$	5.6?	RES	1.43					[3]
Co^{2+}	25	10 mM NaNO$_3$, 4 mM Mg(NO$_3$)$_2$	7.04	RES[i]		4.89			7.32	[36]
Co^{2+}	25	10 mM NaNO$_3$	4.0	RES[i]	3.59					[36]
Cu^{2+}	25	100 mM NaNO$_3$	5.6?	RES	4.41				7.14	[3]
Cu^{2+}	25	10 mM NaNO$_3$	5.6?	RES	4.40					[3]
Cu^{2+}	25	1 mM NaNO$_3$	5.6?	RES	4.18					[3]
Cu^{2+}	25	0.1 mM NaNO$_3$	5.6?	RES	1.77					[3]
Cu^{2+}	25	10 mM NaNO$_3$, 4 mM Mg(NO$_3$)$_2$	7.04	RES[i]		4.70				[36]
Cu^{2+}	25	10 mM NaNO$_3$	4.0	RES[i]	3.82					[36]
Mn	25	10 mM NaNO$_3$, 4 mM Mg(NO$_3$)$_2$	7.04	RES[i]		4.07			7.12	[36]
Mn	25	10 mM NaNO$_3$	4.0	RES[i]	3.52					[36]

(*cont.*)

Table 3.2 (*cont.*)

Analyte	$T°C$	Medium	pH	Gel	D^g DC	D^g DGT	Ligand	D^g M–L	D^w	Ref
Mo(VI)	25	10 mM $NaNO_3$, 4 mM $Mg(NO_3)_2$	7.04	RES^i		4.40			*9.91*	[36]
Mo(VI)	25	10 mM $NaNO_3$	4.0	RES^i	3.96					[36]
Ni^{2+}	25	100 mM $NaNO_3$	5.6?	RES	3.87				7.05	[3]
Ni^{2+}	25	10 mM $NaNO_3$	5.6?	RES	4.28					[3]
Ni^{2+}	25	1 mM $NaNO_3$	5.6?	RES	4.01					[3]
Ni^{2+}	25	10 mM $NaNO_3$, 4 mM $Mg(NO_3)_2$	7.04	RES^i		4.69				[36]
Ni^{2+}	25	10 mM $NaNO_3$	4.0	RES^i	3.68					[36]
Pb^{2+}	25	100 mM $NaNO_3$	5.6?	RES	5.39				9.45	[3]
Pb^{2+}	25	10 mM $NaNO_3$	5.6?	RES	6.14		DGA	3.35		[3]
Pb^{2+}	25	10 mM $NaNO_3$	5.6?	RES	6.14		NTA	2.96	7.3	[3]
Pb^{2+}	25	10 mM $NaNO_3$	5.6?	RES	6.14		AFA	0.52	2.7^f	[3]
Pb^{2+}	25	10 mM $NaNO_3$	5.6?	RES	6.14		AHA	0.18	2.4^f	[3]
Pb^{2+}	25	1 mM $NaNO_3$	5.6?	RES	5.58					[3]
Pb^{2+}	25	10 mM $NaNO_3$, 4 mM $Mg(NO_3)_2$	7.04	RES^i		5.89				[36]
Pb^{2+}	25	10 mM $NaNO_3$	4.0	RES^i	4.99					[36]
Sb(V)	25	10 mM $NaNO_3$, 4 mM $Mg(NO_3)_2$	7.04	RES^i		4.48			8.25	[36]
Sb(V)	25	10 mM $NaNO_3$	4.0	RES^i	4.14					[36]
W(VI)	25	10 mM $NaNO_3$, 4 mM $Mg(NO_3)_2$	7.04	RES^i		4.21			9.23	[36]
W(VI)	25	10 mM $NaNO_3$	4.0	RES^i	2.98					[36]
Cd^{2+}	20	100 mM $NaNO_3$	6	AGE	4.74				7.19	[2]
Cd^{2+}	20	10 mM $NaNO_3$	6	AGE	4.62					[2]
Cd^{2+}	20	1 mM $NaNO_3$	6	AGE	2.53					[2]
Cd^{2+}	20	0.1 mM $NaNO_3$	6	AGE	0.99					[2]
Cu^{2+}	20	100 mM $NaNO_3$	6	AGE	5.42				7.14	[2]
Cu^{2+}	20	10 mM $NaNO_3$	6	AGE	4.59					[2]
Cu^{2+}	20	1 mM $NaNO_3$	6	AGE	4.34					[2]
Cu^{2+}	20	0.1 mM $NaNO_3$	6	AGE	2.30					[2]

Hg^{2+}	25	10 mM NaNO$_3$	6?	AGE		9.07	8.47	[50]
Hg^{2+}	21.5	10 mM NaNO$_3$	6?	AGE[h]		4.41		[45]
Hg^{2+}	21.5	10 mM NaNO$_3$	6?	AGE[h]		3.86		[45]
Hg^{2+}	20	10 mM NaNO$_3$	?	AGE		8.44		[51]
CH$_3$Hg$^+$	25	10 mM NaNO$_3$	6?	AGE		9.06		[50]
CH$_3$Hg$^+$	20	30 mM NaNO$_3$	?	AGE		5.1		[52]
C$_2$H$_5$Hg$^+$	25	10 mM NaNO$_3$	6?	AGE		6.87		[50]
C$_6$H$_5$Hg$^+$	25	10 mM NaNO$_3$	6?	AGE		3.86		[50]
Pb^{2+}	20	100 mM NaNO$_3$	6	AGE	4.03		9.45	[2]
Pb^{2+}	20	10 mM NaNO$_3$	6	AGE	3.12			[2]
Pb^{2+}	20	1 mM NaNO$_3$	6	AGE	1.48			[2]
Pb^{2+}	20	0.1 mM NaNO$_3$	6	AGE	0.25			[2]
Tl$^+$	20	100 mM NaNO$_3$	6	AGE	9.63		19.9	[2]
Tl$^+$	20	10 mM NaNO$_3$	6	AGE	10.2			[2]
Tl$^+$	20	1 mM NaNO$_3$	6	AGE	7.09			[2]
Tl$^+$	20	0.1 mM NaNO$_3$	6	AGE	4.11			[2]
AHA	20	100 mM NaCl	6	AGE	1.89[c]			[2]
PHA	20	100 mM NaCl	6	AGE	1.93[c]			[2]

[a] The cross-linker for these earlier measurements may have produced a more open gel (higher D than later, see [3].

[b] The medium for measurement of the lanthanoid complexes with quin2 was 100 mM KNO$_3$, 20 mM MOPS buffer at pH 7.

[c] Measured using photo correlation spectroscopy.

[d] A 0.82-mm diffusive gel plus a 0.14-mm PES membrane filter was used for DGT and DC.

[e] D^w is higher for HPO$_4^{2-}$ (7.6) and H$_2$PO$_4^-$ (9.6).

[f] These are estimates for pH 6 at I = 5 mM. See [3] for pH and ionic strength dependence.

[g] A 0.4-mm diffusive gel plus a 0.1-mm nylon filter membrane.

[h] A 0.5-mm diffusive gel plus a 0.1-mm nylon filter membrane.

[i] A 0.8-mm diffusive gel plus a 0.1-mm cellulose nitrate filter membrane.

[j] A 0.8-mm diffusive gel plus a 0.14-mm PES filter membrane.

synthetic solutions with known analyte concentrations. The measured mass divided by the known concentration of analyte then gives the slope and equation 3.1 is used to derive D^{mbl}. This requires knowledge of A_E, t, D^w, δ^{mdl} and δ^{dbl}. As early (and some more recent) applications of this approach did not always recognise the need for including the DBL and A_E, rather than the physical area of the exposure window, A_p, the values of D^{mbl} obtained may be approximate. When 0.8-mm thick diffusive gels were used, the errors due to using A_p and disregarding the DBL are likely to have cancelled, as discussed in Chapter 2. Using DGT to derive diffusion coefficients can only provide an aggregated value for the gel and filter membrane together. However, this is not a problem when the APA gel is used with a PES membrane because diffusion coefficients of simple cations have been shown to be indistinguishable in the two materials [3, 12]. For example, DGT measurements using only the PES membrane for the diffusion layer provided consistent results using the established value of D^g for the APA gel [54]. Quite early in the development of DGT, Garmo et al. [55] used this approach to obtain diffusion coefficients for trace metals and the lanthanoids in a 0.8-mm APA gel plus a 0.12-mm cellulose nitrate filter membrane.

Ideally diffusions coefficients in the materials of the diffusion layer should be measured independently of the DGT measurement. Then, if subsequent measurements of c_{DGT} or c_{DGTe} are made in simple solutions using the independently determined values of either D^g and D^f or D^{mdl}, agreement with the known c^{soln} provides a test of both DGT and the measurements of diffusion coefficients.

Simple diaphragm diffusion cells have been used for most independent measurements of diffusion coefficients in the diffusion layer materials commonly used in DGT [3, 56]. Two compartments, which can be vigorously stirred, are separated by the material (usually gel, but sometimes gel and membrane filter) in which diffusion is being measured. The same volume of carrier solution containing an electrolyte is introduced into both compartments, but for the source compartment the carrier solution contains a low concentration of the analyte under investigation. Sub-samples, representing a very small fraction of the total volume, are taken from both compartments at time intervals (typically 5 minutes for simple divalent metal ions, longer for larger complexes) and analysed to obtain the mass of analyte in each compartment.

Calculation of the diffusion coefficient is based on Fick's Law: the flux, J, is given by the product of the diffusion coefficient and the concentration gradient (equation 3.3).

$$J = M/A_p t = D^g c/\delta^g \tag{3.3}$$

Here c should represent the concentration difference, between source and receiving solutions. However, if the duration of the experiment is such that there is minimal depletion of analyte in the source solution, c is effectively constant and equal to the concentration in the source solution. The slope of a plot of the mass accumulated in the receiving solution against time can be used with equation 3.4 to calculate the diffusion coefficient in the gel.

$$D^g = \text{slope} \times \delta^g/A_p c \tag{3.4}$$

Such experiments should be undertaken at constant temperature and there needs to be verification that the combined thickness of the DBLs in the system is negligible. Scally et al. [3] did this by establishing that the same value of D^g was obtained when different thicknesses of gel were used. They also found that diffusion coefficients of metal ions were indistinguishable when measurements were made with a 0.8-mm thick APA gel with and without the presence of a 0.14-mm thick 0.45-μm polysulphone filter membrane. A similar experiment found no difference in the diffusion coefficient of Cs within a 0.8-mm APA gel and a 0.14-mm cellulose acetate membrane filter [12]. These results verify that it is reasonable to set $D^f = D^g$ in a DGT experiment.

Whereas it is acceptable to assume $D^f = D^g$ when the common APA gel is used, the difference should ideally be taken into account when other gels are used. That is the value of D^f should be taken to be the value of D^g in an APA gel. However, with a 0.8-mm thick restricted gel, the error in assuming that the diffusion coefficient of a simple solute in the filter is the same as that in the RES gel would typically only be 5% [3].

Table 3.2 presents values of diffusion coefficients for a range of analytes. The list is not exhaustive, with values being selected on the basis of their consensual acceptance if there is an abundance of data. Values are shown at the measurement temperature unless they were corrected to 25°C by the authors, in which case this value is shown.

The diffusion coefficients of Cd, Cu, Ni and Pb in the APA gel are, to a good approximation 85%, of their values in water [3], which has prompted the suggestion that this percentage will apply to other divalent cations [57]. The tabulated values, based on this premise, provided by DGT Research Limited of a wide range of cations, have been widely validated by the test deployments of DGT in known solutions providing the predicted answer. Recent measurements of diffusion coefficients have also been in broad agreement [36]. The Appendix of this book lists parameters for simple quadratic equations that can be used to obtain diffusion coefficients of a range of ions at any temperature. Cations diffuse predictably more slowly in the restricted gel, with D^g typically having a value 0.62–0.68 [3] or 0.62–0.72 [36] of the value in the APA gel [3]. The diffusion of oxyanions in gels is generally more retarded than cations, with diffusion coefficients reported to be on average 62% and 49% of the value in water for APA and RES gels, respectively [36].

Values for some analytes remain contentious. For example, a research group of the Czech Republic [50] have consistently reported higher values of D for Hg ions and compounds than a group based in Canada (Table 3.2) [45, 46]. Careful diffusion cell measurements are needed to resolve this issue. However, caution needs to be exercised in the use of diffusion cells or DGT to determine D. Cheng [16] found that reported systematic [58, 59] and erratic [60] pH dependencies of the diffusion coefficient of U were most likely due to adsorption of U to the vessels. After taking precautions to prevent adsorption, his DGT measurements agreed with diffusion cell measurements and were independent of pH in the tested range of 3.6–7.2.

It is important when comparing measurements to recognise any differences in solution conditions, particularly pH. Even considering such effects, several papers have reported systematic differences in the values of D^g measured using a diffusion cell and DGT [36, 41].

More confidence can be placed in values where the two approaches agree as shown recently for oxyanions by Price et al. [38] and earlier for cations by Scally et al. [3] whose diffusion cell measurements agreed very well (within a few %) with the DGT measurements of Warnken et al. [10, 54]. Where there is a discrepancy further investigative experiments should be undertaken to establish its cause. Failure to estimate δ^{dbl} in either diffusion cell or DGT, and to account for the effective area of DGT, will introduce an error. It is tempting to simply accept the value obtained by the DGT experiment. While this is likely to provide the most appropriate calibration for DGT, it is important to recognise that, where there is doubt, it can no longer be claimed to be a diffusion coefficient.

Different oxidation states have different diffusion coefficients, as shown in Table 3.2 for As. The inclusion of several datasets for As(V) illustrates that it is often difficult to arrive at a consensus value.

Whereas complexation of metal ions by simple inorganic species is unlikely to affect significantly their diffusion coefficients, they are lowered by complexation with larger organic molecules. Diffusion coefficients of Pb complexed by nitrilotriacetic acid and diglycolic acid were 70–75% in the APA gel and 45–55% in the restricted gel of the value of uncomplexed Pb [3]. Complexation by fulvic acid (FA) and humic acid (HA) have greater effects. Diffusion coefficients in the APA gel of lead complexed with FA and HA extracted from the Suwannee River were about 20% and 8%, respectively, of the value of the uncomplexed metal [3]. The proportional decrease in the restricted gel was much greater, with values of 6% and 2%, respectively.

Diffusion coefficients of organic compounds, measured within the structure of agarose gels at microscopic scales using photocorrelation spectroscopy (PCS) agree with macroscale measurements made using a diffusion cell [61, 62]. Labille et al. [62] were able to show that there is a $\sim$100-μm thick layer at the surface of the agarose gel, which has a lower porosity than the bulk and where diffusion coefficients of solutes are lower. The flux through the whole gel of simple solutes is unlikely to be significantly affected by this effect for gel thicknesses normally used in DGT ($\geq$0.4 mm), but there may be an effect for larger molecules, with substantially lower diffusion coefficients. Measurements of simple solutes using 0.16-mm APA diffusive gels were consistent with simple theory using macroscopic diffusion coefficients [54], suggesting that similar surface differences may not occur in polyacrylamide gels, because of their different structures and preparation procedures. Nonetheless it would be good to have PCS measurements to confirm this.

3.4.3 Effect of Gel Charge on Transport Through the Material Diffusion Layer

The transport of ions through gels and other materials is affected when they are charged. The principles relevant to DGT were described by Yezek and van Leeuwen [63, 64], but Fatin-Rouge et al. [2] had earlier considered quantitatively how specific and electrostatic binding affects diffusion in agarose gels. Conceptually the effect on the diffusion of cations through a gel is quite simple. If a positively charged gel is in contact with a solution, cations will

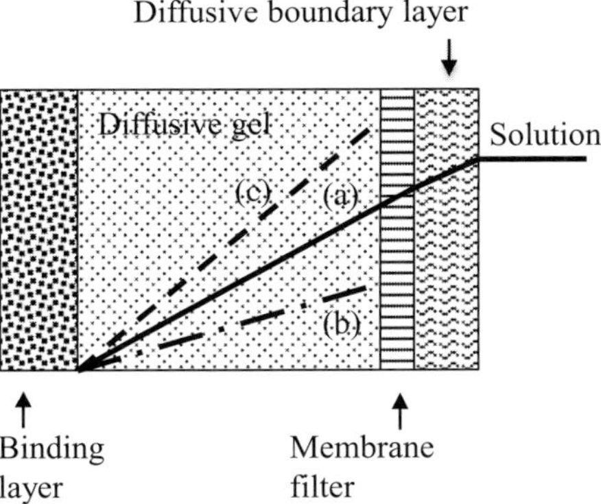

Figure 3.4 Schematic diagram of the concentration gradient of a cation at steady state through a DGT device when the diffusive gel (a) has no charge, (b) is positively charged, (c) is negatively charged. The filter membrane is assumed to be uncharged.

become electrostatically diminished at the surface of the gel. Their concentrations on either side of the gel-solution interface are then unequal. Figure 3.4 illustrates the consequence of this on the concentration gradient through a DGT device, where the filter membrane is assumed to have no charge, so that the gel–solution interface is effectively that between the gel and the porewater of the membrane filter. For a positively charged diffusive gel, the gradient of the cation concentration (b) through the gel is lower than it would be for an uncharged gel (a). Consequently the flux to the binding layer is diminished, causing c_{DGT} to be less than it would be if the gel was uncharged. If a gel is negatively charged, cations are associated with the gel surface, causing the cation concentration to be elevated on the gel side of the gel–solution interface. The concentration gradient ((c) in Figure 3.4) is then elevated through the gel, causing c_{DGT} to be higher than it would be for an uncharged gel. When measuring diffusion coefficients with a diffusion cell or DGT using their simple theories for uncharged systems these charge effects will appear to change the diffusion coefficient. However, it is the concentration gradient rather than the diffusion coefficient that is affected.

Quantification of this effect is based on the charge creating a Donnan potential, ψ, between the gel and solution. If the charge density of the gel, ρ^{g}, is measured or estimated, as in Yesek and van Leeuwen [25] or Fatin Rouge et al. [2], ψ can be calculated using equation 3.5. It provides directly an estimate of the Donnan partition factor, Π, which is simply the ratio of $c^{\mathrm{g}}/c^{\mathrm{soln}}$ at the gel–solution interface (equation 3.6).

$$\psi = \left(\frac{RT^{\mathrm{K}}}{z_{\mathrm{SE}}F}\right) \operatorname{arsinh}\left(\frac{\rho^{\mathrm{g}}}{2z_{\mathrm{SE}}Fc}\right) \tag{3.5}$$

$$\frac{c^{\mathrm{g}}}{c^{\mathrm{soln}}} = \Pi = e^{-z_{\mathrm{C}}F\psi/RT^{\mathrm{K}}} \tag{3.6}$$

The valence of the cation is z_C and of the supporting electrolyte, of concentration, c, is z_{SE}. The Faraday constant is F, R is the gas constant and T^K is the absolute temperature.

3.5 Lessons for DGT Measurements

This chapter has explained what is known about the diffusive gels commonly used in DGT in terms of specific binding of analytes and ligands, charge-induced binding, diffusion of solutes and charge effects on transport. These properties are not unique to gels and are likely to apply to any material used for the MDL, as demonstrated by the slight negative charge of the PES membrane (Figure 3.4) and the more general specific binding of trace metals to filter membranes [5, 28]. In most situations, these effects do not impair measurements made using DGT, but we are fortunate in having this binding information so that we can assess if, and under what circumstances, it should be considered.

3.5.1 Effects of Binding to the Gel

Where there is no binding of analyte to the components of the MDL, the initial transient time before steady state is reached is dependent on the diffusion coefficient of the analyte, as demonstrated in Chapter 2. However, if there is binding of the analyte to the gel, it takes longer to reach steady state, due to the analyte progressively accumulating on the gel. Once the gel binding sites are filled, the steady state flux through the solution phase of the gel is unaffected. Therefore, longer deployment times are required to ensure that, to a good approximation, the simple steady state equations for DGT are applicable. The effect is particularly marked at low concentrations of analyte because diffusional supply to the finite set of binding sites is slower, due to the lower diffusion gradient.

For the APA gel, the most significant binding for trace metals occurs for Cu and Pb (Figure 3.3). Table 3.3, adapted from Garmo et al. [65], provides a simulation of c_{DGT}/c_{soln}, based on the measured binding capacity of the APA gel for Cu. It shows that for strong binding metals, such as Cu and Pb, there are advantages in using a thinner diffusive gel layer. For $\delta^g = 0.4$ mm, errors should be negligible for deployments ≥ 24 h, irrespective of the concentration of metal. Generally, in systems contaminated with metals there should be no errors associated with specific binding. For most metals other than Cu and Pb, the very limited specific binding (Figure 3.3) should not affect DGT deployments in excess of a day at any environmental concentration.

An exception may be Hg, but more data are needed to establish whether this is the case. According to Docekalova and Divis [66], Hg^{2+} binds strongly to the APA gel, with $c^{gel}/c^{soln} = 700$, but Fenandez-Gomeza et al. [45] found no evidence of binding. Conflicting results have also been reported for the binding of Hg to agarose [45]. However, rather surprisingly, both types of gel have been successfully used in DGT for the measurement of Hg ions and compounds [45, 46, 50].

The small but similar values of c^f/c^{soln} for Cd, Co and Ni in the PES membrane exposed to synthetic freshwater (Figure 3.3), suggest it has a negative charge, with these ions

Table 3.3 *Simulated values of c_{DGT}/c_{soln} for DGT measurements of Cu using an APA diffusive gel of three different thicknesses, δ^g, deployed for various times, t, in four solutions containing different concentrations of Cu. The data are taken from Garmo et al. [65], who used $D^g = 5.5 \times 10^{-6}$ cm^2 s^{-1} and obtained the concentration of binding sites on the gel by fitting experimental data. Column N provides simulated c_{DGT}/c_{soln} when there is no binding of Cu to the gel. Adapted with permission from Ref. [65]. Copyright (2010) American Chemical Society.*

δ^g (mm)	t (h)	0.1 nM	1 nM	10 nM	100 nM[a]	N
0.5	1	0.37	0.39	0.49	0.81	0.98
0.5	2	0.61	0.63	0.71	0.90	0.99
0.5	4	0.79	0.80	0.85	0.95	0.99
0.5	8	0.89	0.90	0.93	0.98	1.00
0.5	16	0.94	0.95	0.96	0.99	1.00
0.5	32	0.97	0.97	0.98	0.99	1.00
0.5	64	0.99	0.99	0.99	1.00	1.00
0.5	128	0.99	0.99	1.00	1.00	1.00
0.5	256	1.00	1.00	1.00	1.00	1.00
1.0	1	0.05	0.05	0.06	0.32	0.92
1.0	2	0.17	0.18	0.23	0.61	0.96
1.0	4	0.38	0.40	0.49	0.81	0.98
1.0	8	0.61	0.63	0.71	0.90	0.99
1.0	16	0.79	0.80	0.85	0.95	0.99
1.0	32	0.89	0.90	0.93	0.98	1.00
1.0	64	0.95	0.95	0.96	0.99	1.00
1.0	128	0.97	0.97	0.98	0.99	1.00
1.0	256	0.99	0.99	0.99	1.00	1.00
2.0	1	0.00	0.00	0.00	0.01	0.69
2.0	2	0.01	0.01	0.01	0.08	0.83
2.0	4	0.05	0.05	0.06	0.32	0.92
2.0	8	0.17	0.18	0.23	0.61	0.96
2.0	16	0.38	0.40	0.49	0.81	0.98
2.0	32	0.61	0.63	0.71	0.90	0.99
2.0	64	0.79	0.80	0.85	0.95	0.99
2.0	128	0.89	0.90	0.93	0.98	1.00
2.0	256	0.95	0.95	0.96	0.99	1.00

binding electrostatically. The absence of binding at high ionic strength (seawater) supports this contention. There also appears to be some specific binding of Cu and Pb. As PES membranes are only 0.14 mm thick, the effect of this binding on DGT measurements is likely to be negligible compared to the effect of binding to the diffusive gel.

Like the membrane filter, agarose can bind trace metals both electrostatically and specifically. As the binding is quite weak, it is unlikely to affect appreciably measurements made using DGT equipped with agarose gel, but for many analytes there are no definitive data. Binding of Pb to the restricted gel is similar to that observed for the APA gel, while binding of Cu is enhanced, especially in the presence of fulvic acid (Figure 3.3). This suggests that, for DGT equipped with the restricted gel, deployment times for the measurement of Cu should be even longer than those derived for the APA gel from inspection of Table 3.3. To our knowledge, no direct test of this requirement is available.

As there is no evidence for binding of simple inorganic ligands, such as chloride, sulphate and phosphate (Table 3.1), they should not affect DGT measurements. Interpretation of DGT measurements (APA gel) in the presence of simple organic ligands, including glycine, NTA, EDTA and DPTA have not needed to invoke binding effects [42, 67, 68]. Measurement of polar organic compounds by DGT is in its infancy, but for the antibiotics and bisphenols studied to date it appears that agarose is a more appropriate gel than APA, although evidence for binding of these compounds to the APA gel is not clearly demonstrated [30–32]. The need to use polyfluoroethylene rather than PES filters for measurement of bisphenols is a good illustration of the importance of testing during method development [32].

As trace metals are bound to varying extents by dissolved organic matter (DOM) in natural waters, it is important to know (a) whether DOM binds to the diffusive gel and (b) how this might affect DGT measurements of trace metals. There is little evidence that naturally occurring DOM binds appreciably to the APA gel, as DOM in surface waters and the porewaters of soils was not measurably adsorbed [19] and fulvic and humic acid extracted from natural waters bind only slightly ($c^{gel}/c^{soln} = 1$–3) [18, 19]. The more pronounced binding of humic acid extracted from humic-rich soils ($c^{gel}/c^{soln} = 11$–53) [18, 19], including its selective accumulation in a 15-μm thick surface gel layer [20, 21], suggests that it is mainly the soil bound humic substances, which are released by the extraction process, that bind to the gel. As DGT devices deployed in situ in soils only respond to the solution-phase humic substances, the measurements are unlikely to be affected, as borne out by the many sensible interpretations of DGT measurements in soils which have not invoked binding of humic substances to the gel [69, 70]. However, an experiment that pre-accumulated DOM onto diffusive gels before their use in DGT demonstrated that this pre-exposure could attenuate by up to 50% the mass of Cu accumulated [19]. Cu measurements were unaffected when devices were exposed to waters with low DOM, and Cd measurements were unaffected even after exposure to high DOM waters. As deployment times were only 24 h, the authors considered that there must be a low level of bound DOM that was interacting with Cu, which probably extended the time required to reach steady state. Yasadi et al.'s [24] micro electrode measurements of Cd within the gel confirmed that the equilibrium concentration of Cd^{2+} is unaffected by the binding to the gel or bound humic acid, suggesting that the DGT measurement will be unaffected by these processes provided deployment times are sufficiently long. As van Leeuwen [71] has suggested in an assessment of the effect of humic acid binding on DGT measurements, it is likely that deployments of several days will be needed for accurate measurement of Cu and Pb, but experimental verification is needed. Figure 3.3 shows some evidence for fulvic acid enhancing the binding of metals to the RES gel more than it enhances the binding of metals to the APA gel, especially for Cu and Pb. Although the RES gel has been successfully used in speciation studies involving DOM [72], direct assessment of possible binding affects on DGT measurements is required to establish whether there is a problem.

3.5.2 Charge and Transport

As shown earlier with Figure 3.4, gel charge can modify the interfacial concentration and hence the concentration gradient through the gel. If this effect is substantial, the simple DGT equation is no longer valid because the concentration in solution or, more accurately, within the PES membrane, no longer represents the concentration within the surface of the gel. As cations are elevated within the surface of a negatively charged gel, the concentration gradient is elevated, and the accumulated mass is greater than expected by application of the simple theory. Before the effect of gel charge was fully appreciated, some publications reported that DGT, equipped with an APA gel [73] or a bis cross-linked polyacrylamide gel [26], overestimated concentrations of metals in waters with low ionic strength. We now know that the APA gel has a negative charge if it is not washed thoroughly [10], which may have been the case in this early work. If the gel is exhaustively washed, it can in fact have a slight positive charge. This does not affect DGT measurements in solutions with ionic strengths ≥ 1 mmol L^{-1}, but at $I = 0.1$ mmol L^{-1} c_{DGT}/c^{soln} for Cu and Cd was 0.5 [10]. This Donnan effect is also apparent in diffusion cell measurements, which gave a 50% lower value of D at $I = 0.1$ mmol L^{-1} (Table 3.2) [3].

In summary, for most natural waters, these charge effects should not affect DGT measurements using APA gels provided they have been thoroughly washed. As there is little evidence for any appreciable charge associated with the RES gel (see, e.g. Figure 3.3), it is unlikely that there will be any charge effects, but experiments that directly verify this are lacking.

The general accumulation of cations within both agarose and the PES membrane (Figure 3.3) is due to them having a small negative charge [2, 29]. According to the mechanism of Yesek and van Leeuwen [63, 64] (Figure 3.4), the Donnan potential of 2–3 mV at $I = 1$ mmol L^{-1} and 20–25 mV at $I = 0.1$ mmol L^{-1} would be expected to imperceptibly affect the DGT accumulated mass at $I = 1$ mmol L^{-1}, while approximately doubling it for cations at $I = 0.1$ mmol L^{-1}. Indirect estimates of the effective diffusion coefficient for transition metals indicated the opposite effect, but the authors acknowledged potential problems in the estimations [2]. Again, it is clear that more data are needed on the ability of DGT equipped with agarose gel to measure ions accurately at very low ionic strengths.

The above discussion of the effects of ionic strength on DGT measurements has been restricted to simple solutions where the ionic analyte of interest reacts rapidly with the binding layer without penetration. Uncharged analytes, like many organic compounds, are unaffected by any charge on the diffusive gel and therefore their measurement should be unaffected by ionic strength. If metal ions are present in complexed form, charge-related ionic strength effects may occur due to the penetration of partially labile charged species into the binding layer which, for Chelex, carries a substantial negative charge [23]. This effect is dealt with in more detail in Chapter 5.

3.5.3 Will Colloids and Nanoparticles be Measured by DGT?

The size of analyte that can be measured by DGT has yet to be established unequivocally, as already discussed in some detail [6]. There are four strands of evidence that contribute to our current knowledge and understanding:

(1) For DGT to be sufficiently sensitive to an analyte for it to be measured satisfactorily it must diffuse reasonably quickly. When using DGT to measure a metal in natural water, the value of D^g for the free metal ion in an APA gel would usually be used in the DGT equation. If the metal ion is attached to a larger particle, which can diffuse and release the metal, so that it is measured by DGT, the lower diffusion coefficient will effectively reduce the sensitivity of the measurement. The sensitivity of DGT to a metal complexed with humic acid extracted from peat, with a diameter of less than 4 nm, is <6% of the sensitivity to uncomplexed metal ion because of the low diffusion coefficient of the metal bound to humic acid (Figure 3.5). It is clear from Figure 3.5 that, due to even slower diffusion, the sensitivity to metal attached to a 5-nm particle will be at least a factor of 2 or 3 lower than this. In effect colloidal or nanoparticulate metal with a diameter greater than 5 nm will not be measured by DGT because its diffusion is too slow compared to that of any free metal that may be present.

(2) Careful ultra-filtration and in situ DGT measurements of metals in the Baltic Sea suggested that the DGT metal corresponded to the <1-kDa fraction [74] (see Chapter 6]. However, in freshwater contaminated by water draining mine tailings, more metal was measured by DGT than by ultrafiltration (<1 kDa) [75]. In such a dynamic environment where iron oxyhydroxides are freshly precipitating, there are likely to be small colloids, which straddle the 1-kDa size range, whereas such particle sizes may have been removed by aggregation in the relatively stable waters of the Baltic. Similar information was obtained for soils [76]. Slightly more of the radio-labelled P present in soil water extracts was measured by DGT in samples that were not ultrafiltered (3 kDa). It was speculated that the colloids were humic/fulvic-Fe/Al-P complexes. When zinc oxide, carbon-coated copper and copper oxide particles were added to four different fresh waters and a wastewater, the concentrations measured by DGT were very similar to the ultrafiltered fraction (3 kDa) [77]. Collectively these data indicate that the DGT measurement can include particles of up to 5–10 kDa, indicating diameters approaching 3–4 nm.

(3) It might be expected that DGT measurements of synthetic nanoparticles would provide definitive information, but this is not the case. Santner et al. [78] observed that a small fraction of the P attached to nanoparticles of Al_2O_3 with an average size of 40 nm was taken up by DGT. According to the reasoning of (1) above, only the smallest particles in this size range would be measured. Odzak et al. [79] examined the dissolution of nanoparticles with two nominal sizes (nm) of ZnO (4.5, 27), carbon-coated Cu (7.5, 25)

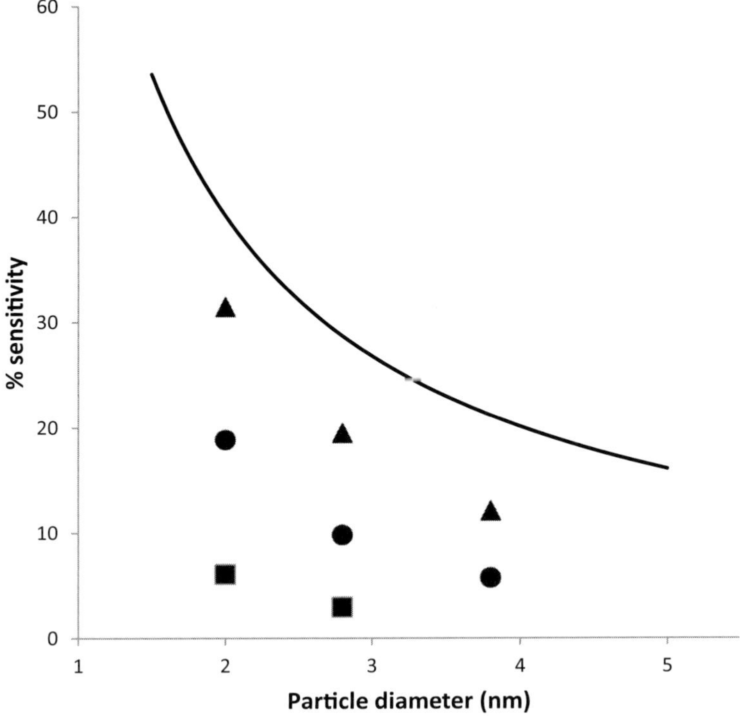

Figure 3.5 Dependence of the sensitivity of the DGT measurement on particle size of the analyte. Percentage sensitivity was calculated by comparing diffusion coefficients of the particles in water or gel to the diffusion coefficients of free Ni or Mn ions in water. Symbols refer to diffusion in water (solid line), in agarose gel (triangles), in APA gel (circles) and RES gel (squares). Data are based on the diffusion coefficients and molecular weights of fulvic acid and humic acid extracted from both water and peat [3, 53, 56].

and CuO (7.5, 45) in synthetic solutions resembling natural waters. They used DGT, ultrafiltration (3 kDa) and dialysis (100–500 molecular weight cut off membrane) to measure the 'dissolved' fraction. In all cases the DGT and ultrafiltration measurement agreed well, but dialysis gave lower values, which was attributed to adsorption of nanoparticles slowing the approach to equilibrium. There was some evidence that labile Pb bound to latex spheres with diameters of 81 nm could contribute to the DGT measurement [80]. The sensitivity was more than 2 orders of magnitude lower than that of the free ion, which is reasonable for this diameter.

(4) Although there are no reliable measurements of the pore size of the APA gel, information exists for the more open structured agarose [61]. Fluorescence correlation spectroscopy of defined particles or molecules spanning a range of sizes indicated that agarose had a mean pore diameter of 74 nm and a more open structure in a 100-μm thick surface layer, but, because of the distribution of pore sizes, only particles <60 nm were thought to be able to diffuse through the gel [61, 62].

With the evidence so far, it is sensible to consider that for most natural waters, DGT will only respond to solution species and very small colloids or nanoparticles (<5 nm). Its practical use as a tool to define the true solution component is therefore a valid approach in many situations [81]. Pouran et al. [82] established that ZnO nanoparticles that were measured by DGT with an APA diffusive gel were not be measured if a 1kDa dialysis membrane was inserted between the diffusive gel and membrane filter. They suggested that these two measurements could be used together to provide estimates of concentrations of nanoparticles, which is certainly an appealing prospect. However, as, according to the above, only nanoparticles less than 5 nm would be measured the practical use of such a procedure is likely to be limited. Moreover, such measurements assume that metals can be readily released from nanoparticles, which may not always be the case (see Chapter 5).

References

1. T. Tanaka, Gels, *Scient. Amer.* 244 (1981), 124–138.
2. N. Fatin-Rouge, A. Milon, J. Buffle, R. R. Goulet and A. Tessier, Diffusion and partitioning of solutes in agarose hydrogels: The relative influence of electrostatic and specific interactions, *J. Phys. Chem. B* 107 (2003), 12126–12137.
3. S. Scally, W. Davison and H. Zhang, Diffusion coefficients of metals and metal complexes in hydrogels used in diffusive gradients in thin films, *Anal. Chim. Acta* 558 (2006), 222–229.
4. H. Zhang and W. Davison, Direct in situ measurements of labile inorganic and organically bound metal species in synthetic solutions and natural waters using diffusive gradients in thin films, *Anal. Chem.* 72 (2000), 4447–4457.
5. H. Brandl and K. W. Hanselmann, Evaluation and application of dialysis porewater samplers for microbiological studies at sediment-water interfaces, *Aquatic Sci.* 53 (1991), 55–73.
6. W. Davison and H. Zhang, Progress in understanding the use of diffusive gradients in thin-films – Back to basics, *Environ. Chem.* 9 (2012), 1–13.
7. M. P. Harper, W. Davison and W. Tych, Temporal, spatial, and resolution constraints for in situ sampling devices using diffusional equilibration: dialysis and DET, *Environ. Sci. Technol.* 31 (1997), 3110–3119.
8. R. J. G. Mortimer, M. D. Krom, P. O. J. Hall, S. Hall and H. Stahl, Use of gel probes for the determination of high resolution solute distributions in marine and estuarine pore waters, *Mar. Chem.* 63 (1998), 119–129.
9. H. Zhang, W. Davison and C. Ottley, Remobilisation of major ions in freshly deposited lacustrine sediment at overturn, *Aquat. Sci.* 61 (1999), 354–361.
10. K. W. Warnken. H. Zhang and W. Davison, Trace metal measurements in low ionic strength synthetic solutions by diffusive gradients in thin films, *Anal. Chem.* 77 (2005), 5440–5446.
11. O. A. Garmo, W. Davison and H. Zhang, Interactions of trace metals with hydrogels and filter membranes used in DET and DGT techniques, *Environ. Sci. Technol.* 42 (2008), 5682–5687.
12. L. Chang, W. Davison, H. Zhang and M. Kelly, Performance characteristics for the measurement of Cs and Sr by diffusive gradients in thin films (DGT), *Anal. Chim. Acta* 368 (1998), 243–253.

13. H. Ernstberger, H. Zhang and W. Davison, Determination of chromium speciation in natural systems using DGT, *Anal. Bioanal. Chem.* 373 (2002), 873–879.

14. W. Davison, H. Zhang and G. W. Grime, Performance characteristics of gel probes used for measuring the chemistry of pore waters, *Environ. Sci. Technol.* 28 (1994), 1623–1632.

15. G. R. Fones, W. Davison and G. W. Grime, Development of constrained DET for measurements of dissolved iron in surface sediments at sub-mm resolution, *Sci. Tot. Environ.* 221 (1998), 127–137.

16. Cheng, H. Investigation of diffusion and binding properties for extending applications of the DGT technique, PhD Thesis, University of Lancaster, UK, 2014.

17. M. D. Krom, P. Davison, H. Zhang and W. Davison, High resolution pore water sampling using a gel sampler, *Limnol. Oceanogr.* 39 (1994), 1967–1972.

18. L. R. van der Veeken, P. Chakraborti and H. P. van Leeuwen, Accumulation of humic acid in DET/DGT gels, *Environ. Sci. Technol.* 44 (2010), 4253–4257.

19. W. Davison, C. Lin, Y. Gao and H. Zhang. Effect of gel interactions with dissolved organic matter on DGT measurements of trace metals, *Aquatic Geochem.* 21 (2015), 281–293

20. K. Zielinska, R. M. Town, K. Yasadi and H. P. van Leeuwen, Partitioning of humic acids between aqueous solution and hydrogel: concentration profiling of humic acids in hydrogel phases, *Langmuir* 30 (2014), 2084–2092.

21. K. Zielinska, R. M. Town, K. Yasadi and H. P. van Leeuwen, Partitioning of humic acids between aqueous solution and hydrogel. 2. Impact of physicochemical conditions, *Langmuir* 31 (2015), 283–291.

22. H. Docekalova, O. Clarisse, S. Salomon and M. Wartel, Use of constrained DET probe for a high-resolution determination of metals and anions distribution in the sediment pore water, *Talanta* 57 (2002), 145–155.

23. J. Puy, J. Galceran, S. Cruz-Gonzalez et al., Measurements of metals using DGT: Impact of ionic strength and kinetics of dissociation of complexes in the resin domain, *Anal. Chem.* 86 (2014), 7740–7748.

24. K. Yasadi, J. P. Pinheiro, K. Zielinska, R. M. Town and H. P. van Leeuwen, Partitioning of humic acids between aqueous solution and hydrogel. 3. Dynamic speciation analysis of free and bound humic metal complex in the gel phase, *Langmuir* 31 (2015), 1737–1745.

25. L. P. Yezek and H. P. van Leeuwen, An electrokinetic characterization of low charge density cross-linked polyacrylamide gels, *J. Coll. Int. Sci.* 278 (2004), 243–250.

26. M. R. Sangi, M. J. Halstead and K. A. Hunter, Use of the diffusion gradient thin film method to measure trace metals in fresh waters at low ionic strength, *Anal. Chim. Acta* 456 (2002), 241–251.

27. J. Morford, L. Kalnejais, W. Martin, R. Francois and I-M. Karle, Sampling marine porewaters for Mn, Fe, U, Re, and Mo: modifications on diffusional gradients in thin film gel probes, *J. Exp. Mar. Biol. Ecol.* 285 (2003), 85–103.

28. L. Weltje, W. den Hollander and H. Wolterbeek, Adsorption of metals to membrane filters in view of their speciation in nutrient solution, *Environ. Toxicol. Chem.* 22 (2003), 265–271.

29. M. Ulbricht, O. Schuster, W. Ansorge, M. Ruetering and P. Steiger, Influence of strongly anisotropic cross-section morphology of a novel polyethersulfone microfiltration membranes onto filtration performance, *Separ. Purif. Techn.* 57 (2007), 63–73.

30. C. Chen, H. Zhang and K. Jones, A novel passive water sampler for in situ sampling of antibiotics, *J. Environ. Monit.* 14 (2012), 1523–1530.

31. C-E. Chen, H. Zhang, G-G. Ying and K. C. Jones, Evidence and recommendations to support the use of a novel passive water sampler to quantify antibiotics in wastewaters, *Environ. Sci. Technol.* 47 (2003), 13587–13593.

32. J-L. Zheng, D-X Guan, J. Luo et al., Activated charcoal based diffusive gradients in thin films for in situ monitoring of bisphenols in waters, *Anal. Chem.* 87 (2015), 801–807.

33. Y. Li and S. Gregory, Diffusion of ions in seawater and in deep-sea sediments, *Geochim. Cosmochim. Acta* 38 (1974) 703–714.

34. V. K. Henry (Ed.) *CRC handbook of thermophysical and thermochemical data* (Boca Raton: CRC Press Inc., 1994).

35. J. Buffle, Z. Zhang and K. Startchev, Metal flux and dynamic speciation at (bio) interfaces. Part 1: Critical evaluation and compilation of physico-chemical parameters for complexes with simple ligands and fulvic/humic substances, *Environ. Sci. Technol.* 41 (2007), 7609–7620.

36. A. H. Shiva, P. R. Teasdale, W. W. Bennett and D. T. Welsh, A systematic determination of diffusion coefficients of trace elements in open and restricted diffusive layers used by diffusive gradients in thin film technique, *Anal. Chim. Acta* 888 (2015), 146–154.

37. J. G. Panther, K. P. Stillwell, K. J. Powell and A. J. Downard, Development and application of the diffusive gradients in thin films technique for the measurement of total dissolved inorganic arsenic in waters, *Anal. Chim. Acta* 622 (2008), 132–142.

38. H. L. Price, P. R. Teasdale and D. F. Jolley, An evaluation of ferrihydrite- and Metsorb-DGT techniques for measuring oxyanion species (As, Se, V, P): Effective capacity, competition and diffusion coefficients, *Anal. Chim. Acta* 803 (2013), 56–65.

39. J. Luo, H. Zhang, J. Santner and W. Davison, Performance characteristics of diffusive gradients in thin films equipped with a binding gel layer containing precipitated ferrihydrite for measuring arsenic(v), selenium(vi), vanadium(v), and antimony(v), *Anal. Chem.* 82 (2010), 8903–8909.

40. H. Osterlund, S. Chlot, M. Faarinen et al., Simultaneous measurements of As, Mo, Sb, V and W using a ferrihydrite diffusive gradients in thin films (DGT) device, *Anal. Chim. Acta* 682 (2010), 59–65.

41. J. G. Panther, R. R. Stewart, P. R. Teasdale et al., Titanium dioxide-based DGT for measuring dissolved As(V), V(V), Sb(V), Mo(VI) and W(VI) in water, *Talanta* 105 (2013), 80–86.

42. E. Unsworth, H. Zhang and W. Davison, Use of DGT to measure cadmium speciation in solutions with synthetic and natural ligands: Comparison with model predictions, *Environ. Sci. Technol.* 39 (2005), 624–630.

43. Y. Pan, D-X. Guan, D. Zhao et al., Novel speciation method based on diffusive gradients in thin-films for in situ measurement of Cr(VI) in aquatic systems, *Environ. Sci. Technol.* in press DOI: 10.1021/acs.est.5b03742.

44. S. Scally, W. Davison and H. Zhang, In situ measurements of dissociation kinetics and labilities of metal complexes in synthetic solutions using DGT, *Environ. Sci. Technol.* 37 (2003), 1379–1384.

45. C. Fenandez-Gomeza, B. Dimock, H. Hintelmann and S. Diez, Development of the DGT technique for Hg measurement in water: Comparison of three different types of samplers in laboratory assays, *Chemosphere* 85 (2011) 1452–1457.

46. O. Clarisse and H. Hintelmann, Measurements of dissolved methylmercury in natural waters using diffusive gradients in thin-film (DGT), *J. Environ. Monit.* 8 (2006), 1242–1247.

47. S. Mason, R. Hamon, A. Nolan, H. Zhang and W. Davison, Performance of a mixed binding layer for measuring anions and cations in a single assay using the diffusive gradients in thin films technique, *Anal. Chem.* 77 (2005), 6339–6346.

48. H. Zhang, W. Davison, R. Gade and T. Kobayashi, In situ measurement of phosphate in natural waters using DGT, *Anal. Chim. Acta* 370 (1998), 29–38.

49. Ø. A. Garmo, N. J. Lehto, H. Zhang, W. Davison, O. Røyset and E. Steinnes, Dynamic aspects of DGT as demonstrated by experiments with lanthanide complexes of a multidentate ligand, *Environ. Sci. Technol.* 40 (2006), 4754–4760.

50. P. Pelcova, H. Docekalova and A. Kleckerova, Development of the diffusive gradient in thin films technique for the measurement of labile mercury species in waters, *Anal. Chim. Acta* 819 (2014), 42–48.

51. Y. Gao, E. De Canck, M. Leermakers, W. Baeyens and P. van der Voort, Synthesized mercaptopropyl nanoporous resins in DGT probes for determining dissolved mercury concentrations, *Talanta* 87 (2011), 262–267.

52. Y. Gao, S. De Craemer and W. Baeyens, A novel method for the determination of dissolved methylmercury concentrations using diffusive gradients in thin films technique, *Talanta* 120 (2014), 470–474.

53. J. R. Lead, K. J. Wilkinson, K. Startchev and S. Canonica, Determination of diffusion coefficients of humic substances by fluorescence correlation spectroscopy: Role of solution conditions, *Environ. Sci. Technol.* 34 (2000), 1365–1369.

54. K. W. Warnken, H. Zhang and W. Davison, Accuracy of the diffusive gradient in thin-films technique: Diffusion boundary layer and effective sampling area considerations, *Anal. Chem.* 78 (2006), 3780–3787.

55. O. A. Garmo, O. Roysett, E. Steinnes and T. P. Flaten, Performance study of diffusive gradients in thin films for 55 elements, *Anal. Chem.* 75 (2003), 3573–3580.

56. H. Zhang and W. Davison, Diffusional characteristics of hydrogels used in DGT and DET techniques, *Anal. Chem. Acta* 398 (1999), 329–340.

57. K. W. Warnken, W. Davison, H. Zhang, J. Galceran and J. Puy, In situ measurements of metal complex exchange kinetics in freshwater, *Environ. Sci. Technol.* 41 (2007), 3179–3185.

58. W. Li, J. Zhao, C. Li, S. Kiser and R. J. Cornett, Speciation measurements of uranium in alkaline waters using diffusive gradients in thin films technique, *Anal. Chim. Acta* 575 (2006), 274–280.

59. G. S. C. Turner, G. A. Mills, P. R. Teasdale et al., Evaluation of the DGT technique for measuring inorganic uranium species in natural waters: Interferences, deployment time and speciation, *Anal. Chim. Acta* 739 (2012), 37–46.

60. C. M. Hutchins, J. G. Panther, P. R. Teasdale et al., Evaluation of a titanium dioxide-based DGT technique for measuring inorganic uranium species in fresh and marine waters, *Talanta* 97 (2012), 550–556.

61. N. Fatin-Rouge, K. Startchev and J. Buffle, Size effects on diffusion processes within agarose gels, *Biophysical J.* 86 (2004), 2710–2719.

62. J. Labille, N. Fatin-Rouge and J. Buffle, Local and average diffusion of nanosolutes in agarose gel: The effect of the gel/solution interface structure, *Langmuir* 23 (2007), 2083–2090.

63. L. P. Yezek and H. P. van Leeuwen, Donnan effects in the steady-state diffusion of metal ions through charged thin films, *Langmuir* 21 (2005), 10342–10347.

64. L. P. Yesek, L. R. van der Veeken and H. P. van Leeuwen, Donnan effects in metal speciation analysis by DET/DGT, *Environ. Sci. Technol.* 42 (2008), 9250–9254.

65. O. A. Garmo, W. Davison and H. Zhang, Effects of binding of metals to the hydrogel and filter membrane on the accuracy of the diffusive gradients in thin films technique, *Anal. Chem.* 80 (2008), 9220–9225.

66. H. Docekalova and P. Divis, Application of diffusive gradient in thin films technique (DGT) to the measurement of mercury in aquatic systems, *Talanta* 65 (2005), 1174–1178.

67. S. Mongin, R. Uribe, J. Puy et al., Key role of the resin layer thickness in the lability of complexes measured by DGT, *Environ. Sci. Technol.* 45 (2011), 4869–4875.

68. M. H. Tusseau-Vuillemin, R. Gilbin, E. Bakkaus and J. Garrie, Performance of diffusion gradient in thin films to evaluate the toxic fraction of copper to Daphnia magna, *Environ. Toxicol. Chem.* 23 (2004), 2154–2161.

69. C. Oporto, E. Smolders, F. Degryse, L. Verheyen and C. Vandecasteele, DGT-measured fluxes explain the chloride-enhanced cadmium uptake by plants at low but not at high Cd supply, *Plant Soil* 318 (2009), 127–135.

70. H. Ernstberger, H. Zhang, A. Tye, S. Young and W. Davison, Desorption kinetics of Cd, Zn and Ni measured in intact soils by DGT, *Environ. Sci. Technol.* 39 (2005), 1591–1597.

71. H. P. van Leeuwen, Revisited: DGT speciation analysis of metal-humic acid complexes, *Environ. Chem.* 13 (2016), 84–88.

72. K. W. Warnken, W. Davison and H. Zhang, Interpretation of in situ speciation measurements of inorganic and organically complexed trace metals in freshwater by DGT, *Environ. Sci. Technol.* 42 (2008), 6903–6909.

73. M. C. Alfaro-Del la Torre, P. Y. Beaulieu and A. Tessier, In situ measurement of trace metals in lakewater using the dialysis and DGT techniques, *Anal. Chim. Acta* 418 (2000), 53–68.

74. J. Forsberg, R. Dahlqvist, J. Gelting-Nyström and J Ingri, Trace metal speciation in brackish water using diffusive gradients in thin films and ultrafiltration: Comparison of techniques, *Environ. Sci. Technol.* 40 (2006), 3901–3905.

75. B. Ohlander, J. Fosberg, H. Osterlund et al., Fractionation of trace metals in a contaminated freshwater stream using membrane filtration, ultrafiltration, DGT and transplanted aquatic moss, *Geochem. Explor. Environ. Anal.* 12 (2012), 303–312.

76. D. Montalvo, F. Degryse and M. J. McLaughlin, Natural colloidal P and its contribution to plant P uptake, *Environ. Sci. Technol.* 49 (2015), 3427–3434.

77. N. Odzak, D. Kistler, R. Behra and L. Sigg, Dissolution of metal and metal oxide nano particles under natural freshwater conditions, *Environ. Chem.* 12 (2015), 138–148.

78. J. Santner, E. Smolders, W. W. Wenzel and F. Degryse, First observation of diffusion-limited plant root phosphorus uptake from nutrient solution, *Plant Cell Environ.* 35 (2012), 1558–1566.

79. N. Odzak, D. Kistler, R. Behra and L. Sigg, Dissolution of metal and metal oxide nanoparticles in aqueous media, *Environ. Pollut.* 191 (2014), 132–138.

80. P. L. R. van der Veeken, J. P. Pinheiro and H. P. van Leeuwen, Metal speciation by DGT/DET in colloidal complex systems, *Environ. Sci. Technol.* 42 (2008), 8835–8840.

81. E. Navarro, F. Piccapietra, B. Wagner et al., Toxicity of silver nanoparticles to Chlamydomonas reinhardtii, *Environ. Sci. Technol.* 42 (2008), 8959–8964.
82. H. M. Pouran, F. L. Martin and H. Zhang, Measurement of ZnO nanoparticles using diffusive gradients in thin films: binding and diffusional characteristics, *Anal. Chem.* 86 (2014), 5906–5913.

4

Binding Layer Properties

WILLIAM W. BENNETT, MAJA ARSIC, JARED G. PANTHER,
DAVID T. WELSH AND PETER R. TEASDALE

4.1 Introduction

The accurate measurement of solutes by DGT relies on the strength of their interaction with the binding layer. Mass transport of the analyte through the diffusive layer can only achieve a steady state when the analyte binds rapidly and irreversibly to the surface of the binding layer, so that the concentration of the solute at the interface between the binding layer and diffusive layer is effectively zero [1]. When this assumption is met, the DGT equation can be used to accurately determine the DGT-labile concentration of the analyte in the bulk solution.

The first binding layer to be used with DGT consisted of a hydrogel containing Chelex 100, a commercially available resin containing iminodiacetic acid functional groups, which strongly bind transition metal ions via chelation [2]. The selectivity of Chelex 100 for transition metals over alkali and alkaline earth metals ensures that the major ions in natural waters (i.e. Na^+, K^+, Mg^{2+}, Ca^{2+}) do not interfere with the measurement of transition metals typically present in much lower concentrations (i.e. trace metals) [3]. Without the high degree of selectivity that Chelex 100 exhibits for trace metals, successful measurement of these analytes in a wide variety of natural waters would not be possible, due to competition for binding sites by major ions.

Natural waters exhibit a variety of major ion concentrations and compositions, as well as a range of hydrogen ion concentrations (pH), which can affect the interaction between the binding layer and the analyte. Competition for binding sites by major ions and H^+, which are often present at several orders of magnitude higher concentrations than trace metals, reduce the effective capacity of the binding layer for the target analyte and thus restrict the length of deployment times that can be used. It is essential, therefore, that binding layers are comprehensively evaluated across a range of pH and ionic concentrations and for deployment times that are representative of actual field deployments, so that measurements in natural systems can be relied on to be both accurate and precise.

This chapter has three primary aims: (1) to provide a detailed description of the different types of binding layers utilized to date and their key features; (2) to outline the steps involved in validating a potential binding layer and to discuss examples in which satisfactory and non-ideal uptake by binding layers has been reported; and (3) to summarize the current validation status of existing binding layers described in the literature and to suggest future developments with regard to DGT binding layers. It is hoped that this chapter will be useful for both researchers looking to develop and evaluate new DGT binding layers and those looking to apply existing DGT techniques for measuring a particular analyte or suite of analytes in the environment.

4.2　Binding Layer Types and Features

A variety of different materials have been described in the literature for application as DGT binding layers (Table 4.1). These layers typically consist of solid resins or powders that are incorporated into a gel matrix (e.g. polyacrylamide) to form a near-homogenous binding layer, although polyacrylamide derivatized with binding functional groups, commercial membranes and liquid (polyelectrolyte) binding layers have also been described [4–6]. The initial focus was on developing DGT for measuring cationic trace metals (e.g. Cd, Cu, Co, Ni, Pb, Zn), which was achieved using the chelating ion-exchange resin, Chelex 100 [2], and expanded through application of other resins such as AG50WX8 (a non-chelating ion exchange resin) for Cs and Sr [7]. Attention then turned to oxyanionic species, such as phosphorus, sulphide and arsenic in which different binding mechanisms were used. Phosphorus and arsenic could be accumulated by adsorption onto metal (hydr)oxide-based binding layers such as ferrihydrite (FeOOH) [8, 9] and Metsorb (TiO_2 – nano-anatase distributed within a polymer backbone) [10, 11]. The measurement of sulphide used a novel binding reaction in which HS^- displaced I^- ions from silver iodide [12], which resulted in a distinct colour change that facilitated two-dimensional measurements (see Chapter 8). The measurement of mercury with DGT was another area of interest, and resulted in the application of selective binding layers containing thiol-functionalized resins, which were later evaluated for the selective measurement of As^{III} [13, 14]. To expand the range of analytes determined in a single DGT deployment, mixed binding layer [15, 16] and multiple binding layer [17] samplers have also been developed. The simple process of encapsulating a solid phase binding agent in a gel matrix has facilitated the development of a great diversity of DGT techniques, and will continue to do so into the future.

To use a solid material as a binding agent within a DGT binding layer, it must be of sufficiently small particle size to allow simple and even incorporation into a gel matrix. The standard thickness of an unhydrated binding gel layer is 250 µm (approx. 400 µm after hydration), so the diameter of solid particles incorporated into the gel must be considerably smaller than this to allow for easy preparation ($<$ 100 µm is recommended). Further details of the practical aspects of preparing binding gel layers are provided in Chapter 10. Many of the existing binding layers utilize commercially available materials as the binding agent, which have the advantage of widespread availability and reproducibility, although laboratory-synthesized materials have also been used successfully [6, 9, 18].

Table 4.1 *Binding agents used in DGT binding layers and their target analyte(s)*

Category	Binding agent	Analyte(s)	Ref
Cation exchange resins	Chelex 100	Al, Ca, Cd, Ce, Co, Cr(III), Cu, Dy, Er, Eu, Fe, Gd, Ho, La, Lu, Mg, Mn, Nd, Ni, Pb, Pu, Pr, Sm, Tb, Tm, U, Y, Yb, Zn	[1, 33, 39, 40]
	AG50W-X8	Cs, Sr	[7]
	Amberlite IRP69	K	[41]
	Suspended particulate reagent-iminodiacetate (SPR-IDA)	Cd, Co, Cu, Ni, Pb	[42]
	Diphonix	U	[43]
Anion exchange resins	Amberlite IRA 910	As(V)	[44]
	Methylthymol blue adsorbed to Dowex 1X8	Cu	[45]
	TEVA resin	^{99}Tc	[46]
Metal (hydr)oxides	Ferrihydrite (FeOOH)	As, Mo, P, Se, Sb, U, V, W	[8, 9, 47, 48]
	Metsorb (TiO_2)	As, Mo, P, Se, Sb, U, V, W	[10, 11, 22, 30]
	ZrO_2	As, Mo, P, Se, Sb, V	[18, 49, 50]
	MnO_2	Ra, U	[51, 52]
Thiol-functionalized resins	3-Mercaptopropyl-functionalized silica	As(III), methyl-Hg	[13, 53]
	Spheron-thiol	Hg	[14]
	Duolite GT73	Hg	[54]
	Iontosorb AV-IM	Hg	[54]
	3-Mercaptopropyl functionalized SBA-15	Hg	[55]
	3-Mercaptopropyl functionalized organosilica	Hg	[55]
	Sumichelate Q10R	Hg	[55]
Liquid binding phases	Poly(4-styrenesulfonate)	Cd, Cu	[5]
	Poly(aspartic acid)	Cd	[56]
	Poly(ethyleneimine)	Cd, Cu, Pb	[57]
	Sodium polyacrylate	Cd, Cu	[58]
	Poly(vinyl alcohol)	Cu	[59]
	Polymer-bound Schiff base	Cd, Cu, Pb	[60]
	Thiol-poly(vinyl alcohol)	Cd	[61]
	Polyquaternary ammonium salt	Cr(VI), P	[62, 63]
Membranes	Whatman P81	Cd, Co, Cu, Hg, Mn, Ni, Pb, Zn	[4, 64, 65]
Modified hydrogels	Poly(acrylamide-co-acrylic acid) (PAM-PAA)	Cd, Cu	[66]

Table 4.1 (*cont.*)

Category	Binding agent	Analyte(s)	Ref
	Poly(acrylamidoglycolic acid-co-acrylamide) (PAAG-PAM)	Cu	[67]
Mixed binding agents	Chelex 100/ferrihydrite	As, Cd, Cu, Mn, Mo, P, Pb, Zn	[15, 68]
	Chelex 100/Metsorb	As, Cd, Co, Cu, Mn, Mo, Ni, P, Pb, Sb, V, W	[16]
	Chelex 100/ZrO_2	As, Fe, P	[69, 70]
	Amberlite IRP69/ferrihydrite	K, P	[71, 72]
	AgI/SPR-IDA	Cd, Co, Cu, Ni, Pb, S^{2-}	[73]
	AgI/ZrO_2	As, Mo, P, S^{2-}, Sb, Se, V	[74]
	AgI/ferrihydrite	As, Mo, P, S^{2-}, Sb, Se, U, V, W As, V	[75]
	ZrO_2/SPR-IDA	As, Cd, Co, Cu, Mo, Ni, P, Pb, Sb, Se, V	[76]
Other binding agents	AgI	S^{2-}	[12]
	Ammonium molybdophosphate (AMP)	Cs	[77]
	Molecularly imprinted polymer	4-Chlorophenol	[78]
	Copper ferrocyanide	Cs	[79]
	8-Hydroxyquinoline-functionalized Spheron-Oxin	U	[80]
	Activated carbon	Bisphenols, Au	[29, 81]
	Montmorrilonite	Cd, Cr, Cu, Mn, Ni, Pb, Zn	[82]
	XAD18	Antibiotics (sulfonamides, fluoroquinolones, macrolides, ionophores, diaminopyimidines, aminocoumarins, lincosamides)	[27, 28]
	Saccharomyces cerevisiae	Cd	[38]

As shown by the wide variety of binding agents in Table 4.1, there are often numerous options when considering which binding layer to use for a particular purpose. The suitability of a binding layer is dependent on the matrix in which it will be deployed, with many binding layers only applicable to freshwaters due to the high concentrations of potential competing ions present in more complex matrices like seawater [11]. In fact, of the forty-six binding layers described in Table 4.1, the majority have not been tested in seawater.

There are four key characteristics when considering a potential binding layer:

1. The binding strength, which is represented by a relevant equilibrium constant for each analyte (e.g. stability or solubility constant). Binding constants for analytes are often available when a commercial resin is used but can also be determined experimentally [19]. Generally, DGT techniques with a high binding strength for their target analyte(s) will be more suitable for making measurements in diverse matrices.

2. The *intrinsic binding capacity* of the binding layer, which relates to the number or density (concentration) of sites able to interact with analyte species at the interface of the diffusive and binding layers, and is independent of the binding strength. Determining the binding layer capacity is an important step of the validation process. Higher binding capacities will be particularly important for DGT techniques in which binding strengths are fairly low or where competition is likely to be an issue.

3. The presence of competition effects. Competition can arise from the presence of other ions with similar or stronger binding strengths, and may be important in matrices with high concentrations of several analytes where a sizeable fraction of the total binding sites are occupied. However, competition effects are more likely for some binding layers when DGT devices are deployed in saline and/or non-circumneutral pH (<5.5 or >7.4) waters. In these instances, major ions, H^+ and/or OH^- interact more weakly with the binding layer, but may be present at much higher concentrations than analytes. Competition has the effect of decreasing the concentration of binding sites available to interact with the analyte and may therefore result in non-ideal accumulation (i.e. not described by the DGT equation) by the binding layer. An *effective binding capacity* should be determined in the presence of competing analytes or in matrices in which competition is likely (see [20] as an example). Binding layers that rely on adsorption or ion exchange mechanisms are generally more prone to competition effects compared with those that rely on complexation.

Kinetic limitation may occur if binding of an analyte is insufficiently fast to outpace diffusional supply or dissociation of a complex (see Chapter 5) and therefore maintain a negligible concentration at the binding layer surface. Kinetic limitations may not be apparent during equilibrium characterisation experiments with the binding layer alone, but could be observed during laboratory DGT deployments. The presence of kinetic limitations can be determined directly, but this has been done infrequently. Some kinetic effects have been observed due to increased competition for binding sites. For instance, uptake of Mn, Co and Cd at pH 4 and Mn at pH 5 by Chelex-DGT was non-ideal [21]. A proportion of the metals was observed to penetrate into the binding layer before being bound, due to a low rate of binding, which was considered to be most likely due to competition from H^+ resulting in a lower concentration of binding sites.

To determine the suitability of a particular binding layer for measuring an analyte, a comprehensive laboratory validation should be undertaken, as described in the following section. Typical experiments in which the above characteristics are determined will be highlighted as well as instances of results that demonstrate satisfactory and non-ideal performance characteristics.

4.3 Laboratory Validation

The process of developing and evaluating a new DGT binding layer involves a series of well-defined laboratory-based experiments. This section provides a general description of each of the recommended validation steps, as well as discussion of the common performance limitations that may be encountered. Comprehensive laboratory validation is essential to ensure new DGT binding layers perform accurately and reproducibly when deployed in the field.

4.3.1 Uptake and Elution

The first step in evaluating a DGT binding layer is to determine that the binding material is capable of accumulating the analyte of interest (uptake). Some researchers also complete comprehensive kinetic studies as an extension of the uptake measurement [15]. This is useful to establish, early on in the validation process, whether the target analyte is suitable for the binding layer. Then it is necessary to find a suitable eluent solution that can quantitatively recover the analyte(s) from the binding layer. The elution step is necessary because the majority of analytical techniques used to quantify analytes require an aqueous sample (e.g. inductively coupled plasma mass spectrometry (ICP-MS)). Solid phase measurement of DGT binding layers is discussed in Chapter 8.

In typical uptake experiments replicate binding layer gel discs are exposed to relatively small volumes of solution (5–10 mL) containing a concentration of analyte that is easily measurable by the relevant analytical technique, while ensuring that the mass of analyte accumulated on the binding layer is typical of masses accumulated in field deployments. These experiments are normally done over 24 h in 0.01 M $NaNO_3$ or NaCl to limit competition effects and adsorption losses to container walls. Following the uptake experiment the binding layer is removed from the analyte solution, rinsed thoroughly, and then placed into a small volume of eluent (typically 1 mL) for a further 24 h. Suitable eluents will vary depending on the binding layer and the mechanism by which the analyte is bound. For example, metals are removed from the Chelex 100 binding layer by a 1 M nitric acid solution [1], whereas oxyanions such as phosphate are removed from the Metsorb binding layer by a 1 M sodium hydroxide solution [11]. With the ferrihydrite binding layer, analytes are quantitatively eluted by dissolving ferrihydrite in HNO_3 or other acids [8, 9, 20]. Some analytes require more complex eluents to remove them from particular binding layers, such as the combination of sodium hydroxide and hydrogen peroxide to elute antimony from Metsorb [22]. Following analysis of the eluent by a suitable analytical technique (e.g. ICP-MS for metals, colorimetry for phosphate), typically after dilution to a matrix concentration suitable for analysis, the uptake and elution factors (often referred to as 'efficiencies' in the literature) can be calculated. The uptake factor, f_u, is calculated from the initial mass of analyte in the solution, M_i, and the mass of analyte remaining in solution after the experiment, M_f (equation 4.1). There is no specific value that is deemed acceptable for f_u, but values of $>85\%$ (Table 4.2) are typical of successful DGT methods.

$$f_u = \frac{M_i - M_f}{M_i} \tag{4.1}$$

The elution factor, f_e, is calculated after measuring the mass of analyte in the eluent solution, M_e, using equation 4.2.

$$f_e = \frac{M_e}{M_i - M_f} \tag{4.2}$$

Elution factors typically vary from 75% to 100% (Table 4.2). Although the reproducibility of the elution procedure is more critical to the accuracy of DGT measurements, higher values of f_e are preferred. For new binding layers, more than one eluent should be tested initially to determine an optimal elution procedure [23].

4.3.2 Linear Mass Accumulation Over Time

Following successful validation of uptake and elution factors, the potential binding layer should be tested as part of an assembled DGT sampler. The fundamental assumption of the DGT equation, of linear accumulation of the target analyte over time, should be met to ensure accurate measurements of solutes by the new technique. Typical experiments involve exposing replicate DGT samplers to a known concentration of analyte(s), for various deployment times (e.g. 6, 12, 18 and 24 h), in a solution of moderate ionic strength (0.01 M $NaNO_3$ or $NaCl$) without interfering ions. The solution should be well stirred at a reproducible and constant rate during such experiments to minimize the diffusive boundary layer thickness [1]. Ideally, a small volume of solution should be collected whenever a set of DGT samplers is removed to monitor any changes in analyte concentration over time.

Following elution of the binding layers and analysis by an appropriate analytical technique, the linearity of the mass accumulation over time can be evaluated by a simple linear regression (Figure 4.1). Such regression lines should have $R^2 > 0.95$. Where the analyte diffusion coefficients are already known, a plot of mass versus time should be compared with a theoretical mass line calculated using the DGT equation. The experimental masses, corrected for f_e, should be very close to this theoretical line (i.e. the slope and intercept of the regression line should be very similar to the theoretical line). If the diffusion coefficients are not known, then these data can be used to calculate the effective diffusion coefficient of the tested analyte through the diffusive gel and filter membrane (D^{mdl}) by simple rearrangement of the DGT equation (see Chapter 3 for details).

After an initial mass vs. time validation experiment, with the time periods described above, it is recommended to do longer term experiments in the same solution for time

Table 4.2 *Summary of performance characteristics of commonly used binding layers. f_u and f_e are uptake and elution efficiencies, respectively. 3-MFS = 3-mercaptopropyl-functionalized silica*

Binding layer	Analyte	Uptake and elution	Validated pH and ionic strength ranges	Linear mass accumulation over time	Representative matrix validation	Comments	Ref
Chelex 100	Al	$f_u > 85\%$ $f_e = 84 \pm 3\%$ (1 M HNO$_3$)	pH 4.0–8.0 0.001–0.7 M NaNO$_3$	96 h in 0.002 M NaNO$_3$	Synthetic freshwater (96 h, ~10 µg L^{-1}, pH 5.4 and 6.1)	Not quantitative at pH > 8 in 0.01 M NaNO$_3$ Not quantitative at pH 7.7–8.1 in synthetic freshwater Not quantitative in synthetic seawater	[30]
	Ca	$f_u > 96\%$ $f_e = 98.3 \pm 5.6\%$ (1 M HNO$_3$)	pH 5.0–6.7	No data	Natural freshwater (6 h, 2.2 mg L^{-1}, pH not reported)	Not quantitative at pH < 5	[39]
	Cd	f_u – No data $f_e = 83.9 \pm 2.7\%$ (1 M HNO$_3$) $f_e = 94.9 \pm 4.7\%$ (1 M HNO$_3$) $f_e = 98.7 \pm 1.1$ (conc. HNO$_3$)	pH 5.0–10.0 10 nM–1 M NaNO$_3$	72 h in 0.01 M NaNO$_3$	No data	Not quantitative at pH < 5 in 0.01 M NaNO$_3$	[1, 16, 25, 33]
	Co	f_u – No data $f_e = 93.0 \pm 2.0\%$ (1 M HNO$_3$) $f_e = 98.6 \pm 1.2$ (conc. HNO$_3$)	pH 3.5–10.0 No competitive effects from Ca^{2+}	72 h in 0.01 M NaNO$_3$	No data		[16, 25, 33]
	Cr(III)	f_u (pH 4.4) = 99 $\pm$ 2% f_u (pH 5.3) = 94 $\pm$ 5% f_e (pH 4.4) = 81 $\pm$ 7% (1 M HNO$_3$) f_e (pH 5.3) = 79 $\pm$ 6% (1 M HNO$_3$)	pH 2.9–5.0	10 h in 0.01 M NaNO$_3$ (pH 5.0)	No data	Not quantitative at pH 7.5 in 0.01 M NaNO$_3$	[83]

(*cont.*)

Table 4.2 (*cont.*)

Binding layer	Analyte	Uptake and elution	Validated pH and ionic strength ranges	Linear mass accumulation over time	Representative matrix validation	Comments	Ref
	Cu	f_u – No data $f_e = 79.3 \pm 6.4\%$ (1 M HNO$_3$) $f_e = 91.8 \pm 1.2\%$ (1 M HNO$_3$) $f_e = 97.7 \pm 1.0\%$ (conc. HNO$_3$)	pH 2.0–10.0 (0.01 M NaNO$_3$) No competitive effects from Ca^{2+}	72 h in 0.01 M NaNO$_3$	No data		[1, 16, 25, 33]
	Fe	f_u – No data $f_e = 69.7 \pm 5.0\%$ (1 M HNO$_3$) $f_e = 80.8 \pm 10.1$ (conc. HNO$_3$)	No data	No data	No data	Not quantitative at pH 4.7–5.9	[1, 33]
	Mg	$f_u > 96\%$ $f_e = 95.4 \pm 6.5\%$ (1 M HNO$_3$)	pH 4.0–6.7	24 h in deionized water	Natural freshwater (6 h, 1.1 mg L^{-1}, pH not reported)	Not quantitative at pH < 4	[39]
	Mn	$f_u > 99\%$ $f_e = 81.4 \pm 2.2\%$ (1 M HNO$_3$) $f_e = 90.7 \pm 9.1\%$ (1 M HNO$_3$) $f_e = 98.3 \pm 1.1$ (conc. HNO$_3$)	pH 3.5–10.00.001–0.7 M NaNO$_3$	24 h in 0.01 M NaNO$_3$/ 0.004 M Mg(NO$_3$)$_2$, pH 6.06	Synthetic freshwater (96 h, 10.9 μg L^{-1}, pH 7.48) Synthetic seawater (52 h, 12.7 μg L^{-1}, pH 8.02)	Sensitivity is reduced at pH 5–6. Competitive effects under high iron concentrations (5 mM) can be avoided by using short deployment times (<12 hours)	[1, 16, 24, 25]
	Ni	f_u – No data $f_e = 81.4 \pm 2.2\%$ (1 M HNO$_3$) $f_e = 92.0 \pm 2.1\%$ (1 M HNO$_3$) $f_e = 93.5 \pm 0.8$ (conc. HNO$_3$)	pH 4.7–6.0	24 h in 0.01 M NaNO$_3$/0.004 M Mg(NO$_3$)$_2$	No data		[9, 16, 33]
	Pb	f_u – No data $f_e = 93.5 \pm 2.7\%$ (1 M HNO$_3$) $f_e = 97.4 \pm 0.5$ (conc. HNO$_3$)	pH 4.7–6.0	24 h in 0.01 M NaNO$_3$/0.004 M Mg(NO$_3$)$_2$	No data		[16, 33]

	U	$f_u > 80\%$ $f_e = 80.0 \pm 6\%$ (2 M HNO$_3$) $f_e = 89.0 \pm 0.8\%$ (1 M HNO$_3$) $f_e = 96.2 \pm 1.3$ (conc. HNO$_3$)	pH 5.0–9.0 0.01–1 M NaNO$_3$	120 h in 0.01 M NaNO$_3$	Not quantitative under tested conditions	Not quantitative at pH 4.7 – 5.9 Strong competitive effects from PO$_4^{3-}$ and HCO$_3^-$ Poor uptake in synthetic freshwater (76%) Not quantitative in synthetic seawater	[23, 33, 51, 80]
	Zn	f_u – No data $f_e = 80.3 \pm 5.5\%$ (1 M HNO$_3$) $f_e = 91.1 \pm 2.8\%$ (1 M HNO$_3$) $f_e = 96.3 \pm 1.6$ (conc. HNO$_3$)	pH 3.5–10.0	24 h in 0.01 M NaNO$_3$/ 0.004 M Mg(NO$_3$)$_2$	No data	No competitive effects from Ca^{2+}	[9, 16, 25, 33]
Ferrihydrite	As(III)	$f_u = 100\%$ $f_e = 89 \pm 0.5\%$ (3.2 M HNO$_3$) $f_e = 100 \pm 5\%$ (conc. HCl)	pH 3.0–8.0 0.001–0.2 M NaNO$_3$ 5 mg L^{-1} fulvic acid	72 h in 0.01 M NaNO$_3$	Synthetic freshwater (41 h, 90 µg L^{-1}, pH 8.1) Natural seawater (24 h, 100 µg L^{-1}, pH ~8)	No competitive effects on uptake from As(V), PO$_4^{3-}$, V(V), Se(VI), Se(IV)	[8, 13, 20, 47]
	As(V)	$f_u > 95\%$ $f_u = 100\%$ $f_e = 78 \pm 0.48\%$ (1 M HNO$_3$) $f_e = 88 \pm 0.3\%$ (3.2 M HNO$_3$) $f_e = 94\%–104\%$ (1.4 M HNO$_3$, 0.1 M HF) $f_e = 100 \pm 5\%$ (conc. HCl)	pH 3.0–8.0 0.001–0.2 M NaNO$_3$ 5 mg L^{-1} fulvic acid	10 h in 0.01 M NaNO$_3$ 72 h in 0.01 M NaNO$_3$	Synthetic freshwater (93 h, 15 µg L^{-1}, pH 7.2) Synthetic seawater (92 h, 15 µg L^{-1}, pH 8.3)	Lower adsorption in presence of colloidal iron (III) hydroxide (1–5 mg L^{-1}) No competitive effects on uptake from As(III), PO$_4^{3-}$, V(V), Se(VI), Se(IV) in synthetic fresh or seawater	[8, 10, 20, 22, 47, 48]

(cont.)

Table 4.2 (*cont.*)

Binding layer	Analyte	Uptake and elution	Validated pH and ionic strength ranges	Linear mass accumulation over time	Representative matrix validation	Comments	Ref
	PO_4^{3-}	$f_u > 96\%$ f_u: 100% $f_e = 81 \pm 0.7\%$ (3.2 M HNO_3) $f_e = 92 \pm 5\%$ (1 M NaOH) $f_e = 101.4 \pm 2.1\%$ (0.25 M H_2SO_4)	pH 2.0–10.0 0.0001–1.0 M $NaNO_3$	55 h in 0.01 M $NaNO_3$	Synthetic freshwater (96 h, 47.5 µg L^{-1}, pH 7.2) Synthetic seawater (72 h, 95 µg L^{-1}, pH 8.1)	Not quantitative > 48 hours at 2.9–5.7 mM HCO_3^- No competitive effects from Cl^- or SO_4^{2-} No competitive effects on uptake from As(III), As(V), V(V), Se(VI), Se(IV) in synthetic fresh or seawater Lower recovery at pH 1 (10.5%)	[9, 11, 20, 31]
	Se(IV)	f_u – No data $f_e = 81 \pm 0.7\%$ (3.2 M HNO_3)	pH 4.0–8.0 0.001–0.2 M $NaNO_3$	72 h in 0.01 M $NaNO_3$	Synthetic freshwater (41 h, 79 µg L^{-1}, pH 8.1) Synthetic seawater (72 h, 40 µg L^{-1}, pH 8.1)	Competition for uptake by As, V and PO_4^{3-}	[20, 47]
	Sb(V)	$f_u > 95\%$ $f_e = 85 \pm 0.33\%$ (1 M HNO_3) $f_e = 94\%–104\%$ (1.4 M HNO_3, 0.1 M HF)	pH 3.0–8.0 0.001–0.1 M $NaNO_3$	72 h in 0.01 M $NaNO_3$	Synthetic freshwater (6 h, 4 µg L^{-1}, pH 7.2)	Slightly slower and weaker adsorption at pH > 8 Lower binding affinity than V, As, Mo and W Not quantitative in synthetic seawater Not quantitative in synthetic freshwater for deployment times > 6 h	[22, 47, 48]
	Mo(VI)	$f_u > 95\%$ $f_e = 94\% – 104\%$ (1.4 M HNO_3, 0.1 M HF)	pH 4.0–6.2	24 h in 0.01 M $NaNO_3$	Synthetic freshwater (6 h, 30 µg L^{-1}, pH 7.2)	Slightly slower and weaker adsorption at pH > 8 Not quantitative in synthetic seawater Not quantitative in synthetic freshwater for deployment times > 6 h	[22, 48]

	V(V)	$f_u > 95\%$ $f_e = 63 \pm 0.17\%$ (1 M HNO_3) $f_e = 91 \pm 0.4\%$ (3.2 M HNO_3) $f_e = 94\% - 104\%$ (1.4 M HNO_3, 0.1 M HF)	pH 5.95–7.820.001–0.1 M $NaNO_3$	72 h 0.01 M $NaNO_3$	Synthetic freshwater (93 h, 12 µg L^{-1}, pH 7.2) Synthetic seawater (92 h, 13 µg L^{-1}, pH 8.3)	Competition effects on uptake by As(V) and PO_4^{3-} at typical marine pH ($\sim$8) Reduced uptake ($< 80\%$) at pH 3–4.4	[20, 22, 47, 48]
	W(VI)	$f_u > 95\%$ $f_e = 94\% - 104\%$ (1.4 M HNO_3, 0.1 M HF)	pH 4.0–8.0	24 h in 0.01 M $NaNO_3$	Synthetic freshwater (93 h, 17 µg L^{-1}, pH 7.2)	Not quantitative in synthetic seawater	[22, 48]
Metsorb	Al	f_u (pH 4.8) $> 98\%$ f_u (pH 8.4) $> 85 \pm 5\%$ $f_e = 85 \pm 6\%$ (1 M NaOH)	pH 5.0–8.5 0.001–0.7 M $NaNO_3$ (pH 5.0 and 8.4)	96 h in 0.002 M $NaNO_3$ (pH 5.05 and 8.35)	Synthetic freshwater (96 h, 10.2–22.6 µg L^{-1}, pH 5.4–8.1) Synthetic seawater (96 h, 20 µg L^{-1}, pH 8.15)	Not quantitative at pH < 5	[30]
	As(III)	$f_u = 100\%$ $f_e = 81.2 \pm 1.1\%$ (1 M NaOH) $f_e = 99 \pm 13\%$ (1 M NaOH)	pH 3.5–8.5 0.0001–0.75 M $NaNO_3$	24 h in 0.01 M $NaNO_3$	Synthetic freshwater (41 h, 90 µg L^{-1}, pH 8.1) Natural seawater (72 h, 50 µg L^{-1}, pH $\sim$8)	No competitive effects observed with As(V) or PO_4^{3-}	[10, 13, 20]
	As(V)	$f_u > 98\%$ $f_u = 100\%$ $f_e = 75.2 \pm 1.5\%$ (1 M NaOH) $f_e = 92.1 \pm 2.5\%$ (1 M NaOH) $f_e = 100 \pm 2\%$ (1 M NaOH)	pH 3.5–8.5 0.0001–0.75 M $NaNO_3$	24 h in 0.01 M $NaNO_3$	Synthetic freshwater (93 h, 15 µg L^{-1}, pH 7.2) Synthetic seawater (92 h, 15 µg L^{-1}, pH 8.3)	No competitive effects observed with As(III) or PO_4^{3-}	[10, 20, 22]
	PO_4^{3-}	$f_u > 98\%$ $f_e = 61 \pm 9\%$ (1 M NaOH) $f_e = 92 \pm 5.0\%$ (1 M NaOH)	pH 3.9–8.2 0.0001–1.0 M $NaNO_3$	55 h in 0.01 M $NaNO_3$	Synthetic freshwater (96 h, 47.5 µg L^{-1}, pH 7.2) Synthetic seawater (96 h, 45 µg L^{-1}, pH 8.22)	No observed competition effects with HCO_3^-, Cl^- or SO_4^{2-}	[11, 20, 31]

(cont.)

Table 4.2 (*cont.*)

Binding layer	Analyte	Uptake and elution	Validated pH and ionic strength ranges	Linear mass accumulation over time	Representative matrix validation	Comments	Ref
	Se(IV)	$f_u = 96.5 \pm 0.5\%$ (1 M NaOH) $f_e = 86.2 \pm 5.6\%$ (1 M NaOH) $f_e = 94 \pm 3\%$ (1 M NaOH)	pH 3.5–8.5 0.0001–0.75 M NaNO$_3$	24 h in 0.01 M NaNO$_3$	Synthetic freshwater (41 h, 79 µg L^{-1}, pH 8.1) Synthetic seawater (48 h, 40 µg L^{-1}, pH 8.1)	Lower adsorption affinity compared to As(V), V(V) and PO$_4^{3-}$	[10, 20]
	Sb(V)	$f_u > 98\%$ $f_e = 90.3 \pm 4.6\%$ (1 M NaOH/1 M H$_2$O$_2$)	pH 4.0–8.2 0.001–0.7 M NaNO$_3$	24 h in 0.01 M NaNO$_3$	Synthetic freshwater (93 h, 4 µg L^{-1}, pH 7.2) Synthetic seawater (92 h, 3 µg L^{-1}, pH 8.3)		[22]
	Mo(VI)	$f_u > 98\%$ $f_e = 94.8 \pm 3.6\%$ (1 M NaOH)	pH 4.0–8.2 0.001–0.7 M NaNO$_3$	24 h in 0.01 M NaNO$_3$	Synthetic freshwater (93 h, 30 µg L^{-1}, pH 7.2) Synthetic seawater (4 h, 15 µg L^{-1}, pH 8.3)	Not quantitative > 4 h in synthetic seawater	[22]
	V(V)	$f_u > 98\%$ $f_e = 90 \pm 5\%$ (1 M NaOH) $f_e = 90.8 \pm 2.3\%$ (1 M NaOH)	pH 4.0–8.2 0.001–0.7 M NaNO$_3$	24 h in 0.01 M NaNO$_3$	Synthetic freshwater (93 h, 12 µg L^{-1}, pH 7.2) Synthetic seawater (92 h, 13 µg L^{-1}, pH 8.3)	Possible competition effects after 24 h from As(V) and PO$_4^{3-}$ in synthetic seawater matrix	[20, 22]
	U	$f_u > 90\%$ $f_u > 100 \pm 3\%$ $f_e = 83 \pm 3\%$ (1 M HNO$_3$/1 M H$_2$O$_2$) $f_e = 95.2 \pm 5\%$ (1 M NaOH/1 M H$_2$O$_2$)	pH 3.0–9.0 0.001–1 M NaNO$_3$	120 h in 0.01 M NaNO$_3$	Synthetic freshwater (96 h, 14.7 µg L^{-1}, pH 8.25) Natural seawater (8 h, 26.3 µg L^{-1}, pH 7.86)	Competition affects uptake in synthetic seawater for deployments > 8 h Strong competitive effects from SO$_4^{2-}$, PO$_4^{3-}$ or HCO$_3^{-}$ at high concentrations Addition of Ca^{2+} increased f_u	[23, 51]

Substrate	Analyte	Uptake / elution	Conditions	Equilibration	Sample type	Notes	Ref.
	W(VI)	$f_u > 98\%$ $f_e = 97.9 \pm 1.6\%$ (1 M NaOH)	pH 4.0–8.2 0.001–0.7 M NaNO$_3$	24 h in 0.01 M NaNO$_3$	Synthetic freshwater (93 h, 17 µg L^{-1}, pH 7.2) Synthetic seawater (92 h, 17 µg L^{-1}, pH 8.3)		[22]
ZrO$_2$	As	$f_u = 100\%$ As(III) $f_e = 86.9 \pm 2.5\%$ (1 M NaOH) As(V) $f_e = 88.4 \pm 2.0\%$ (1 M NaOH) As (inorganic) $f_e = 96.9 \pm 1.8\%$ (0.5 M NaOH)	pH 2.0–9.1 0.00001–0.75 M NaNO$_3$	72 h in 0.01 M NaNO$_3$	Synthetic and natural freshwater (96 h, 50 µg L^{-1}, pH 7.2–7.6) Synthetic seawater (96 h, 200 µg L^{-1}, pH 8.2)	$f_e = 87.7\%$ used for total inorganic As(III + V) elution efficiency No competitive effects from phosphate (0.25–10 mg L^{-1})	[49, 50]
	Mo(VI)	$f_u > 85\%$ within 30 mins $f_e = 98.9 \pm 1.7\%$ (0.5 M NaOH)	pH 3.8–7.8 0.0001–0.5 M NaNO$_3$	72 h in 0.01 M NaNO$_3$	No data		[50]
	PO$_4{}^{3-}$	$f_u = 100\%$ $f_e = 95.3 \pm 2.1\%$ (1 M NaOH) $f_e = 96.2 \pm 1.8\%$ (0.5 M NaOH)	pH 3.0–10.0 0.00001–0.5 M NaNO$_3$	72 h in 0.01 M NaNO$_3$	No data	No interferences from Cl$^-$, SO$_4{}^{2-}$, As(V) and As(III)	[18, 32, 50]
	Sb(V)	$f_u > 85\%$ within 30 mins $f_e = 101.8 \pm 0.7\%$ (0.5 M NaOH; 0.5 M NaOH/0.5 M H$_2$O$_2$)	pH 3.8–7.8 0.0001–0.5 M NaNO$_3$	72 h in 0.01 M NaNO$_3$	No data		[50]
	Se(VI)	$f_u > 85\%$ $f_e = 95.3 \pm 5.2\%$ (0.5 M NaOH)	pH 3.8–7.8 0.001–0.1 M NaNO$_3$	72 h in 0.01 M NaNO$_3$	No data	Uptake affected at low (0.0001 M NaNO$_3$) and high (0.5 M NaNO$_3$) ionic strengths	[50]
	V(V)	$f_u > 85\%$ $f_e = 98.2 \pm 1.0\%$ (0.5 M NaOH)	pH 3.8–7.8 0.0001–0.5 M NaNO$_3$	72 h in 0.01 M NaNO$_3$	No data		[50]

(*cont.*)

Table 4.2 (*cont.*)

Binding layer	Analyte	Uptake and elution	Validated pH and ionic strength ranges	Linear mass accumulation over time	Representative matrix validation	Comments	Ref
3-MFS	As(III)	$f_u > 99\%$ $f_e = 85.6 \pm 1.7\%$ (1 M HNO$_3$, 0.01 M KIO$_3$)	pH 3.5–8.5 0.0001–0.75 M NaNO$_3$	24 h in 0.01 M NaNO$_3$	Natural seawater (72 h, 100 μg L^{-1}, pH ~8)		[13]
	Hg	f_u – No data f_e = No data (5% thiourea) f_e = No data (aqua regia)	pH 5.0	32 h in 0.01 M NaCl	No data	No interference observed with Al, V, Mn, Cr, Fe, Co, Ni, Cu, Zn, Ag, Sn and Pb Quantitative in presence of DOM	[14, 55, 84]
	CH$_3$Hg	f_u – No data $f_e = 91 \pm 3\%$ (1.31 mM thiourea in 0.1 M HCl)	pH 3.0–9.0	32 h in 0.01 M NaCl	No data		[53]
AG50W-X8	Cs	$f_u = 100\%$ $f_e = 91 \pm 0.1\%$ (2 M HNO$_3$)	pH 4.0–9.0 1 μM–1 mM NaCl,	No data	Synthetic freshwater (20 h, 0.03 μg L^{-1}, pH 7.5)	Na competes effectively at >1 mM NaCl	[7]
	Sr	$f_u = 100\%$ $f_e = 100 \pm 0.2\%$ (2 M HNO$_3$)	pH 3.0–9.0 1 μM–1 mM NaCl,	No data	Natural freshwater (25 h, 16.5 μg L^{-1}, pH 7.5)	Na competes effectively at >1 mM NaCl	[7]
XAD18	SMX	f_u – No data $f_e = 97 \pm 2\%$ (methanol)	pH 6.0–9.0 0.001–0.01 M NaCl	48 h in 0.01 M NaCl	No data	Partial field validation in wastewater conducted for 40 antibiotic compounds	[27]

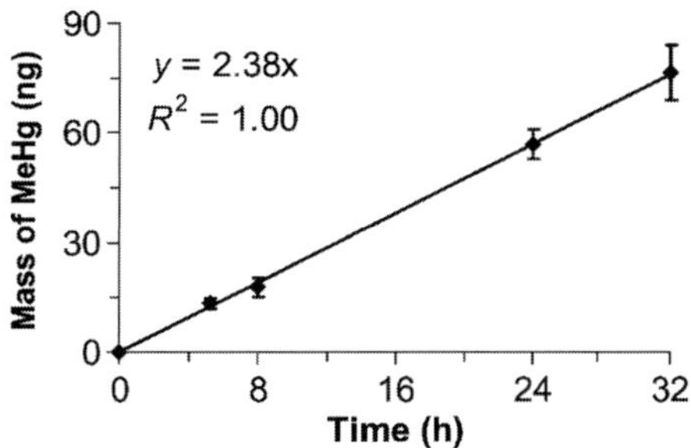

Figure 4.1   Linear accumulation of methylmercury over time in a DGT experiment where the binding layer contains 3-mercaptopropyl-functionalized silica. Reproduced from Ref 53 with permission of The Royal Society of Chemistry.

periods likely to be used for field deployments as this may help to identify non-ideal behaviour. Non-ideal accumulation of an analyte can arise due to several reasons, which are numerically simulated and discussed in Chapter 5. If the rate of binding at and within the binding layer is slow, the accumulation can still be represented by steady state conditions, but the accumulation at any given time will be less than expected for rapid binding. In this case the mass accumulated still increases linearly with time but with a reduced slope compared to the theoretical case. Such non-ideal behaviour has been observed, but the most common examples of non-ideal behaviour are characterised by non-linear plots of mass versus time (Figure 4.2) [16, 24]. They are easily identified by an $R^2 < 0.95$, in which deviation from linearity increases over time (other patterns of non-linearity are likely to indicate analytical issues). This non-linearity essentially signifies that steady state conditions do not apply, which may occur for two reasons.

The accumulated mass is a sizeable fraction of the intrinsic binding capacity (see Section 4.3.3). Effectively, the concentration of the analyte species at the interface between the binding and diffusive layers is progressively increasing and the flux into the DGT samplers is decreasing over time.

Competition effects by other ions in solution may also promote non-linearity of mass versus time plots by decreasing the effective binding capacity through a reduction in available binding sites. Competition effects in laboratory solutions can be minimised by using a single analyte and/or changing the solution matrix or concentration.

4.3.3 Capacity

The capacity of the binding layer for the target analyte(s) needs to be determined, as established above. The *intrinsic binding capacity* is most efficiently determined by exposing binding layer discs to a high concentration of analyte, until equilibrium is reached, in the absence of any competition effects [11]. If multiple analytes are determined together then there will inevitably be competition between analytes as the capacity is approached. The

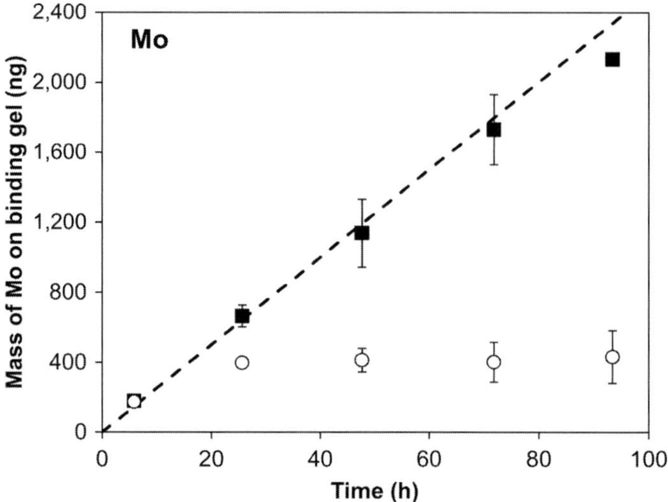

Figure 4.2   Non-linear accumulation of molybdenum over time by ferrihydrite DGT samplers ($\circ$), compared to linear accumulation by Metsorb DGT samplers ($\blacksquare$). Reprinted from *Talanta*, Vol 105, J. G. Panther, R. R. Stewart, P. R. Teasdale, W. W. Bennett, D. T. Welsh and H. Zhao, Titanium dioxide-based DGT for measuring dissolved As(V), V(V), Sb(V), Mo(VI) and W(VI) in water, 80–86, Copyright (2013), with permission from Elsevier.

capacity measurement can also be made with assembled DGT samplers rather than just the binding layers [18]; the obtained values should be quite similar, regardless of the approach used, although the DGT measurement is slower and more expensive. The effective binding capacity is determined in a similar way to the intrinsic binding capacity, but in the presence of potentially competing ions (e.g. HCO_3^-, H^+, SO_4^{2-}, etc.), which can compete for binding sites and reduce the number available for interacting with the analyte of interest [20]. A comparison between the intrinsic and effective binding capacities for a particular analyte could be helpful in revealing the susceptibility of a binding layer to competition effects in various matrices.

A distinction should be made between the intrinsic and effective capacities of the binding layer and the mass at which linear accumulation is no longer observed under particular measurement conditions [13]. This *linear accumulation capacity* is determined by deploying DGT samplers in a high concentration of analyte for various deployment times, so that deviation from the predicted linear uptake is observed. The linear accumulation capacity will be lower in the presence of competing ions, which occupy binding sites on the binding layer, the extent of which is determined by their concentration in solution and/or their affinity for the binding sites. This measurement is therefore highly operational and depends largely upon competition effects in the sample matrix; it provides an operationally useful estimate of conditions under which DGT can be used reliably. As DGT is a kinetic passive sampler, which is not designed to approach equilibrium during deployment, knowledge of the linear accumulation capacity measured in this way will be more useful practically than the intrinsic capacity determined simply by making measurements in a range of solution concentrations. Unfortunately, modelling studies (Chapter 5) demonstrate that there is no simple rule that can be applied to determine the proportion of a binding layer's capacity that should not be exceeded to ensure theoretical linear analyte accumulation by a DGT sampler. For this reason we recommend a thorough investigation of the linear accumulation performance of new DGT binding layers in representative matrices, and for representative deployment times, as discussed in detail in Section 4.3.5.

4.3.4 Ionic Strength and pH

Potential binding layers must be tested to ensure accurate performance in a range of environmental conditions. Ionic strength and pH are two critical chemical parameters that vary significantly amongst natural systems, and that can impact the strength of the interaction between the analyte and the binding layer. The typical approach for evaluating the effect of ionic strength and pH on a binding layer involves exposure of replicate DGT samplers to simple synthetic solutions containing a range of ionic concentrations (e.g. 0.0001–0.7 M $NaNO_3$) or pH (e.g. pH 3–8.5) that are typical of natural systems. The DGT-measured analyte concentration (c_{DGT}) is then compared to the analyte concentration measured directly in the solution by an appropriate analytical technique (c^{soln}); a ratio of c_{DGT} to c^{soln} close to unity (0.9–1.1) indicates that the DGT technique is accurately measuring the analyte concentration in the tested solution.

Hydrogen ions can compete with cationic species for binding sites, as evidenced by the poor performance of the Chelex 100 binding layer at pH < 4 for some analytes (Figure 4.3) [25]. Fortunately, the majority of aquatic systems have a pH > 4 and thus the Chelex 100 binding layer has been widely applied to measuring cationic trace metals in the environment. Ionic strength has minimal impact on the accumulation of most cationic trace metals by the Chelex 100 binding layer due to its high selectivity, and thus it generally performs well in matrices like seawater that contain high concentrations of competing ions.

The main caveat of this approach to assessing the impact of pH and ionic strength on a binding layer is that the experiments are typically done for no longer than 24 h. While this may be sufficient to observe significant deviations from the predicted performance, most DGT field deployments are for multiple days. It is, therefore, strongly recommended that potential binding layers are further tested in representative matrices (e.g. synthetic fresh or seawater) for deployment times similar to those that are to be used for actual field deployments, as described in the following section.

4.3.5 Extended Deployment in Representative Matrices

The final step in the laboratory validation of a potential DGT binding layer is to test its performance in synthetic solutions that are representative of natural systems (e.g. synthetic

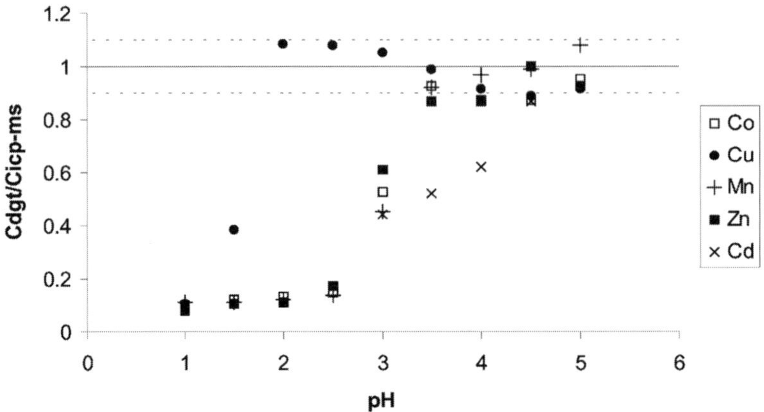

Figure 4.3   Performance of Chelex 100 DGT samplers for the measurement of Co, Cu, Mn, Zn and Cd over the pH range 1–5, as measured by comparison of the DGT-measured concentration (c_{DGT}) to the concentration measured in solution by ICP-MS (c^{soln}) [25]. A value close to one indicates that the DGT method is accurately measuring the analyte concentration at the tested pH. Reprinted from *Anal. Chim. Acta*, Vol 448, J. Gimpel, H. Zhang, W. Hutchinson and W. Davison, Effect of solution composition, flow and deployment time on the measurement of trace metals by the diffusive gradient in thin films technique, 93–103, Copyright (2001), with permission from Elsevier.

freshwater and synthetic seawater), for deployment times that reflect those commonly used in DGT field deployments (typically < 1 week). For deployments greater than one to two weeks, it may not be feasible to perform such laboratory validation experiments, although other factors such as biofouling of the samplers may limit the use of in situ DGT deployments for these extended times [26].

Panther and co-workers [11] evaluated Metsorb and ferrihydrite DGT for measuring dissolved reactive phosphorus in both synthetic freshwater and synthetic seawater, for deployment times of up to four days. This experiment revealed that while both techniques accurately measured DRP in freshwater over a four-day deployment time, only Metsorb DGT was capable of accurately measuring DRP in seawater after four days (ferrihydrite DGT underestimated the DRP concentration by 31%). This was due to competition for binding sites by bicarbonate anions (HCO_3^-), which are present at relatively high concentrations in seawater. Similar experiments [16] showed that the measurement of Mn by Chelex 100 DGT experienced similar limitations in synthetic seawater, underestimating the Mn concentration by 49% after four days. Interestingly, accurate measurement of Mn was possible with deployment times of up to 48 h, suggesting that the short-term validation experiments often reported in the literature (typically < 48 h) may not reveal the limitations of binding layers, which only become apparent after longer deployment times. Tankéré-Muller and co-workers [24] provide a detailed explanation of the susceptibility of Mn measurements with Chelex 100 DGT to competitive interference from Ca^{2+} and Mg^{2+} ions present at high concentrations, as in seawater.

4.4 Performance Characteristics of Existing Binding Layers and Future Research Requirements

Over forty different DGT binding layers have been described in the literature to date for approximately fifty elements (Table 4.1). This demonstrates that the DGT technique is very well suited to elemental analysis. However, only a few studies have looked at a range of organic substances [27–29]. This is clearly an emerging area of application, the implications of which are discussed further below. The extent of validation for each binding layer varies considerably, with many only being partially validated for some analytes. As it is not feasible to present the performance characteristics of every binding layer reported in the literature, only the binding layers that have found widespread, multi-analyte, application will be addressed in detail. These are Chelex 100, ferrihydrite, Metsorb, zirconium dioxide (ZrO_2), 3-mercaptopropyl functionalized silica, AG50W-X8 and XAD18. A summary of the performance characteristics for these binding layers, determined by laboratory validations as described in Section 4.2, is presented in Table 4.2.

The effectiveness of Chelex 100 as a DGT binding layer, in selectively measuring twenty-nine elements, is immediately apparent from Table 4.2. The success of this first commercial resin in DGT measurements was clearly instrumental in attracting other researchers to the technique. However, inspection of Table 4.2 indicates that there are many analytes for which validation data in representative matrices is missing for Chelex 100, compared to newer binding layers. Perhaps there was such confidence in the performance of Chelex-DGT that the research community moved quickly onto interesting and important applications – the nature of scientific funding undoubtedly had a role in this. To some extent the validation expectations have developed over time, as new challenges were encountered with other binding layer types.

More recently developed DGT techniques, especially those for oxyanionic species, have undergone a more comprehensive validation, particularly with respect to longer term deployments and the identification of competing ions [20, 30–32]. This has identified limitations with some analytes and binding layers, illustrating the need for this rigour. With further research some limitations have been observed for Chelex 100 too, for instance for Al and Mn [24, 30]. Although not included in Table 4.2, some of these limitations have been overcome by using mixed binding layers, for instance Chelex 100 with either ferrihydrite or Metsorb [15, 16]. These mixed binding layers have also increased the number of analytes able to be determined from single DGT binding layers, which is an important practical development and will likely make the technique more attractive to environmental consultants and regulatory users. The summary in Table 4.2 will allow researchers to establish whether sufficient validation has been done for a particular application of a DGT technique and perhaps direct research to address some of the gaps still present. Clearly, application of any DGT technique in matrices with high salinity or beyond the typical pH range will require careful validation.

Despite the number of elements able to be determined by DGT, there are some important analytes missing from the list, such as dissolved inorganic nitrogen species. There are emerging contaminants such as platinum, palladium and rare earth elements, which have only been looked at in a few studies [33–35]. The potential of the DGT technique for determination of polar organic contaminants has been demonstrated as a proof of concept but with limited detailed validation to date [27, 28]. This emerging research direction requires careful consideration, as there are actually many other passive samplers developed for various classes of organic substances. Organic contaminants are often present at ultra-trace concentrations in waters that have required other passive samplers to be deployed for longer times than are preferred for DGT deployments (due to the difficulties associated with complex biofilm formation that affect all passive samplers) and to have sampler configurations that maximize the rates of analyte uptake. Given that many DGT techniques have been developed for application in water initially, these factors may explain why there are relatively few examples of DGT being used for organic contaminants to date. However, DGT samplers have a configuration that is inherently flexible and may be further optimized for organic analytes, for instance by using a larger surface area and/or minimal thickness diffusive layer. For some applications, such as in industrial wastewater [28] and contaminated soils and sediments [36, 37], the small size, the relatively low cost of DGT samplers and the fact that the same apparatus can be used to determine elemental contaminants may be advantageous. Furthermore, deployment in sediments and soils can make use of the extensive theory and models developed for metals [36, 37].

Various approaches have been used to prepare DGT binding layers, as described in Section 4.2. It is apparent from Table 4.2 that incorporating binding materials into a polyacrylamide hydrogel has been the most popular and successful approach. In our experience most materials (e.g. resins and powders) can readily be incorporated into polyacrylamide hydrogels. But it is worthwhile for the DGT research community to consider if there are other reasons for the dominance of this approach. Does it relate to the preferences of the most active research groups? Is it because many research groups new to the DGT field, and some perhaps with less background or interest in chemistry, prefer to purchase their binding layers commercially? Such considerations are particularly useful when we consider future directions for the DGT technique – it may be that other approaches to preparing binding layers will provide new opportunities in the future. For instance, may liquid binding layers eventually be combined with real-time measurements for in situ sensors? Might the use of functionalized polyacrylamide gels or membranes be useful for ultra-high resolution measurements in soils and sediments? An interesting binding layer application has been the use of *Saccharomyces cerevisiae* [38] – will incorporation of microorganisms open up interesting new applications of DGT? Will some of the new range of selective nanomaterials be useful as DGT binding layers? Can the DGT technique be improved with regard to sustainability practices? We encourage the DGT research community to consider these matters – it is apparent that there remains plenty of opportunity for research in the future.

References

1. H. Zhang and W. Davison, Performance characteristics of diffusion gradients in thin films for the in situ measurement of trace metals in aqueous solution, *Anal. Chem.* 67 (1995), 3391–3400.
2. W. Davison and H. Zhang, In situ speciation measurements of trace components in natural waters using thin-film gels, *Nature* 367 (1994), 546–548.
3. S. C. Pai, P. Y. Whung and R. L. Lai, Pre-concentration efficiency of Chelex-100 resin for heavy metals in seawater: Part 1. Effects of pH and salts on the distribution ratios of heavy metals, *Anal. Chim. Acta* 211 (1988), 257–270.
4. B. L. Larner and A. J. Seen, Evaluation of paper-based diffusive gradients in thin film samplers for trace metal sampling, *Anal. Chim. Acta* 539 (2005), 349–355.
5. W. Li, H. Zhao, P. R. Teasdale and F. Wang, Trace metal speciation measurements in waters by the liquid binding phase DGT device, *Talanta* 67 (2005), 571–578.
6. W. Li, H. Zhao, P. Teasdale, R. John and S. Zhang, Synthesis and characterisation of a polyacrylamide–polyacrylic acid copolymer hydrogel for environmental analysis of Cu and Cd, *React. Funct. Polym.* 52 (2002), 31–41.
7. L. Y. Chang, W. Davison, H. Zhang and M. Kelly, Performance characteristics for the measurement of Cs and Sr by diffusive gradients in thin films (DGT), *Anal. Chim. Acta* 368 (1998), 243–253.
8. J. G. Panther, K. P. Stillwell, K. J. Powell and A. J. Downard, Development and application of the diffusive gradients in thin films technique for the measurement of total dissolved inorganic arsenic in waters, *Anal. Chim. Acta* 622 (2008), 133–142.
9. H. Zhang, W. Davison, R. Gadi and T. Kobayashi, In situ measurement of dissolved phosphorus in natural waters using DGT, *Anal. Chim. Acta* 370 (1998), 29–38.
10. W. W. Bennett, P. R. Teasdale, J. G. Panther, D. T. Welsh and D. F. Jolley, New diffusive gradients in a thin film technique for measuring inorganic arsenic and selenium (IV) using a titanium dioxide based adsorbent, *Anal. Chem.* 82 (2010), 7401–7407.
11. J. G. Panther, P. R. Teasdale, W. W. Bennett, D. T. Welsh and H. Zhao, Titanium dioxide-based DGT technique for in situ measurement of dissolved reactive phosphorus in fresh and marine waters, *Environ. Sci. Technol.* 44 (2010), 9419–9424.
12. P. R. Teasdale, S. Hayward and W. Davison, In situ, high-resolution measurement of dissolved sulfide using diffusive gradients in thin films with computer-imaging densitometry, *Anal. Chem.* 71 (1999), 2186–2191.
13. W. W. Bennett, P. R. Teasdale, J. G. Panther, D. T. Welsh and D. F. Jolley, Speciation of dissolved inorganic arsenic by diffusive gradients in thin films: selective binding of As^{III} by 3-mercaptopropyl-functionalized silica gel, *Anal. Chem.* 83 (2011), 8293–8299.
14. H. Dočekalová and P. Diviš, Application of diffusive gradient in thin films technique (DGT) to measurement of mercury in aquatic systems, *Talanta* 65 (2005), 1174–1178.
15. S. Mason, R. Hamon, A. Nolan, H. Zhang and W. Davison, Performance of a mixed binding layer for measuring anions and cations in a single assay using the diffusive gradients in thin films technique, *Anal. Chem.* 77 (2005), 6339–6346.
16. J. G. Panther, W. W. Bennett, D. T. Welsh and P. R. Teasdale, Simultaneous measurement of trace metal and oxyanion concentrations in water using diffusive gradients in thin films with a chelex-metsorb mixed binding layer, *Anal. Chem.* 86 (2014), 427–434.
17. C. Naylor, W. Davison, M. Motelica-Heino, G. A. Van Den Berg and L. M. Van Der Heijdt, Simultaneous release of sulfide with Fe, Mn, Ni and Zn in marine harbour

sediment measured using a combined metal/sulfide DGT probe, *Sci. Total Environ.* 328 (2004), 275–286.

18. S. Ding, D. Xu, Q. Sun, H. Yin and C. Zhang, Measurement of dissolved reactive phosphorus using the diffusive gradients in thin films technique with a high-capacity binding phase, *Environ. Sci. Technol.* 44 (2010), 8169–8174.

19. W. Li, P. R. Teasdale, S. Zhang, R. John and H. Zhao, Application of a poly(4-styrenesulfonate) liquid binding layer for measurement of Cu^{2+} and Cd^{2+} with the diffusive gradients in thin-films technique, *Anal. Chem.* 75 (2003), 2578–2583.

20. H. L. Price, P. R. Teasdale and D. F. Jolley, An evaluation of ferrihydrite- and Metsorb^{TM}-DGT techniques for measuring oxyanion species (As, Se, V, P): Effective capacity, competition and diffusion coefficients, *Anal. Chim. Acta* 803 (2013), 56–65.

21. J. L. Levy, H. Zhang, W. Davison, J. Puy and J. Galceran, Assessment of trace metal binding kinetics in the resin phase of diffusive gradients in thin films, *Anal. Chim. Acta* 717 (2012), 143–150.

22. J. G. Panther, R. R. Stewart, P. R. Teasdale et al., Titanium dioxide-based DGT for measuring dissolved As(V), V(V), Sb(V), Mo(VI) and W(VI) in water, *Talanta* 105 (2013), 80–86.

23. C. M. Hutchins, J. G. Panther, P. R. Teasdale et al., Evaluation of a titanium dioxide-based DGT technique for measuring inorganic uranium species in fresh and marine waters, *Talanta* 97 (2012), 550–556.

24. S. Tankéré-Muller, W. Davison and H. Zhang, Effect of competitive cation binding on the measurement of Mn in marine waters and sediments by diffusive gradients in thin films, *Anal. Chim. Acta* 716 (2012), 138–144.

25. J. Gimpel, H. Zhang, W. Hutchinson and W. Davison, Effect of solution composition, flow and deployment time on the measurement of trace metals by the diffusive gradient in thin films technique, *Anal. Chim. Acta* 448 (2001), 93–103.

26. E. Uher, H. Zhang, S. Santos, M. H. Tusseau-Vuillemin and C. Gourlay-Francé, Impact of biofouling on diffusive gradient in thin film measurements in water, *Anal. Chem.* 84 (2012), 3111–3118.

27. C. E. Chen, H. Zhang and K. C. Jones, A novel passive water sampler for in situ sampling of antibiotics, *J. Environ. Monit.* 14 (2012), 1523–1530.

28. C. E. Chen, H. Zhang, G. G. Ying and K. C. Jones, Evidence and recommendations to support the use of a novel passive water sampler to quantify antibiotics in wastewaters, *Environ. Sci. Technol.* 47 (2013), 13587–13593.

29. J. L. Zheng, D. X. Guan, J. Luo et al., Activated charcoal based diffusive gradients in thin films for in situ monitoring of bisphenols in waters, *Anal. Chem.* 87 (2015), 801–807.

30. J. G. Panther, W. W. Bennett, P. R. Teasdale, D. T. Welsh and H. Zhao, DGT measurement of dissolved aluminum species in waters: comparing Chelex-100 and titanium dioxide-based adsorbents, *Environ. Sci. Technol.* 46 (2012), 2267–2275.

31. J. G. Panther, P. R. Teasdale, W. W. Bennett, D. T. Welsh and H. Zhao, Comparing dissolved reactive phosphorus measured by DGT with ferrihydrite and titanium dioxide adsorbents: Anionic interferences, adsorbent capacity and deployment time, *Anal. Chim. Acta* 698 (2011), 20–26.

32. Q. Sun, Y. Chen, D. Xu, Y. Wang and S. Ding, Investigation of potential interferences on the measurement of dissolved reactive phosphate using zirconium oxide-based DGT technique, *J. Environ. Sci.* 25 (2013), 1592–1600.

33. Ø. A. Garmo, O. Røyset, E. Steinnes and T. P. Flaten, Performance study of diffusive gradients in thin films for 55 elements, *Anal. Chem.* 75 (2003), 3573–3580.

34. K. Andersson, R. Dahlqvist, D. Turner et al., Colloidal rare earth elements in a boreal river: changing sources and distributions during the spring flood, *Geochim. Cosmochim. Acta* 70 (2006), 3261–3274.

35. Ø. A. Garmo, N. J. Lehto, H. Zhang et al., Dynamic aspects of DGT as demonstrated by experiments with lanthanide complexes of a multidentate ligand, *Environ. Sci. Technol.* 40 (2006), 4754–4760.

36. C. E. Chen, W. Chen, G. G. Ying, K. C. Jones and H. Zhang, In situ measurement of solution concentrations and fluxes of sulfonamides and trimethoprim antibiotics in soils using o-DGT, *Talanta* 132 (2015), 902–908.

37. C. E. Chen, K. C. Jones, G. G. Ying and H. Zhang, Desorption kinetics of sulfonamide and trimethoprim antibiotics in soils assessed with diffusive gradients in thin-films, *Environ. Sci. Technol.* 48 (2014), 5530–5536.

38. A. A. Menegário, P. S. Tonello and S. F. Durrant, Use of Saccharomyces cerevisiae immobilized in agarose gel as a binding agent for diffusive gradients in thin films, *Anal. Chim. Acta* 683 (2010), 107–112.

39. R. Dahlqvist, H. Zhang, J. Ingri and W. Davison, Performance of the diffusive gradients in thin films technique for measuring Ca and Mg in freshwater, *Anal. Chim. Acta* 460 (2002), 247–256.

40. R. Cusnir, P. Steinmann, F. Bochud and P. Froidevaux, A DGT technique for plutonium bioavailability measurements, *Environ. Sci. Technol.* 48 (2014), 10829–10834.

41. S. Tandy, S. Mundus, H. Zhang et al., A new method for determination of potassium in soils using diffusive gradients in thin films (DGT), *Environ. Chem.* 9 (2012), 14–23.

42. K. W. Warnken, H. Zhang and W. Davison, Performance characteristics of suspended particulate reagent-iminodiacetate as a binding agent for diffusive gradients in thin films, *Anal. Chim. Acta* 508 (2004), 41–51.

43. G. S. C. Turner, G. A. Mills, J. L. Burnett and S. Amos, Evaluation of diffusive gradients in thin-films using a Diphonix® resin for monitoring dissolved uranium in natural waters, *Anal. Chim. Acta* 854 (2015), 78–85.

44. A. M. C. M. Rolisola, C. A. Suarez, A. A. Menegario et al., Speciation analysis of inorganic arsenic in river water by Amberlite IRA 910 resin immobilized in a polyacrylamide gel as a selective binding agent for As(V) in diffusive gradient thin film technique, *Analyst* 139 (2014), 4373–4380.

45. R. W. McGifford, A. J. Seen and P. R. Haddad, Direct colorimetric detection of copper(II) ions in sampling using diffusive gradients in thin-films, *Anal. Chim. Acta* 662 (2010), 44–50.

46. M. A. French, H. Zhang, J. M. Pates, S. E. Bryan and R. C. Wilson, Development and performance of the diffusive gradients in thin-films technique for the measurement of technetium-99 in seawater, *Anal. Chem.* 77 (2005), 135–139.

47. J. Luo, H. Zhang, J. Santner and W. Davison, Performance characteristics of diffusive gradients in thin films equipped with a binding gel layer containing precipitated ferrihydrite for measuring arsenic(V), selenium(VI), vanadium(V), and antimony(V), *Anal. Chem.* 82 (2010), 8903–8909.

48. H. Österlund, S. Chlot, M. Faarinen and A. Widerlund, Simultaneous measurements of As, Mo, Sb, V and W using a ferrihydrite diffusive gradients in thin films (DGT) device, *Anal. Chim. Acta* 682 (2010), 59–65.

49. Q. Sun, J. Chen, H. Zhang et al., Improved diffusive gradients in thin films (DGT) measurement of total dissolved inorganic arsenic in waters and soils using a hydrous zirconium oxide binding layer, *Anal. Chem.* 86 (2014), 3060–3067.

50. D. X. Guan, P. N. Williams, L. Jun and J. L. Zheng, Novel precipitated zirconia-based DGT technique for high-resolution imaging of oxyanions in waters and sediments, *Environ. Sci. Technol.* 49 (2015), 3653–3661.

51. G. S. C. Turner, G. A. Mills, P. R. Teasdale et al., Evaluation of DGT techniques for measuring inorganic uranium species in natural waters: Interferences, deployment time and speciation, *Anal. Chim. Acta* 739 (2012), 37–46.

52. M. Leermakers, Y. Gao, J. Navez et al., Radium analysis by sector field ICP-MS in combination with the diffusive gradients in thin films (DGT) technique, *J. Anal. At. Spectrom.* 24 (2009), 1115–1117.

53. O. Clarisse and H. Hintelmann, Measurements of dissolved methylmercury in natural waters using diffusive gradients in thin film (DGT), *J. Environ. Monit.* 8 (2006), 1242–1247.

54. P. Diviš, R. Szkandera and H. Dočekalová, Characterization of sorption gels used for determination of mercury in aquatic environment by diffusive gradients in thin films technique, *Cent. Eur. J. Chem.* 8 (2010), 1105–1109.

55. Y. Gao, E. De Canck, M. Leermakers, W. Baeyens and P. Van Der Voort, Synthesized mercaptopropyl nanoporous resins in DGT probes for determining dissolved mercury concentrations, *Talanta* 87 (2011), 262–267.

56. Y. Liu, H. Chen, F. Bai, J. Gu and L. Liu, Application of sodium poly(aspartic acid) as a binding phase in the technique of diffusive gradients in thin films, *Chem. Lett.* 41 (2012), 1471–1472.

57. D. P. Sui, H. T. Fan, J. Li et al., Application of poly(ethyleneimine) solution as a binding agent in DGT technique for measurement of heavy metals in water, *Talanta* 114 (2013), 276–282.

58. H. Fan, T. Sun, W. Li et al., Sodium polyacrylate as a binding agent in diffusive gradients in thin-films technique for the measurement of Cu and Cd in waters, *Talanta* 79 (2009), 1228–1232.

59. H. Fan, Y. Bian, D. Sui, G. Tong and T. Sun, Measurement of free copper(II) ions in water samples with polyvinyl alcohol as a binding phase in diffusive gradients in thin-films, *Anal. Sci.* 25 (2009), 1345–1349.

60. H. T. Fan, J. X. Liu, D. P. Sui et al., Use of polymer-bound Schiff base as a new liquid binding agent of diffusive gradients in thin-films for the measurement of labile Cu, Cd and Pb, *J. Hazard. Mater.* 260 (2013), 762–769.

61. H. T. Fan, T. Sun, M. Z. Xue, G. F. Tong and D. P. Sui, Preparation of thiol-polyvinyl alcohol and its application to diffusive gradients in thin-films technique, *Chinese J. Anal. Chem.* 37 (2009), 1379–1381.

62. H. Chen, Y. Y. Zhang, K. L. Zhong et al., Selective sampling and measurement of Cr (VI) in water with polyquaternary ammonium salt as a binding agent in diffusive gradients in thin-films technique, *J. Hazard. Mater.* 271 (2014), 160–165.

63. H. Chen, H. Chen, M. H. Zhang, J. L. Gu and G. Zhao, Measurement of dissolved reactive phosphorus in water with polyquaternary ammonium salt as a binding agent in diffusive gradients in thin-films technique, *J. Agric. Food Chem.* 62 (2014), 12112–12117.

64. C. D. Colaço, L. N. M. Yabuki, A. M. Rolisola et al., Determination of mercury in river water by diffusive gradients in thin films using P81 membrane as binding layer, *Talanta* 129 (2014), 417–421.

65. E. De Almeida, V. F. Do Nascimento Filho and A. A. Menegário, Paper-based diffusive gradients in thin films technique coupled to energy dispersive X-ray fluorescence spectrometry for the determination of labile Mn, Co, Ni, Cu, Zn and Pb in river water, *Spectrochim. Acta, Part B* 71–72 (2012), 70–74.

66. W. Li, H. Zhao, P. R. Teasdale, R. John and S. Zhang, Synthesis and characterisation of a polyacrylamide–polyacrylic acid copolymer hydrogel for environmental analysis of Cu and Cd, *React. Funct. Polym.* 52 (2002), 31–41.

67. W. Li, H. Zhao, P. R. Teasdale and R. John, Preparation and characterisation of a poly(acrylamidoglycolic acid-co-acrylamide) hydrogel for selective binding of Cu^{2+} and application to diffusive gradients in thin films measurements, *Polymer* 43 (2002), 4803–4809.

68. T. Huynh, H. Zhang and B. Noller, Evaluation and application of the diffusive gradients in thin films technique using a mixed-binding gel layer for measuring inorganic arsenic and metals in mining impacted water and soil, *Anal. Chem.* 84 (2012), 9988–9995.

69. Q. Sun, L. Zhang, S. Ding et al., Evaluation of the diffusive gradients in thin films technique using a mixed binding gel for measuring iron, phosphorus and arsenic in the environment, *Environ. Sci.: Processes Impacts* 17 (2015), 570–577.

70. D. Xu, Y. Chen, S. Ding et al., Diffusive gradients in thin films technique equipped with a mixed binding gel for simultaneous measurements of dissolved reactive phosphorus and dissolved iron, *Environ. Sci. Technol.* 47 (2013), 10477–10484.

71. Y. Zhang, S. Mason, A. Mcneill and M. J. Mclaughlin, Application of the diffusive gradients in thin films technique for available potassium measurement in agricultural soils: Effects of competing cations on potassium uptake by the resin gel, *Anal. Chim. Acta* 842 (2014), 27–34.

72. Y. Zhang, S. Mason, A. Mcneill and M. J. McLaughlin, Optimization of the diffusive gradients in thin films (DGT) method for simultaneous assay of potassium and plant-available phosphorus in soils, *Talanta* 113 (2013), 123–129.

73. Y. Gao, S. Van De Velde, P. N. Williams, W. Baeyens and H. Zhang, Two-dimensional images of dissolved sulfide and metals in anoxic sediments by a novel diffusive gradients in thin film probe and optical scanning techniques, *TrAC, Trends Anal. Chem.* 66 (2015), 63–71.

74. S. Ding, Q. Sun, D. Xu et al., High-resolution simultaneous measurements of dissolved reactive phosphorus and dissolved sulfide: The first observation of their simultaneous release in sediments, *Environ. Sci. Technol.* 46 (2012), 8297–8304.

75. A. Stockdale, W. Davison and H. Zhang, High-resolution two-dimensional quantitative analysis of phosphorus, vanadium and arsenic, and qualitative analysis of sulfide, in a freshwater sediment, *Environ. Chem.* 5 (2008), 143–149.

76. A. Kreuzeder, J. Santner, T. Prohaska and W. W. Wenzel, Gel for simultaneous chemical imaging of anionic and cationic solutes using diffusive gradients in thin films, *Anal. Chem.* 85 (2013), 12028–12036.

77. C. Murdock, M. Kelly, L. Y. Chang, W. Davison and H. Zhang, DGT as an in situ tool for measuring radiocesium in natural waters, *Environ. Sci. Technol.* 35 (2001), 4530–4535.

78. J. Dong, H. Fan, D. Sui, L. Li and T. Sun, Sampling 4-chlorophenol in water by DGT technique with molecularly imprinted polymer as binding agent and nylon membrane as diffusive layer, *Anal. Chim. Acta* 822 (2014), 69–77.

79. W. Li, F. Wang, W. Zhang and D. Evans, Measurement of stable and radioactive cesium in natural waters by the diffusive gradients in thin films technique with new selective binding phases, *Anal. Chem.* 81 (2009), 5889–5895.

80. M. Gregusova and B. Docekal, New resin gel for uranium determination by diffusive gradient in thin films technique, *Anal. Chim. Acta* 684 (2011), 142–146.

81. A. Lucas, A. Rate, H. Zhang, S. U. Salmon and N. Radford, Development of the diffusive gradients in thin films technique for the measurement of labile gold in natural waters, *Anal. Chem.* 84 (2012), 6994–7000.

82. V. E. Dos Anjos, G. Abate and M. T. Grassi, Potentiality of the use of montmorillonite in diffusive gradients in thin film (DGT) devices for determination of labile species of Cu, Cr, Cd, Mn, Ni, Pb, and Zn in natural waters, *Braz. J. Anal. Chem.* 1 (2010), 187–193.

83. H. Ernstberger, H. Zhang and W. Davison, Determination of chromium speciation in natural systems using DGT, *Anal. Bioanal. Chem.* 373 (2002), 873–879.

84. C. Fernández-Gómez, B. Dimock, H. Hintelmann and S. Díez, Development of the DGT technique for Hg measurement in water: Comparison of three different types of samplers in laboratory assays, *Chemosphere* 85 (2011), 1452–1457.

5

Interpreting the DGT Measurement

Speciation and Dynamics

JAUME PUY, JOSEP GALCERAN AND CARLOS REY-CASTRO

5.1 Introduction

Diffusive gradients in thin-films (DGT) is a technique widely used for in situ monitoring a range of analytes, including metal cations [1–3], nanoparticles [4, 5], oxyanions [6–8] and other inorganic and organic [9] components in waters, sediments [10, 11] and soils [12]. The reasons for its success include the simple manipulation of the devices, their optimized design for *in situ* measurements (see Chapter 1 and [13]), the summarizing of the experimental information in a straightforward usable quantity and the variety of the applications developed [14].

Parallel to the experimental development, there has also been a need to develop a theoretical framework for the interpretation of the measurements in terms of the system parameters and for the assessment of the influence of phenomena not considered previously [15–22]. As described in the preceding chapters, a combination of conveniently modified DGT devices, carefully designed experiments and simulation tools have been used to understand the relevant phenomena in DGT measurements and to assess the accuracy of simple analytical expressions to describe the analyte accumulation in the resin. These analytical expressions have mostly been derived under limiting conditions, such as the assumption of perfect sink behaviour for the resin, negligible transient effects, and a common diffusion coefficient for DBL, filter and gel domain.

Chapter 2 discusses the basic principles of DGT measurements in solutions where the speciation of the analyte is not relevant. In the present chapter, we review a DGT framework, which essentially considers the analyte accumulation as arising from the dynamic interplay of diffusion and dissociation of the species present. In general, trace metals in solution interact with a broad range of components, such as small inorganic and organic ligands, macromolecular ligands arising from the decay of organic matter, particles, colloids and surfaces. In most environmental compartments, concentrations of trace metals are significantly smaller than the concentrations of ligands, so that the complexes are the dominant metal species. In such systems, the dynamics of the speciation processes, as well as the mobility of the different species, will influence their availability to biota, as considered in more detail in Chapter 9. When metal ions bind to the DGT resin layer, all these processes tend to buffer the metal ion consumption, contributing in this way to the metal accumulation. The development of a comprehensive framework of metal availability aims

93

to: (i) quantify the species that mostly contribute to the metal accumulation, which can be, then, grouped according to their common chemical traits and (ii) describe and predict the impact of the different physicochemical parameters and solution conditions. An important outcome of this quantitative treatment is that it explains why DGT usually measures as labile most of the metal complexes within a natural system, a result that stems from the finite thickness of the resin-binding disc compared to an ideal binding surface. This key role of the binding layer in the metal accumulation is another indication that, as stated in Chapter 1, DGT is not just a typical "passive sampler."

The simplest case considered in this chapter assumes that the metal M forms just one complex ML with a ligand L:

$$M + L \underset{k_d}{\overset{k_a}{\rightleftharpoons}} ML \tag{5.1}$$

The stability constant K (ratio of the association and dissociation kinetic constants, k_a and k_d) governs the concentration ratio in the bulk (denoted with superscript *) of the sample solution:

$$K = \frac{k_a}{k_d} = \frac{c_{ML}^*}{c_M^* c_L^*} \tag{5.2}$$

The framework developed here for metals can be extended to other analytes targeted by DGT, such as organic pollutants or anions, which may also interact with surfaces, dissolved organic matter, colloids, etc. These interactions influence the accumulation of the analyte in a similar way as the metal case described here.

While some detail will be devoted to dynamic speciation issues, other phenomena will be omitted for the sake of simplicity. For instance, in most of this work, a common diffusion coefficient for a given species will be considered in the diffusive gel, filter and diffusive boundary layer (DBL). The interested reader will be redirected to the relevant literature for further details on specific issues.

5.2 Which Species Contribute to Metal Accumulation in DGT?

5.2.1 Lability Degree: Definition and Measurement

As indicated in the Introduction, metal speciation can include a large number of species in natural media due to the presence of a rich mixture of ligands. DGT accumulation is dependent on the features of this complex mixture, and thus a better appreciation of which species are being measured would greatly increase our understanding of the natural systems [23]. The introduction of the lability degree in this section aims at quantifying the specific contribution of each complex species to the DGT accumulation.

As described in Chapter 1, the DGT device consists of a resin layer (of thickness δ^r), covered with a diffusive gel layer (of thickness δ^g) and a filter in contact with the probed phase (an aqueous solution in the present chapter) where a diffusive boundary layer, DBL (of thickness δ^{dbl}) can develop. When the diffusion coefficients for a given species are the same in gel, filter and DBL, the expressions derived considering diffusion only in the gel domain (e.g. those in this Section 5.2) can be applied to the whole system, just by understanding that δ^g refers to the aggregate thickness of gel, filter and DBL. Let n_M be the number of moles of the analyte (a metal M) accumulated in the resin layer at a given time (t). The flux can be computed as

$$J(t) = \frac{1}{A}\frac{dn_M}{dt} \tag{5.3}$$

where A is the effective area of the gel-solution interface. A detailed plot of n_M as a function of increasing deployment times would reveal four regimes: a transient stage (where the flux is growing), a (quasi) steady-state regime (practically constant flux), a transition to equilibrium (decreasing flux) and a final equilibrium state (null flux) [24–26]. As detailed in Chapter 2, the practical use of DGT has focussed on the steady-state regime (while neglecting the very small effects of the short transient regime), so that the time integral of the flux becomes just

$$n_M = JAt \tag{5.4}$$

Thus, DGT accumulation is also providing a quantification of the flux.

This steady-state flux will result from the contributions of free M and other M-containing species arriving to the sensor. We start with the case of just one complex as given in equation 5.1. If the dissociation rate constant is sufficiently small, the complex does not dissociate and, thus, the flux of free metal, J_{free}, would be the only contribution to the total flux. At the other extreme, for very large values of the dissociation rate constant, metal and complex are in equilibrium at all points, and the resulting flux (J_{labile}) is called *fully labile*. The extent of the actual participation of the complex in the total flux can be quantified by the lability degree ξ [27], which is defined as

$$\xi \equiv \frac{J - J_{free}}{J_{labile} - J_{free}} \tag{5.5}$$

With this definition, $\xi = 0$ and $\xi = 1$ correspond to completely inert and fully labile complexes, respectively.

In most actual trace speciation measurements J_{free} is almost negligible because the concentration of free metal is very low. Under these conditions, and if the diffusion coefficient of the complex is known, one can compute the lability degree as

$$\xi \approx \frac{J}{J_{labile}} \approx \frac{n_M/(At)}{\left(\dfrac{D_{ML}}{D_M}\right) n_M^{\text{no-ligand}}/(At)} = \frac{D_M\, n_M}{D_{ML}\, n_M^{\text{no-ligand}}} \tag{5.6}$$

where the labile flux has been estimated from the accumulation of M in the same system at a fixed deployment time, but without ligand (i.e. just the metal), and a re-scaling factor to take into account the different diffusion coefficients of the complex (D_{ML}) and metal (D_M). In this sense, the lability degree can be interpreted as a normalized flux (where differences in diffusion coefficients are also accounted for). Thus, in principle, the lability degree of a complex can be determined experimentally from simple DGT experiments measuring the accumulation in systems with and without ligand (for a fixed total metal concentration) whenever the diffusion coefficients of metal and complex are known.

5.2.2 Diffusion and Reaction Processes in a DGT Device. Limiting Cases: Inert and Labile Complexes

As explained in Chapter 1, the diffusive gel was introduced in the design of the DGT device to minimize convection effects in the DGT accumulations. Thus, diffusion (of the free metal, metal complex and free ligand species) is the only relevant transport phenomenon in both the diffusive and the resin gels. Even though, in the early days of DGT, it was speculated that some metal complexes were unable to penetrate into the diffusive gel, it is now clear that complexes with organic ligands or DOM can penetrate both the diffusive and resin gels [28, 29]. DGT devices with a stack of resin discs have been used to evidence these phenomena [30]. Together with diffusion, association/dissociation reactions corresponding to the process (5.1) have to be considered in both (resin and diffusive) gel domains. In most cases, complexes with different stoichiometric metal to ligand ratios are formed and ligands are involved in acid-base equilibria. In both cases, the formulation must include the transport of all the chemical species coupled with the corresponding reaction processes. Additionally, the following reaction arises in the resin domain between free metal and resin sites (R) to form occupied sites (MR):

$$\mathrm{M} + \mathrm{R} \underset{k_{d,R}}{\overset{k_{a,R}}{\rightleftharpoons}} \mathrm{MR} \tag{5.7}$$

where $k_{a,R}$ and $k_{d,R}$ are the association and dissociation rate constants for this process.

In general, no ternary species (such as those corresponding to the direct binding of the complexes to the resin sites) are assumed, so that the binding of the metal requires previous dissociation of the complexes. The standard DGT device can be modelled in just one relevant dimension (i.e., edge effects can be neglected [22]). The general mathematical formulation of such systems, for the case of one ligand can be found in references [17, 19, 31, 32], while systems with a set of ligands are discussed in [33].

Different simplifications from this general formulation are of interest, because they are a requisite to move from numerical simulation to analytical solutions. For trace metals, ligands are commonly in excess, so that $c_{ML}^* \ll c_{T,L}^*$, where $c_{T,L}^*$ represents the total ligand concentration. In other cases, the ligand concentration is often buffered by fast acid–base processes [34, 35]. In both cases, the concentration of the ligand is constant throughout the DGT binding and diffusion layers (a flat ligand profile):

$$c_L(x) = c_L^* \tag{5.8}$$

and the corresponding system of equations is conveniently simplified by linearization.

Another important simplification arises when the metal binds so fast and strongly to the resin sites that the free metal concentration can be neglected in the resin domain. This is the so-called "perfect sink" regime (for metal) which, together with conditions of ligand excess, allows approximate analytical expressions for the metal accumulation to be obtained in some cases [36].

Figure 5.1 shows the normalized concentration profiles in a standard DGT device obtained from numerical simulation (as described in ref. [36]) of a generic system with only one metal complex. The overlapping of the normalized profiles of metal and complex indicates local equilibrium conditions between M and ML [25]. The region where M and ML profiles diverge is called the reaction layer and, in DGT, extends to both sides of the resin/diffusive gel interface. Once these profiles have been established, i.e., once the system has reached quasi-steady-state conditions, the metal complexes that diffuse from the solution towards the resin domain do not dissociate until they reach the reaction layer domain. In the reaction layer, the normalized complex profile is above the metal one, indicating net dissociation and production of free M. As shown in ref. [36], the extent of the reaction layer in the resin and diffusive gels (for ligand excess and perfect sink conditions) has a thickness of λ_{ML} (see equation 5.9) and m (see equation 5.10), respectively (whenever $\delta^r > \lambda_{ML}$ and $\delta^g > m$). Thus, the summation $\lambda_{ML} + m$ is a good estimation of the total reaction layer thickness.

$$\lambda_{ML} = \sqrt{\frac{D_{ML}}{k_d}} \tag{5.9}$$

is related to the distance of penetration of ML inside the resin, and it can be seen as the intercept of the tangent line of c_{ML}-profile at $x = \delta^r$ with the abscissae axis (i.e., zero ML concentration); whereas

$$m = \sqrt{\frac{D_{ML}}{k_d\,(1 + \varepsilon K')}} \tag{5.10}$$

is a quantification of the thickness (within the diffusive gel phase) where c_M and c_{ML} are not in equilibrium (i.e., the reaction layer in the gel domain), being

$$K' = K c_L^* = \frac{c_{ML}^*}{c_M^*} \tag{5.11}$$

and

$$\varepsilon = \frac{D_{ML}}{D_M} \tag{5.12}$$

Notice that m (which is typically of the order of microns) corresponds to the total reaction layer thickness in voltammetric sensors, since complexes cannot penetrate into the electrode, whereas in DGT the overall reaction layer can be extended up to $\delta^r = 0.4$ mm in a standard DGT device. Thus, the resin thickness has a relatively large impact on the lability behavior of complexes measured by DGT, rendering them much more labile than in electrochemical sensors. Complex penetration is expected since both the resin disc and the diffusive gel are made of the same polymer, APA, (see Chapter 3) in the standard DGT device.

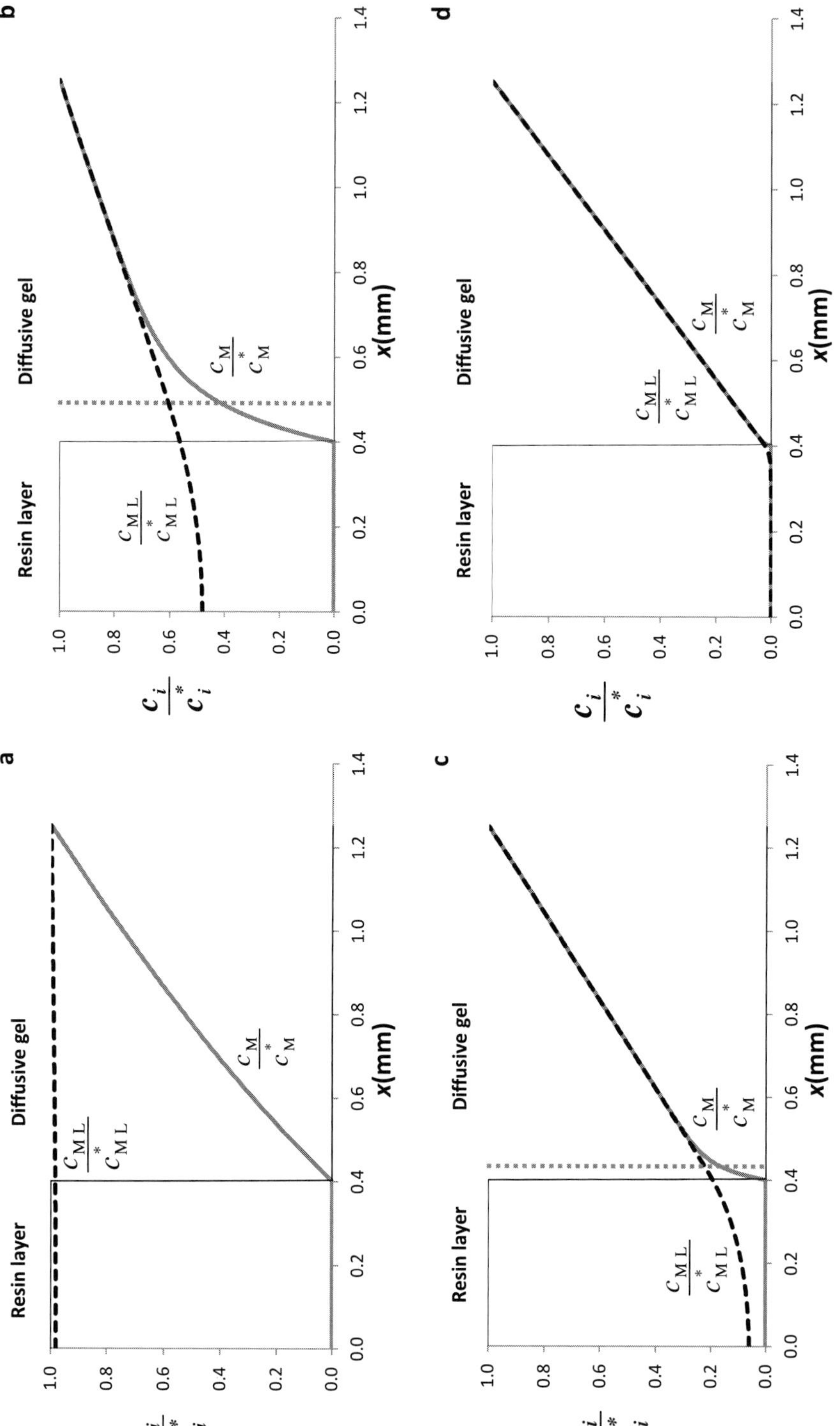

Figure 5.1 Concentration profiles of metal and complex species in a standard DGT device, normalized with respect to their respective values in bulk solution for four different complex dissociation rates: $k_{\mathrm{d}} = 10^{-4}$ s^{-1} (inert case) (a); 10^{-3} s^{-1} (b); 10^{-2} s^{-1} (c); 10 s^{-1} (labile case) (d). Parameters: $\delta^{\mathrm{r}} = 4.00 \times 10^{-4}$ m; $\delta^{\mathrm{g}} = 8.52 \times 10^{-4}$ m; $c_{\mathrm{T,M}} = 0.01$ mol/m^3; $c_{\mathrm{T,L}} = 1$ mol/m^3; $D_{\mathrm{M}} = 7.06 \times 10^{-10}$ m^2 s^{-1}, $D_{\mathrm{ML}} = D_{\mathrm{L}} = 4.80 \times 10^{-10}$ m^2 s^{-1}, and $K = 100$ m^3/mol. Binding to the resin is fast and strong, so that it practically fulfils perfect sink conditions. Saturation effects are also negligible. In panels b and c, the dotted vertical lines indicate the effective position of the reaction layer boundary.

In the inert case, depicted in Figure 5.1a, the complex does not dissociate at all. The ML profile is flat, and only free M is diffusing towards the resin. The steady-state concentration profile of M is linear and thus, according to Fick's first law,

$$J = J_{\text{free}} = D_M \frac{c_M^*}{\delta^g} \tag{5.13}$$

Conversely, in the labile case represented in Figure 5.1d, the normalized metal and complex profiles match each other along the diffusion domain from bulk solution to the resin–gel interface, where both concentrations become negligible. In this case, the dissociation is so fast that the reaction layer has been reduced to the resin–diffusive gel plane, where all the arriving metal and complex instantaneously react with the resin sites. Thus, the flux can be written as

$$J = J_{\text{labile}} = D_M \frac{c_M^*}{\delta^g} + D_{ML} \frac{c_{ML}^*}{\delta^g} \tag{5.14}$$

Equation 5.14 indicates that, in this case, the contribution of the complex reaches its maximum value.

In a general case, by replacing J_{free} and J_{labile} in equation 5.5, the flux of metal bound to the resin can be expressed as

$$J = D_M \frac{c_M^*}{\delta^g} + D_{ML} \frac{c_{ML}^*}{\delta^g} \xi \tag{5.15}$$

indicating that the flux depends on both the mobility (D_M and D_{ML}) and the lability degree, ξ, of the species.

It can be shown [27, 37] that, for systems with only one ligand and vanishing concentration of M at the resin–gel interface ($x = \delta^r$), the lability degree is directly connected to the normalized concentration of ML at that point:

$$\xi = 1 - \frac{c_{ML}^r}{c_{ML}^*} \tag{5.16}$$

where the superscript r indicates that the concentration of ML is taken at $x = \delta^r$.

Thus, for an inert complex, $\xi = 0$ and $c_{ML}^r = c_{ML}^*$, indicating that there has not been complex dissociation at $x = \delta^r$, and the contribution of the complex to the flux is null. Conversely, for a labile complex, $\xi = 1$ and $c_{ML}^r = 0$, indicating that the complex is fully dissociated at $x = \delta^r$, the contribution of the complex is maximum and $J = J_{\text{labile}}$.

5.2.3 Mixtures of Ligands. Interpretation of c_{DGT}

Natural systems are complex mixtures of ligands. In the simplest case of 1:1 stoichiometric ratios, a set of parallel reactions involving h ligands (denoted from ^1L to ^hL) can form h complexes. It can be shown – see Eqn. (SI-21) in the Supporting Information of [33] – that, when M vanishes at $x = \delta^r$, the total flux can be formally written as [38]

$$J = \frac{D_M c_M^*}{\delta^g} + \sum_{j=1}^{h} \left[\frac{D_{M^jL} c_{M^jL}^*}{\delta^g} \xi_j \right] \tag{5.17}$$

where index j scans all the complexes in the system and

$$\xi_j \equiv 1 - \frac{c^r_{M^jL}}{c^*_{M^jL}} \tag{5.18}$$

The previous expressions 5.17 and 5.18 hold even in the absence of ligand excess [37].

The meaning of equation 5.17 is quite simple: the total flux is just the sum of the free metal flux plus a fraction of the maximum contribution of each complex given by its corresponding lability degree.

Notice that expression 5.18 for the lability degree of a complex in a mixture is formally the same as in the case of a single ligand system (equation 5.16). However, the values of $c^r_{M^jL}$ in a mixture will differ from those in the single ligand system, as discussed in the next section.

Equation 5.17 also allows a simple interpretation of c_{DGT} in terms of the composition of the system. The conventional definition of c_{DGT}

$$J = \frac{D_M \, c_{DGT}}{\delta^g} \tag{5.19}$$

indicates that c_{DGT} is the metal concentration of a hypothetical "only-M" system that would yield the same flux as the system under study (where the complexes also contribute to the flux by dissociation). A comparison of equations 5.19 and 5.17 allows to solve for c_{DGT}:

$$c_{DGT} = c^*_M + \sum_j \xi_j \frac{D_{M^jL}}{D_M} c^*_{M^jL} \tag{5.20}$$

which provides an interpretation of c_{DGT} in terms of the species present in the system: *each complex species contributes to c_{DGT} with an "effective concentration" which may be regarded as the hypothetical free metal concentration that would provide the same input to the overall metal flux.*

Notice the presence of two factors in the contribution of each complex to c_{DGT}: one factor depends on the relative mobility of the complex and free metal species, while the other factor stands for the lability of the complex. Both factors can be relevant to understand why c_{DGT} differs from the total metal concentration as discussed with more detail in [26].

If all diffusion coefficients are similar ($D_M \simeq D_{M^jL} \quad \forall j$)

$$c_{DGT} \approx c^*_M + \sum_j \xi_j c^*_{M^jL} \tag{5.21}$$

which *supports the basic notion that DGT is essentially measuring the labile fraction of the metal present in the system* [14, 39–41]. This is an interesting result that has simplified the interpretation and manipulation of the DGT outcome, by allowing it to be referred to as a concentration rather than a flux.

Notice, however, that for a general system with complexes having very different mobilities, c_{DGT} cannot be interpreted as the labile metal concentration present in the system, but as the effective free metal concentration leading to the measured accumulation.

Via c_{DGT}, one essentially quantifies availability (a flux), which is dependent on both lability and mobility. The good correlation between c_{DGT} and availability to biota supports the statement that fluxes are important in determining uptake [42, 43] (See Chapter 9).

5.2.4 Lability Degree in Complex Mixtures

A strategy for dealing with complex mixtures consists in using information obtained from simpler systems. This section aims to discuss whether lability degrees measured in single ligand systems as indicated in Section 1.1, can be used in complex mixtures. A detailed discussion of this topic was given in ref. [33]. Here we will reproduce the main conclusion together with a simple explanation.

Actually, parallel complexes mutually influence their lability degrees, which would suggest that lability degrees measured in single ligand systems cannot be used in a mixture. This is the outcome of studies both on voltammetric sensors [38, 44–46] and on DGT [33].

A simple explanation of the change in individual lability degree follows. In a single ligand system, the normalized profiles of complex and metal almost merge from the bulk solution up to the reaction layer in the diffusive gel (see Figure 5.1b or c). The value of the metal and complex normalized profiles out of the reaction layer depends on the lability degree of the complex. A mixture of two complexes would require the metal profile to converge to two different complex profiles (different lability degrees), which is not possible. So, some re-adaptation is needed. A new steady state situation will be reached where the less labile complex will increase its lability, while for the more labile complex, the situation will be just the opposite. In summary, there is a mutual and opposite influence.

However, the modification of the lability degree due to the "mixture effect" has a limited impact on the DGT flux. Indeed, in DGT, metal accumulation is mainly due to the dissociation processes in the resin domain, where there is no free metal. There, dissociation of each complex proceeds as an independent process provided that the association of the released metal to the resin sites is faster than any other process, i.e., whenever the resin acts as a perfect sink. Under these conditions, the addition of other ligands has only a mild influence on the lability degree of the rest of complexes, this influence coming from the coupling of the dissociation processes taking place in the reaction layer located in the diffusive gel. Additionally, due to the opposite influence on the lability of each member of each couple of complexes, there is a trend to global cancellation.

This result allows an approximate estimation of the metal accumulation in mixtures by using the lability degree available for each complex in a single ligand system [26, 33]

$$J \approx \frac{D_M c_M^* + \sum_j \xi_j^{h=1} D_{M^jL} c_{M^jL}^*}{\delta^g} \tag{5.22}$$

where the superscript $h = 1$ indicates that the lability degree is obtained (either experimentally or theoretically) from a system where there is only ligand j.

Maximum discrepancies of 10 percent in the accumulation flux have been found using these predictions [33]. This discrepancy can be reduced further if thicker resin gels are used, so that dissociation in the resin domain becomes even more dominant.

5.2.5 Simple Analytical Expressions for the Metal Flux and Lability Degree

The availability of theoretical expressions to predict the lability degree in a single ligand system can be useful to: (i) compare with the experimental procedure described in Section 5.2 (see equation 5.6) and (ii) estimate the total flux in systems with several complexes where $\xi_j^{h=1}$ can be a surrogate for the actual lability degree ξ_j (see equation 5.22).

The prediction of the lability degree (and, therefore, the flux) in a system with just one complex, taking into account the kinetics of association/dissociation, was achieved with the recognition that penetration of ML into the resin layer (considered as an ideal one-dimensional domain with binding sites evenly distributed) exerts a major influence on the accumulation, as supported by numerical simulation [17, 19, 36]. On the basis of this key idea, a penetration analytical model (PAM) was developed [26, 36], with the help of additional hypotheses such as: (i) perfect sink for M throughout the resin layer (as backed up by experimental observations [47, 48]) and (ii) excess of ligand L. According to this model, the lability degree becomes

$$\xi = 1 - \frac{c_{ML}^r}{c_{ML}^*} = 1 - \frac{\left(1 + \varepsilon K'\right)}{\varepsilon K' + \dfrac{\delta^g}{m} \coth\left(\dfrac{\delta^g}{m}\right) + \dfrac{\delta^g}{\lambda_{ML}} \left(1 + \varepsilon K'\right) \tanh\left(\dfrac{\delta^r}{\lambda_{ML}}\right)} \quad (5.23)$$

This approach was able to justify the lability of the CdNTA system obtained experimentally with DGT [31].

It is worth noticing that the lability degree is not an intrinsic parameter of a complex. It depends on both the geometric characteristics of the sensor, like δ^g or δ^r, and the characteristics of the media like c_L^*. Expression 5.23 allows assessment of the influence of the different parameters on the lability degree.

The high impact of the thickness of the resin disc on the lability of the complex [31] is linked to the extension of the reaction layer due to complex penetration into the binding layer, as commented in Section 5.1.2. This increase in the lability degree is steepest for δ^r-values of the order of λ_{ML} (see Figure 5.2, where $\lambda_{ML} = 2.5 \times 10^{-7}$ m). When $\delta^r \gg \lambda_{ML}$, the resin is thick enough for the complex to reach full dissociation in the resin layer (in the distance λ_{ML}) and further increases of δ^r do not modify the lability degree.

The condition $\delta^g > m$ applied to equation 5.23 leads to

$$\xi \approx 1 - \frac{1}{1 + \dfrac{\delta^g}{\lambda_{ML}} \tanh\left(\dfrac{\delta^r}{\lambda_{ML}}\right)} \quad (5.24)$$

which indicates that, for given values of δ^r and δ^g, ξ is only dependent on D_{ML} and k_d (*i.e.*, independent of D_M, k_a) Conceptually, the dependence of the lability degree on just D_{ML}

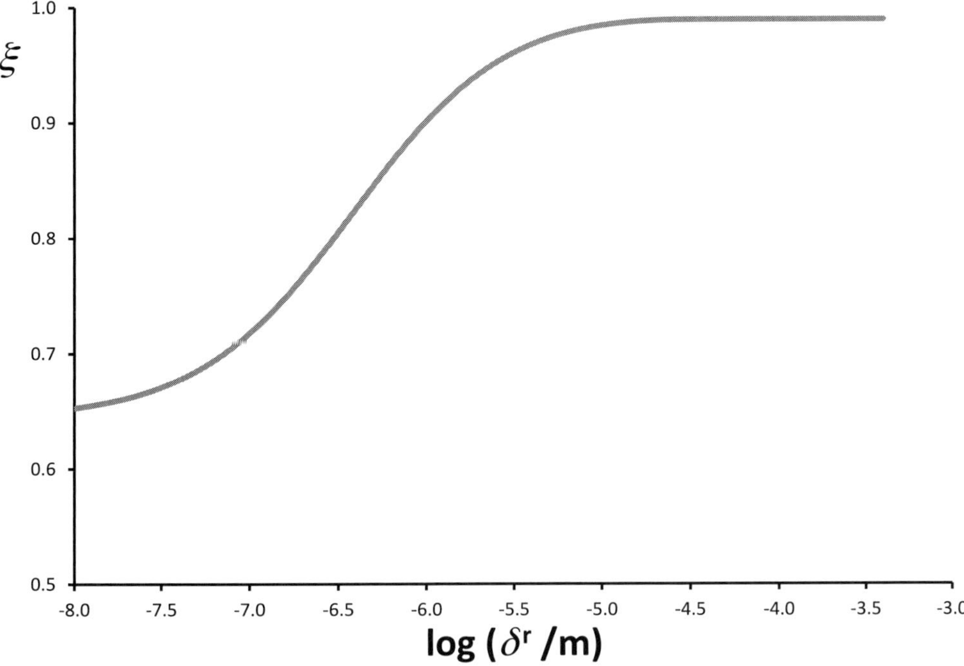

Figure 5.2 Variation of the lability degree, ξ, with the logarithm of the thickness of the resin layer δ^r, as determined by the analytical expression 5.23. Parameters: $D_{ML} = 4.26 \times 10^{-10}$ m^2 s^{-1}; $D_M = 6.09 \times 10^{-10}$ m^2 s^{-1}; $\delta^g = 1.13 \times 10^{-3}$ m; $k_d = 2.76$ s^{-1}.

and k_d is consistent with dissociation and diffusion of the complex in the resin layer being the main mechanism for the metal accumulation.

As discussed in [36, 49], expression 5.24 also allows to evidence that a change in the value of δ^g (around typical values) has a mild impact on ξ. In contrast to the significant influence of the ligand concentration on the lability degree in measurements made with voltammetric sensors [45], equation 5.24 predicts a *mild dependence of the lability degree on the ligand concentration*. This *special feature of DGT sensors* [25, 33] follows from the absence of metal in the resin domain which prevents, for instance, the shift of the complexation process towards association when the ligand concentration is increased.

Equation 5.24 does not consider any influence of pH on the DGT accumulation, since it follows from equations 5.1 and 5.7, which do not involve any reaction with protons. However, an influence of pH is expected from two sources: the participation of the ligand L on acid-base equilibria and the protonation of the resin sites. Actually, the participation of the ligand in acid base processes leads to more involved formulation schemes than indicated in equation 5.1. Under some conditions, the resulting scheme can be reduced to that represented in equation 5.1 by considering effective kinetic and thermodynamic constants dependent on the ligand concentration and pH. In these cases, an indirect influence

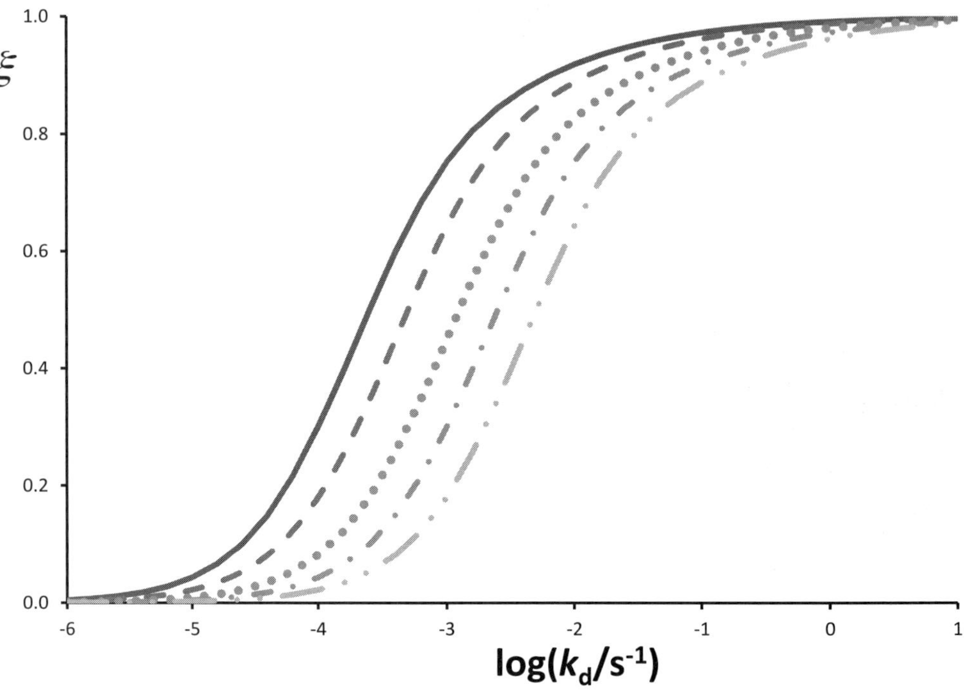

Figure 5.3   Variation of the lability degree ξ in a standard DGT sensor ($\delta^g = 1.13 \times 10^{-3}$ m; $\delta^r = 4 \times 10^{-4}$ m) according to equation 5.24 for different diffusion coefficients of the complex, $D_{ML} = 1.00 \times 10^{-10}$ m^2 s^{-1} (continuous line), 2×10^{-10} m^2 s^{-1} (dashed line), 5.00×10^{-10} m^2 s^{-1} (dotted line), 1.00×10^{-9} m^2 s^{-1} (dash dotted line); and 2.00×10^{-9} m^2 s^{-1} (dash double dotted line)

of pH or $c^*_{T,L}$ on the accumulation is expected via its influence in the "apparent" kinetic dissociation constant [24, 31, 34, 35].

The influence of the pH on the DGT accumulation via the protonation of the resin is discussed in Chapter 4 as a competition effect. In general, as pH decreases the "effective" binding strength of the resin sites for the metal decreases due to the increasing part of the binding energy that has to be expended in extracting the protons. The effects of a decreasing binding strength of the resin are discussed in the next section.

Figure 5.3 shows a series of theoretical estimates of the lability with equation 5.24 as a function of the kinetic dissociation constant for different D_{ML} values [36]. For example, complexes with $D_{ML} = 5 \times 10^{-10}$ m^2 s^{-1} will start showing a reduced lability at $k_d < 0.1$ s^{-1}. As seen in Figure 5.3, the decrease in lability starts at higher k_d values as D_{ML} increases, since the effective time scale for dissociation is reduced. Simple algebra on equation 5.24 can be used to obtain the relationship between the diffusion coefficient of the complex and the kinetic dissociation constant corresponding to a lability degree equal to $1/2$. For δ^r and δ^g corresponding to the standard DGT devices ($\delta^r = 4 \times 10^{-4}$ m and $\delta^g = 9 \times 10^{-4}$ m), the relationship in SI units is $D_{ML} = 3.08 \times 10^{-7} k_d$. This simple expression allows us to assess the approximate k_d value below which complexes are inert for a given value of the complex diffusion coefficient.

5.3 Further Refinements

5.3.1 Does the Resin Act as a Perfect Sink? Kinetic Experiments when Only Metal is Present

As seen in Section 5.1.5, the Penetration Analytical Model assumes that trace metals react instantaneously with the binding phase and the binding is so strong that the resin acts as a perfect sink, i.e., there is no free metal in the resin domain. This assumption was tested using multiple binding layers in stacks, to simulate analysis of different depths of a DGT binding phase [47].

These experiments found that in simple metal solutions (i.e., in absence of ligand) at pH 7, metal penetration to the back layer was very low and similar for many metals (Mn, Co, Ni, Cu, Cd, Pb). As pH decreased down to 4, the percentage of metal accumulated in the back resin layer remained low for Ni, Cu and Pb, but an increase was observed for Mn at pH 4 and 5 (34.2 percent and 8.2 percent of total metal accumulation, respectively), as well as for Cd and Co, whose accumulations increased to 10.9 percent and 25.4 percent at pH 4, respectively. Penetration into the back layers was also significant in tests with only Mn, so the increased proportion in back layers could not be explained by simple competition effects. Saturation effects of the resin (which will be further discussed in Section 5.3.2) were also excluded. This suggested that binding of metals to Chelex-type resins may sometimes be kinetically limited.

A new expression for accumulation in a metal-only solution was developed. With standard assumptions (steady state, excess of resin sites, diffusion of M in the resin layer, common diffusion coefficient of metal in diffusive gel, filter and DBL), the metal flux received by the resin can be written [47] as

$$J = \frac{D_M \, c_M^*}{\delta^g + \lambda_M \coth\left(\dfrac{\delta^r}{\lambda_M}\right)} \tag{5.25}$$

where

$$\lambda_M = \sqrt{\frac{D_M}{k'_{a,R}}} \tag{5.26}$$

and $k'_{a,R}$ is the product of the association rate with the resin $k_{a,R}$ (see equation 5.7) and the concentration of free resin sites. The term $\lambda_M \coth\left(\frac{\delta^r}{\lambda_M}\right)$ is just a measure of the effective distance of penetration of M, i.e., the distance necessary for the metal concentration to drop to zero in the resin domain as calculated by linear extrapolation of the metal profile at the resin/diffusive gel interface.

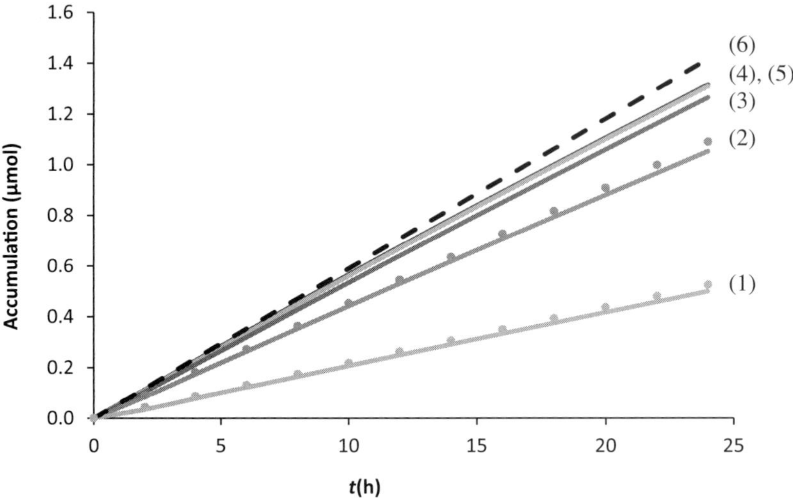

Figure 5.4 Evolution of the accumulation for different values of the kinetic association constant between metal and resin sites. Parameters: $k_{a,R} = 2.55 \times 10^{-5}$ m^3/(mol·s) (1), 2.55×10^{-4} (2), 2.55×10^{-3} (3), 2.55×10^{-2} (4), 2.55 (5). Other parameter values: $K_{MR} = 1000$ m^3/mol, $c_{T,M} = 9 \times 10^{-2}$ mol/m^3, $D_M = 4.94 \times 10^{-10}$ m^2 s^{-1}, $c_{T,R} = 38$ mol/m^3. Dotted lines correspond to the accumulations predicted by equation 5.25 for curves (1) and (2), respectively. Dashed line (6) stands for the perfect sink limit (equation 5.13).

A simple interpretation can be given to equation 5.25: $\delta^g + \lambda_M \coth\left(\frac{\delta^r}{\lambda_M}\right)$ is just the "effective" thickness of a linear diffusion profile of M starting at the bulk solution and eventually dropping to zero within the DGT device.

An interesting application of equation 5.25 is the determination of λ_M from metal accumulation data. However, an alternative independent measurement of λ_M may be obtained using a stack of resin discs. When there are two resin discs with the same thickness, the ratio of the number of moles of metal accumulated in the front layer (n_f) to the number of moles of metal accumulated in the back resin layer (n_b) becomes

$$\frac{n_f}{n_b} = \frac{\sinh(\delta^r/\lambda_M)}{\sinh(\delta^r/(2\lambda_M))} - 1 = 2\cosh\left(\frac{\delta^r}{2\lambda_M}\right) - 1 \qquad (5.27)$$

where δ^r is the total thickness of the resin layer (front plus back layers). This equation allows the determination of the λ_M as well as the operational kinetic constant $k'_{a,R}$.

Maximum λ_M values have been found to be ~ 100 µm for Co, Ni, Cu, Cd and Pb at pH 5 and 7 and for Mn at pH 7. These values are comparable to the diameter of the standard Chelex 100 resin beads. Therefore, for these pH conditions (and in absence of ligand) the resin layer acts almost like a perfect sink, and thus the use of the simple DGT equations is justified.

Figure 5.4 is used here to exemplify the performance of the DGT experiment for slow metal binding to the resin. Continuous lines in this figure have been computed by using

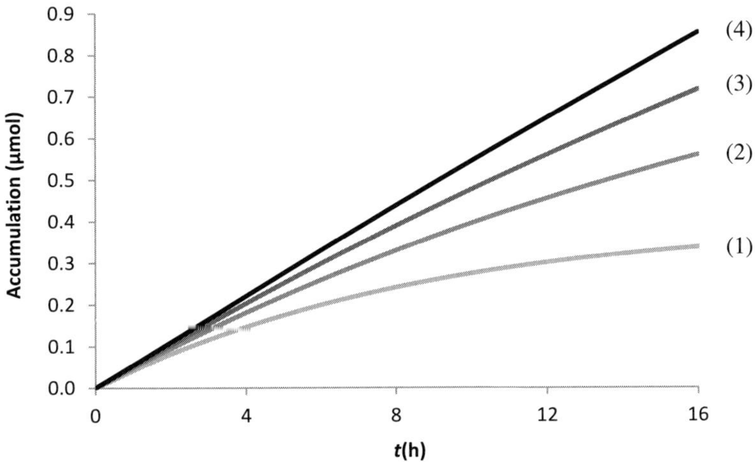

Figure 5.5 Evolution of the accumulation for different values of the stability constant between metal and resin sites. Parameters: $K_{MR} = 1$ m^3/mol (1), 3 m^3/mol (2), 10 m^3/mol (3) and 100 m^3/mol (4). In all cases, $k_{a,R} = 0.1$ m^3/(mol·s) and the rest of parameters as in Figure 5.4.

numerical simulation as described in the SI of [32]. Figure 5.4 shows a decrease of the metal accumulation as $k_{a,R}$ decreases. As explained in Section 5.3.5, this change could experimentally be achieved in some cases by just increasing the ionic strength. Notice that, for the $k_{a,R}$-values in Figure 5.4, and as demonstrated for Mn at pH 5–6 in the presence of high concentrations of Ca and Mg [50], the accumulation is still almost linear, but with a smaller slope than the value expected for the perfect sink case, equation 5.13, depicted as curve (6). The linearity of the accumulation indicates that a steady state is reached where all the metal transported by diffusion binds to the resin and the free metal concentration in the resin domain remains time independent. The smaller slope indicates that the diffusive metal flux to the resin domain is smaller than in the perfect sink case and decreases as $k_{a,R}$ decreases due to the kinetic limitation of the metal binding to the resin. As the equilibrium constant is quite high, under the conditions of the figure, the dissociation of metal bound to the resin is still negligible, especially at short times, so that equation 5.25, derived by neglecting this dissociation, is expected to be a good approximation for the accumulation. Computations of the expected accumulation using equation 5.25 are shown in dotted lines in Figure 5.4 for curves (1) and (2). Both dotted lines appear close to the corresponding continuous lines, especially at short times.

For a sufficiently large amount of bound metal, dissociation from the resin starts to be relevant, and the net metal binding rate decreases as time increases. This situation, which has often been observed experimentally (e.g. for phosphates measured by ferrihydrite DGT [6],), is depicted in Figure 5.5, where we can see that the linearity of the accumulation is lost, indicating an approach to the equilibrium conditions with the bulk metal concentration. Then, the rate of association is close to that of dissociation and the net accumulation rate tends to zero as seen at the rightmost part of curve (1) in Figure 5.5. This situation should

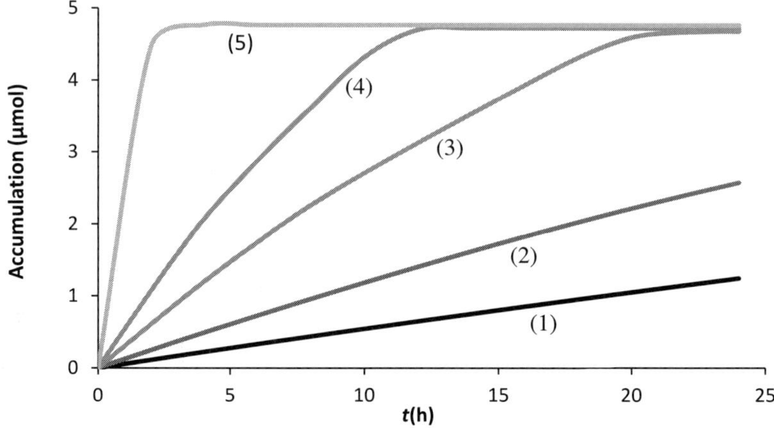

Figure 5.6   Evolution of the accumulation for different values of the total metal concentration in solution. Parameters: $c_{T,M} = 9 \times 10^{-2}$ mol/m^3 (1), 0.2 mol/m^3 (2), 0.5 mol/m^3 (3), 0.9 mol/m^3 (4), 5 mol/m^3 (5). In all cases, $K_{MR} = 100$ m^3/mol, $k_{a,R} = 0.1$ m^3/(mol·s) and the rest of parameters as in Figure 5.4.

not be identified with saturation of the resin. Under equilibrium conditions, if the metal concentration is increased (Figure 5.6), the maximum accumulation increases. Saturation is reached when further increases of the metal concentration in solution do not increase the accumulation, but only reduce the time to reach this maximum accumulation as curves (3) to (5) in Figure 5.6 indicate.

5.3.2 Saturation and Equilibrium Effects in the Presence of Complexes

A fundamental question for the practical application of DGT as dynamic sampling devices in the environmental monitoring of trace metals is the influence (via kinetic and thermodynamic mechanisms) of pH, competing metal cations and dissolved ligands on the accumulation linearity. Protons and other metal cations compete with the probe metal ion for the binding to the DGT resin sites at relatively low pH or high concentration of the competing metals, whereas high affinity dissolved ligands compete with resin sites for the binding of metals. Any of the two phenomena can lead to a departure from the linear accumulation regime (at shorter times than expected for a significant saturation of the resin) and to an underestimation of the actual species concentration in solution. While competing effects due to the presence of other metal cations are described in Chapter 4, the effects of ligands were studied in [24]. The results were reported as a chart to delimitate the range of experimental conditions (pH, ligand concentration and binding affinity) where the linear accumulation regime prevails (see figures 4, SI.5 and SI.6 in [24]) and are briefly commented hereafter. For example, it was concluded that separations from the linear regime within 10 h of deployment were negligible above pH 5 with weak complexation ($\log K' < 0$) or above pH 7 with strong complexation ($\log K' < 3$), where K' is the effective stability constant (the ratio of complexed to free metal ion concentrations in bulk solution, see equation 5.11). These plots can also be approximately used for partially labile systems whenever the time is replaced with the product of lability degree and t.

Can DGT accumulate metals complexed with ligands stronger than Chelex?

The local value of $K' = Kc_L = c_{ML}/c_M$ (corresponding to the metal binding to dissolved ligands) and the local value of $K'_{MR} = K_{MR}c_R = c_{MR}/c_M$ (corresponding to the binding reaction to the resin sites) are the parameters determining the thermodynamic fate of the metal ion. Let us exemplify the application of the thermodynamic criterion for binding (i.e., $K'_{MR} > K'$) to the accumulation of Ni in a solution where the main species is NiNTA. The value reported in the literature for the equilibrium $Ni^{2+} + NTA^{3-} \rightleftharpoons Ni - NTA^-$ is $\log(K(Ni - NTA)/M^{-1}) = 12.79$ ([51]), while the value of Ni-Chelex binding according to the reaction $Ni^2 + +R^{2-} \rightleftharpoons NiR$ can be estimated as $\log(K(Ni-R)/M^{-1}) = 10.04$ (as derived from data provided in references [52–54]). Even though the binding of Ni with NTA is stronger than the binding to Chelex, Ni^{2+} accumulation in the resin disc proceeds. This fact can be justified from the local equilibrium value for the ratio c_{NiR}/c_{NiNTA}.

$$\frac{c_{NiR}}{c_{NiNTA}} = \frac{K_{NiR}c_{Ni}c_R}{K_{NiNTA}c_{Ni}c_{NTA}} = \frac{K'_{NiR}}{K'_{NiNTA}} \tag{5.28}$$

For experimental conditions (at pH 7.5) in ref. [48], the bulk concentration of NTA^{3-} is around 4×10^{-7} M, while the concentration of R^{2-} in the resin disc is around 3.8×10^{-2} M. Using these concentrations as estimates of the actual concentration in the resin domain, equation 5.28 provides an estimation for $K'(Ni - R)/K'(Ni - NTA) = (10^{10.04} \times 3.8 \times 10^{-2})/(10^{12.79} \times 4 \times 10^{-7})$ which is greater than 150, a factor that indicates that the resin is preferred to the NTA for Ni binding. An experimental confirmation of the preferred binding to the R sites was the linearity of the time evolution of the accumulated Ni-NTA and the observation that the accumulation of Ni was well over the value expected for the accumulation due to just the free Ni in solution [24].

However, even though a given ratio $K'_{MR}/K' > 1$ might be sufficient to ensure accumulation of a metal complexed with a ligand in solution proceeds, it may not do so under perfect sink conditions. An illustration of the influence of K'_{MR}/K' on the accumulation is depicted in Figure 5.7. For the concentrations used in the figure, with c_L^* and $c_{T,R}$ (the total concentration of resin sites, also referred to as the intrinsic binding capacity in Chapter 4) as estimates for c_L and c_R in the resin domain, respectively, we have $K = 10^4$ m^3/mol, $K' = Kc_L^* = 900$ and $K'_{MR} = 38K_{MR}$ (where K_{MR} is in m^3 mol^{-1}), which indicate that K'_{MR}/K' ranges from 4.22×10^4 when $\log K_{MR} = 6$ to 4.22 when $\log K_{MR} = 2$. Even though $K'_{MR}/K' > 1$ for all the lines of the figure, the accumulation is less than that for perfect sink behavior (depicted as dashed line). The deviation at 10 h accumulation from the perfect sink behavior is small (3.16 percent) for $\log K_{MR} = 6$, but this underestimation increases as K_{MR} decreases reaching around 92 percent when $\log K_{MR} = 2$. While the accumulation is still linear for $\log K_{MR} = 6$ or $\log K_{MR} = 5$, it bends downwards for lower values of $\log K_{MR}$. When $\log K_{MR} = 2$, there is no further noticeable accumulation after 3 h of deployment, indicating that the free metal has reached a flat concentration profile,

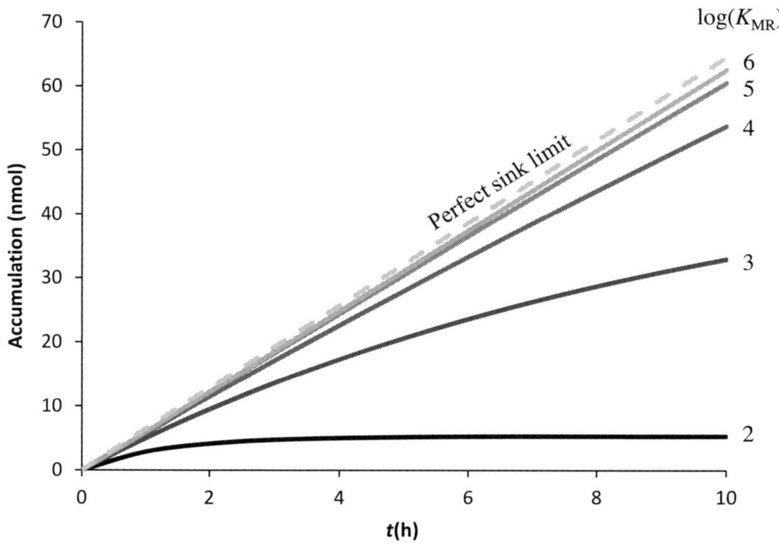

Figure 5.7 Evolution of the accumulation for different values of the stability constant of the metal binding to the resin (K_{MR}). Values of K_{MR} (in $m^3\ mol^{-1}$) are indicated in the figure. $k_{d,R} = 0.1\ s^{-1}$, $c_{T,M} = 0.01\ mol/m^3$, $c_{T,L} = 0.1\ mol/m^3$, $K = 10^4\ m^3/mol$, $k_a = 10^4\ m^3/(mol\cdot s)$, $D_M = 7 \times 10^{-10}\ m^2\ s^{-1}$, $D_{ML} = 0.7 \times D_M$. Rest of parameters as in Figure 5.4.

i.e., the free metal inside the resin layer is c_M^*, which for the concentrations of the figure results $c_M^* = 1.1 \times 10^{-5}\ \frac{mol}{m^3}$, and the metal bound in the resin has reached the equilibrium value with c_M^*. This bound concentration is also known as the effective binding capacity (Chapter 4).

$$\frac{c_{MR}}{c_{T,R}} = \frac{K_{MR}c_M^*}{1 + K_{MR}c_M^*} \tag{5.29}$$

Notice that we are still far from saturation (where $\frac{c_{MR}}{c_{T,R}}$ should be close to 1). Indeed, in line 2 of Figure 5.7, $\frac{c_{MR}}{c_{T,R}} \leq 0.001$ as computed from equation 5.29.

In all cases depicted in Figure 5.7, the rate constants of the metal binding to the resin are high enough for the process 5.7 to reach local equilibrium at any time. However, there could also be kinetic limitations in the metal binding to the resin. These are illustrated in Figure 5.8, which is obtained from the line corresponding to $\log K_{MR} = 5$ in Figure 5.7, but decreasing $k_{a,R}$ and $k_{d,R}$ while keeping their ratio constant. Notice how the accumulation decreases due to the kinetic limitation of the metal binding to the resin, but the accumulation remains linear in all cases. This indicates that the system has reached a steady state where the complex arriving to the resin domain dissociates and the released metal binds to the resin at a speed controlled by the kinetics of the metal–resin binding.

What has been exemplified with the Ni-NTA system can be quite common. The complexation constants of all metals with NTA are higher than the affinity for the iminodiacetic acid or the Chelex resin. But even though the complexation constant is higher than the

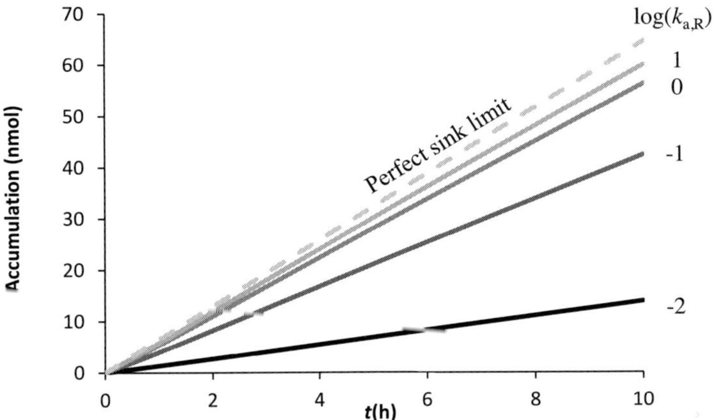

Figure 5.8   Evolution of the accumulation for different values of the kinetic association constant between metal and resin sites. Values of $k_{a,R}$ in $m^3 mol^{-1}s^{-1}$ are indicated in the figure. $\log K_{MR} = 5$ (K_{MR} in $m^3 mol^{-1}$) in all lines. Rest of parameters as in Figure 5.7.

affinity for the resin, $K > K_{MR}$, we could find $K'_{MR} > K'$ in many cases. This is due to the fact that ligands are usually involved in acid base equilibria and the active species for the metal binding (the so-called free ligand L in this chapter) can be, at a given pH, a minor component in the ligand speciation. Additionally, the resin concentration is used in a high enough concentration, so that $c_L \ll c_R$ and $K'_{MR} > K'$.

As seen in this section, linear accumulations arise for labile or partially labile complexes under perfect sink conditions or with kinetic limitations in the metal binding to the resin. In all these cases, a quasi steady state is reached, a sufficient condition for a linear accumulation. Non-linear accumulations arise when a steady state is not reached. For instance, this is the case when saturation effects are non-negligible, but also when the concentration of bound metal starts to be close to the value where it achieves equilibrium with the bulk free metal concentration, i.e., $\frac{c_{MR}}{c_{T,R}} \sim \frac{K_{MR}c_M^*}{1+K_{MR}c_M^*}$.

5.3.3 ADBL: Experiments Varying the Thickness of the Gel Layer. Kinetic Signatures

Although the diffusive gel was designed in DGT to minimize the influence of convection on the metal accumulation [39, 55] (see Chapter 1), it was early recognized that the diffusion domain extended into the solution in the so-called "diffusive boundary layer," DBL. Specific hypotheses considered next are:

(i) Full stirring for $x > \delta^r + \delta^g + \delta^f + \delta^{dbl}$ (i.e., bulk conditions), but quiescent conditions in between $x = \delta^r + \delta^g + \delta^f$ and $x = \delta^r + \delta^g + \delta^f + \delta^{dbl}$.

(ii) The thickness of the DBL is assumed to be independent of the diffusion coefficients of the species present in the system [56].

(iii) Different diffusion coefficients of the metal species in the gel domain and in the water solution. Actually, the case of common diffusion coefficients in the gel and water domains is trivial and can be analyzed by just replacing δ^g with the addition $\delta^g + \delta^f + \delta^{dbl}$ in the expressions reported in the above sections.

Experiments with varying gel thickness have been performed to estimate the physical length of the DBL (δ^{dbl}), as explained in Chapter 2 and references therein. Here we will only revise the influence of the presence of complexes on these experiments taking penetration into account.

With the standard assumptions of PAM (penetration of complexes, perfect sink for metal, excess of ligand, a common diffusion coefficient for each species in gel and filter, but different from that corresponding to the solution) and assuming equilibrium between M and ML in the DBL, one arrives (see [26] and SI of [20]) to

$$J = \cfrac{c_M^*}{\cfrac{\delta^{dbl}}{D_M^w (1 + \varepsilon^w K')} + \cfrac{\delta^g + g_{kin}^p}{D_M^g (1 + \varepsilon^g K')}} \tag{5.30}$$

where

$$g_{kin}^p \equiv \cfrac{\varepsilon^g K' m \tanh \cfrac{\delta^g}{m}}{1 + \varepsilon^g K' m \left(\tanh \cfrac{\delta^g}{m} \right) \left(\cfrac{1}{\varepsilon^g K'} + 1 \right) \cfrac{1}{\lambda_{ML}} \tanh \cfrac{\delta^r}{\lambda_{ML}}} \tag{5.31}$$

Diffusion coefficients in the DBL are indicated with superscript "w," while superscript "g" is added for gels and filter.

g_{kin}^p in equation 5.30 can be seen as the equivalent extra length that would be necessary to add to the physical gel layer thickness to account for the decrease of accumulated metal due to kinetic limitations [20]. For labile complexes, m and λ_{ML} tend to zero and g_{kin}^p becomes negligible, in agreement to the fact that there is no kinetic limitation for these complexes.

As expected, equation 5.30 indicates that the flux decreases when δ^{dbl} increases, because the distance required for the species in solution to reach the resin domain increases. This influence is dominant over the increase in lability, which is due to the increased diffusional "time" available for complex dissociation.

The representation of the inverse of the accumulated mass [20, 39, 40] for varying δ^g can be used to experimentally estimate the thickness of the DBL. For solutions with one homogeneous ligand, taking into account penetration of the complex and no further complications, the equation of $1/n_M$ vs δ^g is expected to be the straight line

$$\frac{1}{n_M} = \left[\frac{\delta^{dbl}}{D_M^w (1 + \varepsilon^w K') c_M^* At} + \frac{g_{kin}^p}{D_M^g (1 + \varepsilon^g K') c_M^* At} \right] + \frac{\delta^g}{D_M^g (1 + \varepsilon^g K') c_M^* At} \tag{5.32}$$

The ratio of the intercept i over the slope s (both obtained from the linear regression) is

$$\frac{i}{s} = \frac{D_M^g \left(1 + \varepsilon^g K' \right)}{D_M^w \left(1 + \varepsilon^w K' \right)} \delta^{dbl} + g_{kin}^p \tag{5.33}$$

which allows to obtain g_{kin}^p. Then, k_d can be obtained from g_{kin}^p using equations 5.31, 5.9 and 5.10. The representation of the so-called apparent diffusive boundary layers (i.e., ADBL $= \frac{i}{s}$) for several metals has been termed the "kinetic signature" of the mixture [20, 40].

This method was applied, for instance, to the determination of the kinetic signature of metal–fulvic acid complexes. Results indicate a substantial kinetic limitation at pH 7 for Cu, Pb and Ni complexes with Suwannee River Fulvic Acid, while none for Cd, Co and Mn. In the absence of kinetic limitation, a common value of ADBL (which corresponds to δ^{dbl}), was found. Other works dealt with pristine river water [57], algal exudates [58] and treated wastewater [59].

Several reasons can explain experimental deviations from the linearity expected for $1/n_M$ vs δ^g, ranging from failure of some of the assumptions to accuracy limitations (see [26]).

5.3.4 Low Ionic Strength and Electrostatic Effects

Measurement of metals by DGT in low salinity media has been carefully studied due to its relevance for freshwaters. Given that the resin and (to a lesser extent) gel phases include fixed charges in their polymeric structures, electrostatic effects may arise [60]. This leads to the development of differences in electrostatic potential between the resin-gel phases and/or between the gel-solution phases, as well as around the resin beads. Electrostatic effects due to charges in the diffusive gel were reviewed in Chapter 3. Due to the high concentration of negatively charged binding sites in the Chelex beads, electrostatic effects of the resin are expected to be particularly relevant when charged complexes are the main metal species present in the system. These effects have been studied in [48] and will be summarized here.

The presence of electrostatic potentials might imply migration terms in the transport equations of charged species, especially at low ionic strengths, where screening of the fixed charges by the salt background is relatively low. Rigorous solution requires, then, the application of the Nernst-Planck transport equations coupled to the Poisson equation. However, a simple situation arises when the electrostatic potential can be considered constant in both the solution and resin phases [56] with potential jumps being restricted to the interfaces (the Donnan potential). This can be a good approximation when the average distance among fixed charges is shorter than the Debye length (the characteristic length scale associated to the electrostatic screening effect of background ions) and the influence of the trace metal accumulation on the Donnan potential is negligible [61, 62]. Although this approximation has been used in the literature in several cases [28, 63], the fulfillment of this hypothesis could be questionable for the typical DGT applications, so that the following results (next paragraphs) must be taken as a first approximation to this problem. More rigorous treatments require an explicit model for the Chelex beads embedded in the resin disc and the distribution of charged sites, in a similar way as described in literature for soft permeable charged particles [64].

The Donnan potential difference can be computed by applying the electroneutrality condition to the resin domain and by taking into account that concentrations outside and inside this domain are related by a Boltzmann factor or partition coefficient. For a symmetrical univalent background electrolyte (such as $NaNO_3$) and assuming a negligible contribution of the trace metals to the overall charge balance (as a consequence of the excess of background salt), the Donnan potential becomes

$$\Psi_{resin} - \Psi_{solution} = \frac{RT^K}{F} \text{arcsinh}\left(\frac{\rho_{charge,resin}}{2I}\right) \tag{5.34}$$

where $\rho_{charge,resin}$ (with its sign) and Ψ_{resin} are, respectively, the charge density (of fixed charges) and the potential in the resin domain. R is the gas constant, T^K the Kelvin temperature, F is the Faraday constant and I the ionic strength. In Chelex devices at usual pH values, $\rho_{charge,resin}$ is negative.

Donnan potential differences lead to partitioning effects, with the consequent increase in the concentration of those ions with opposite charge and decrease of those with the same charge as the resin phase.

For simple solutions of metals where all the complexes are fully labile, the effect of the resin charges is expected to be negligible, since the metal concentration effectively drops to almost zero at the resin-gel interface whenever perfect sink conditions are fulfilled. However, relevant effects will arise for background cations, anions or partially labile charged complexes. In the latter case (and for Chelex binding phases), we may expect decreasing/increasing accumulations as the ionic strength decreases, depending on the negative/positive net charge of the complexes, respectively [48].

The partition coefficient at the resin-gel interface for a generic complex ML with charge z_{ML} can be written as

$$\Pi = \frac{c_{ML}^{gel}}{c_{ML}^{resin}} = \exp\left(\frac{z_{ML}\,F\,(\Psi_{resin} - \Psi_{solution})}{RT^K}\right) \tag{5.35}$$

So, equation 5.23 can be extended to account for the electrostatic effects:

$$J = \frac{D_M c_M^*}{\delta^g} + \frac{D_{ML} c_{ML}^*}{\delta^g}\left[1 - \frac{(1+\varepsilon K')}{\varepsilon K' + \frac{\delta^g}{m}\coth\left(\frac{\delta^g}{m}\right) + \frac{\delta^g}{\lambda_{ML}}\frac{(1+\varepsilon K')}{\Pi}\tanh\left(\frac{\delta^r}{\lambda_{ML}}\right)}\right] \tag{5.36}$$

For a negatively charged Chelex binding layer, equations (5.34) and (5.35) indicate that with decreasing ionic strength, both Π increases, and the concentration of negative charged species in the resin domain decreases with respect to the diffusive gel. These results agree with experimental observations. For instance, accumulation of Ni in solutions with an excess of the ligand nitrilotriacetic acid (NTA) has been found to decrease by a factor 30 when the ionic strength decreases from 0.5M to 10^{-3}M, while the decrease in the Cd accumulation is almost negligible under the same conditions. It must be recalled that complexes of divalent cations with NTA are negatively charged for the 1:1 stoichiometric ratio, so that these results would indicate a labile behavior of CdNTA while a partially labile one for NiNTA. This has been verified by analyzing the percentage of back accumulation using DGT devices with a stack of two resin discs: while the back percentage of Cd is almost negligible throughout the range of ionic strengths studied, the back percentage of Ni is close to 30 percent at

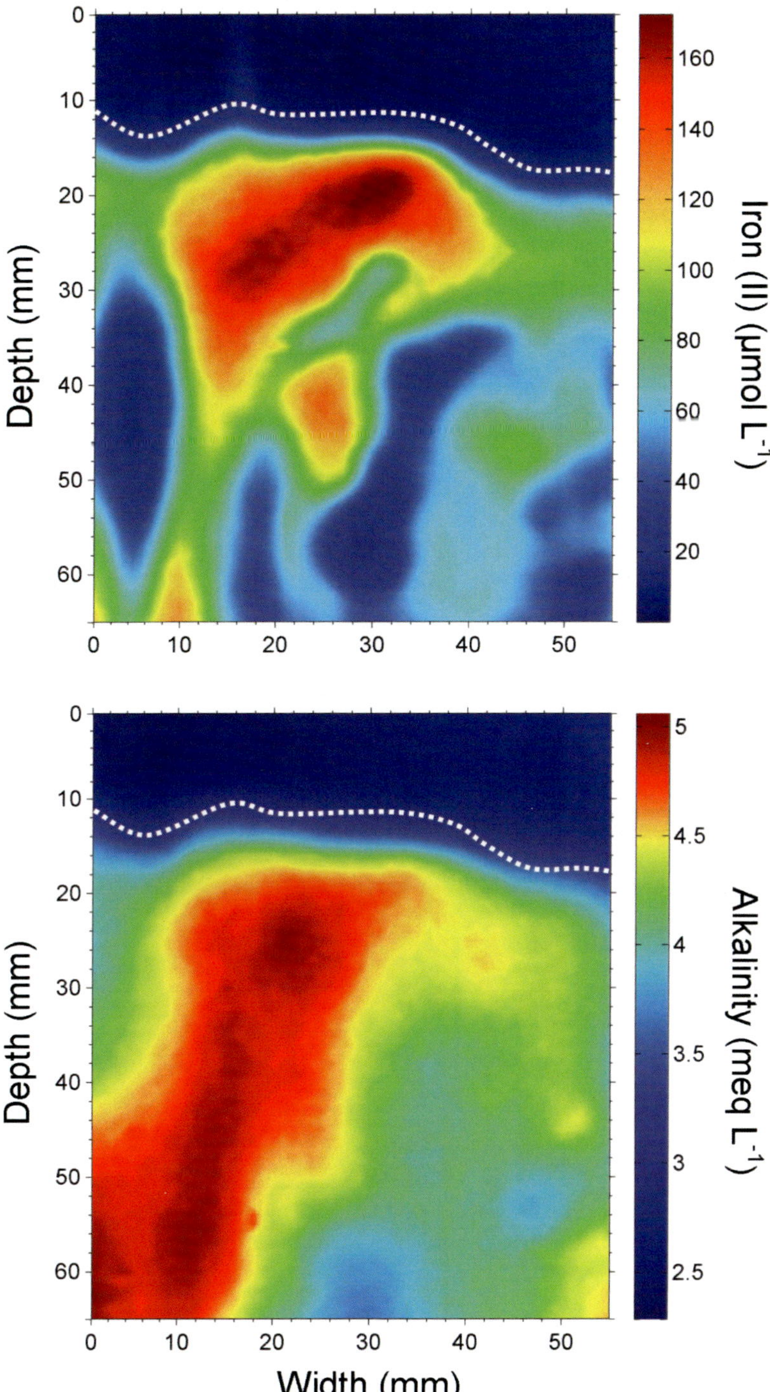

Plate 8.2 Simultaneous colorimetric DET images of the Fe(II) and alkalinity distributions in a coastal marine sediment. This sediment was colonised by the seagrass *Zostera capricorni*. The highly heterogeneous Fe(II) distribution might be linked to oxygen leakage of the seagrass roots, which leads to Fe(II) oxidation and precipitation as Fe(III) oxyhydroxides. The alkalinity distribution is less heterogeneous. In some parts, alkalinity and Fe(II) are clearly linked, possibly because Fe(II) and alkalinity are generated by microbial dissimilatory iron reduction during organic matter mineralisation. The dotted line represents the sediment–water interface. Adapted from Bennett et al. [29], with permission from Elsevier.

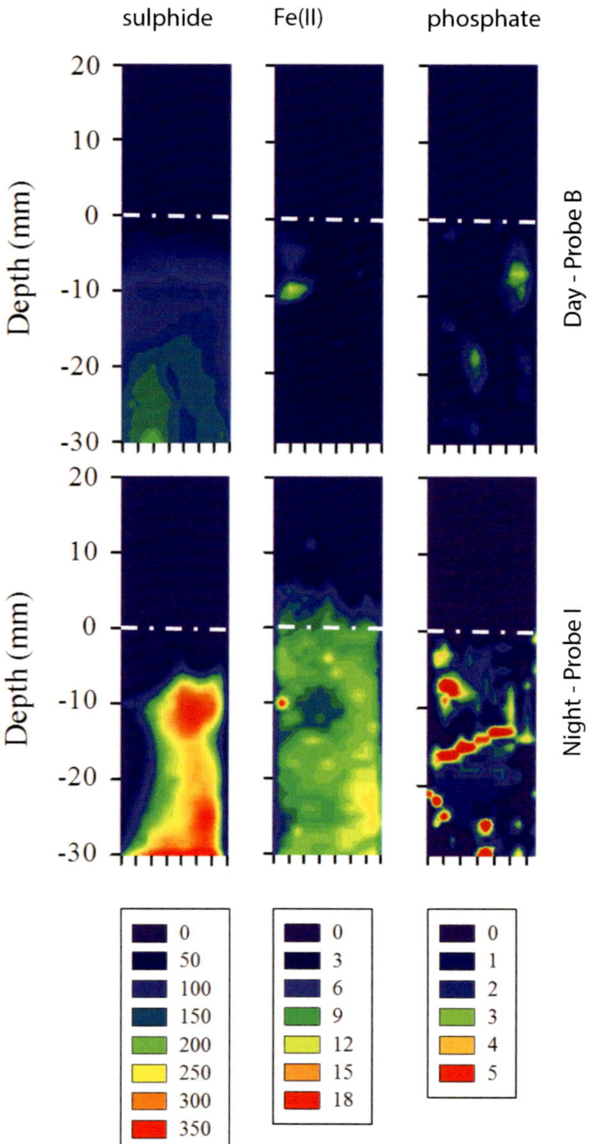

Plate 8.3   Distribution of sulphide, phosphate and Fe(II) in a microbial mat, measured by triple-layer DGT/DET probes. The width of all gels is 14 mm. Distinct differences in solute concentrations and distributions between the day (top) and the night (bottom) deployments are visible, while no clear element co-distribution features can be seen. The dotted line represents the sediment–water interface. Reprinted from Pagès et al. [30], with permission from Elsevier.

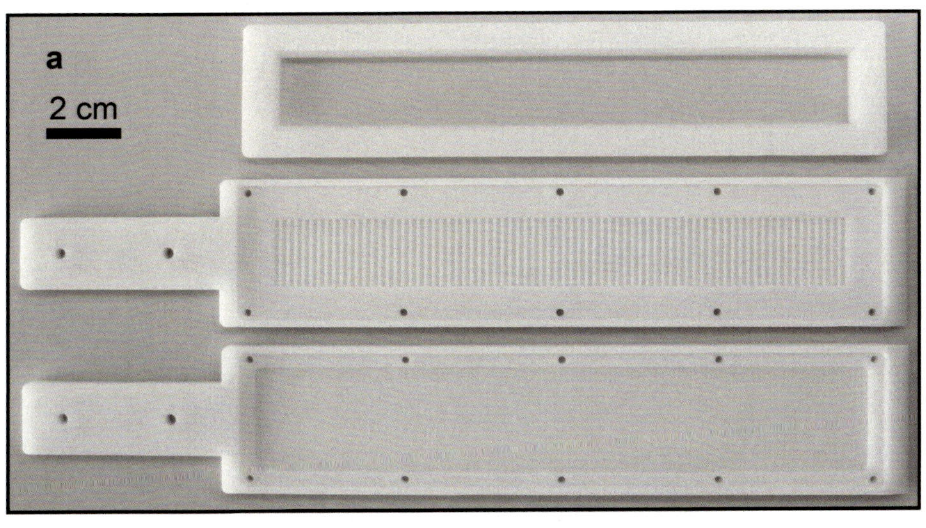

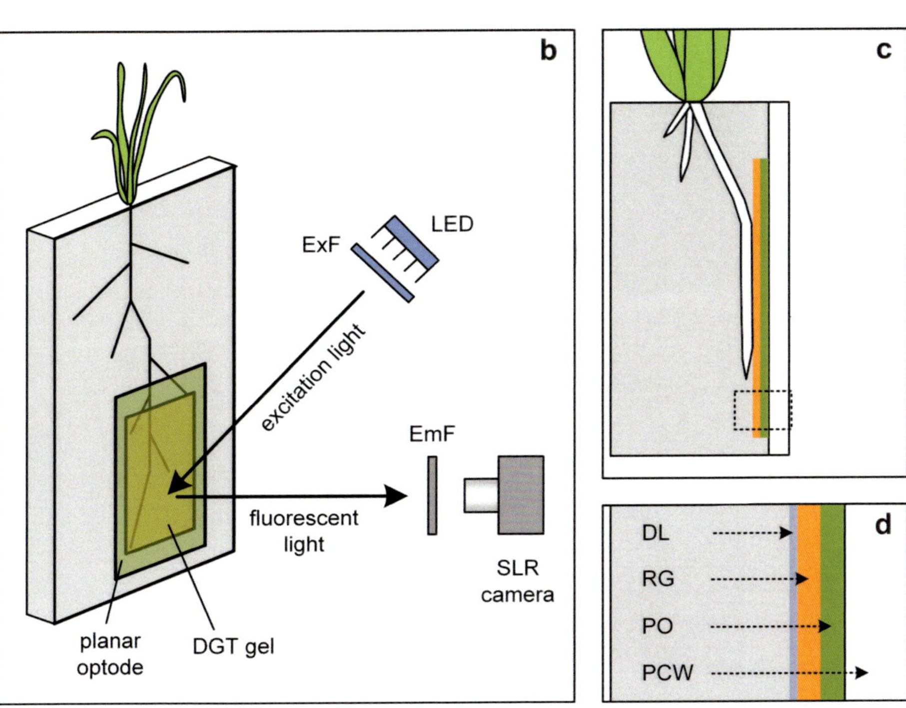

Plate 8.4 Sampling setup designs. (a) Standard size DGT and DET gel probes. Bottom: Backside plate of an open gel probe. The gel(s) and the covering membrane are cut to size, placed onto the backing plate and the assembly is closed with the front part containing the sampling window (top). Middle: Backside plate of a constrained probe. After filling the gel slits, the gel is hydrated, covered with protective membrane and closed with the front part. (b) Schematic of a simultaneous DGT – planar optode application on soil and roots. The DGT gel is overlaid by a planar optode, thereby serving as diffusion layer for the optode sensor. The fluorophore is excited by LED light, the emitted fluorescence light is imaged using a standard single lens reflex camera, ExF: excitation filter. EmF: emission filter. (c) Side view of the DGT-optode assembly inside the growth box. (d) Detail of (c) showing the membrane, gel and optode layers. DL: diffusion layer (membrane), RG: resin gel, PO: planar optode, PCW: plastic container wall. (b–d) reproduced from Santner et al. [11].

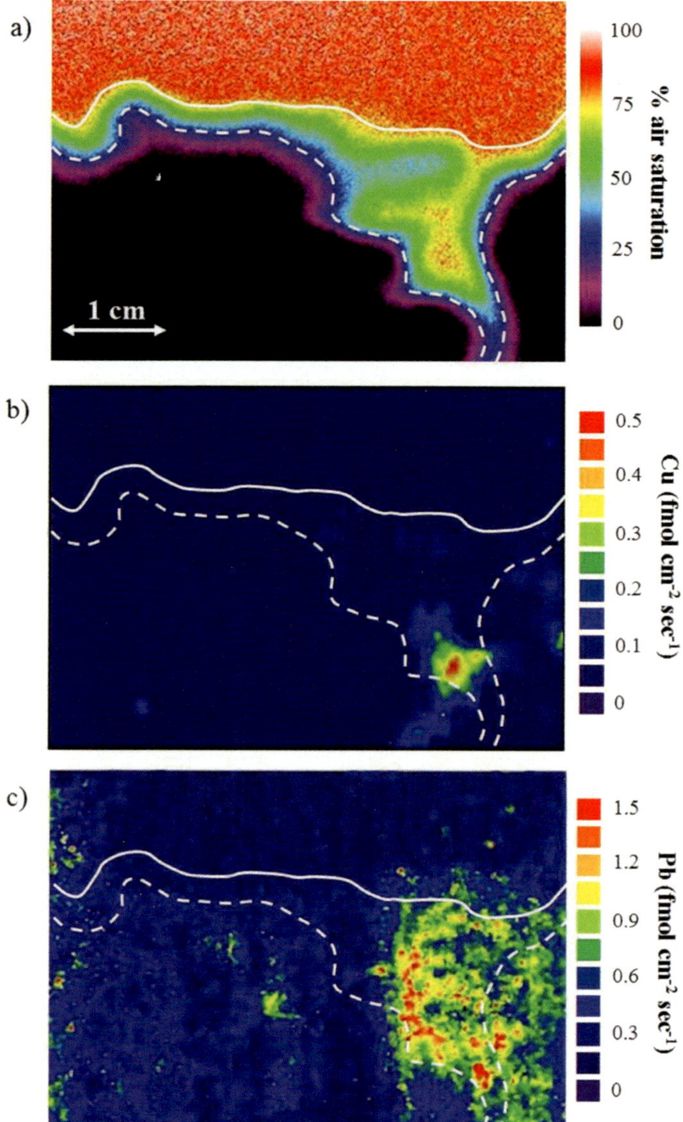

Plate 8.5 Simultaneous high-resolution two-dimensional imaging of Cu, Pb and O$_2$, in polychaete (*Hediste diversicolor*) burrows. Large-scale flume mesocosms were filled with intact harbour sediment containing *H. diversicolor*. The distribution of (a) O$_2$, (b) Cu and (c) Pb inside and adjacent to an individual burrow was imaged with a simultaneous deployment of planar O$_2$ optodes and DGT gels. The majority of localised Cu and Pb hotspots were located within the irrigated burrow. However, some Pb release was also observed in the neighbouring sediment. Movement and burrowing activity of the polychaete, in and on top of the sediment, may have affected the metal distributions during the gel deployment period, which might have caused the differences between the near instantaneous concentration measurement of the optode compared with the integrated flux (12 h) of the DGT. Reproduced from Stahl et al. [95], with permission of John Wiley and Sons.

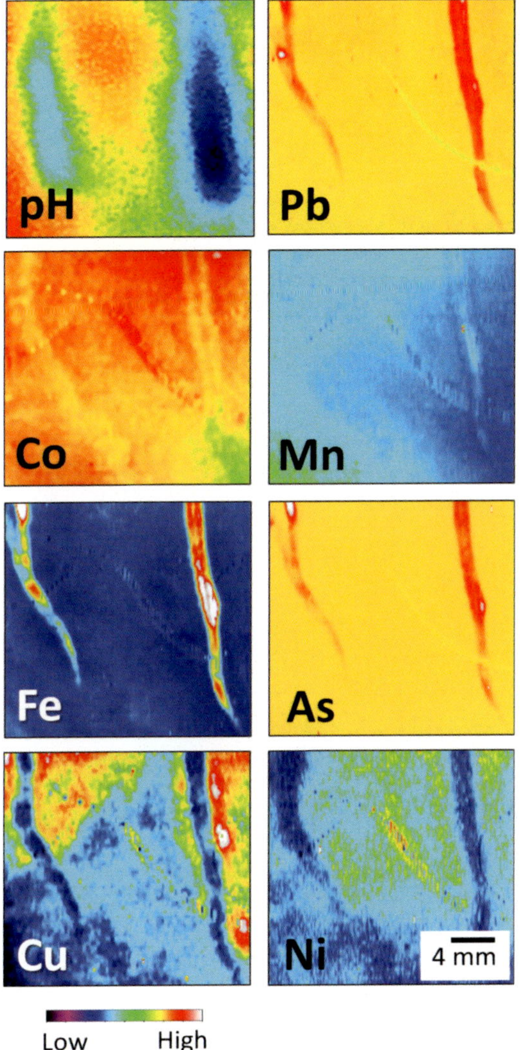

Plate 8.6 Zones of greatly enhanced As, Pb and Fe(II) mobilisation, located immediately adjacent to a set of rice root tips. Rice seedlings were grown for three weeks with deployment of an ultrathin 0.05-mm thick SPR-IDA DGT binding gel, partnered with a planar pH optode, 8 cm below the soil–water interface for 24 h. The blue to white colour scale represents a sequential increase in metal fluxes (f_{DGT}, pg cm^{-2} s^{-1}). Planar optode pH measurements demonstrate a localised acidification around the rice roots of ~0.5 pH units. Adapted from Williams et al. [89].

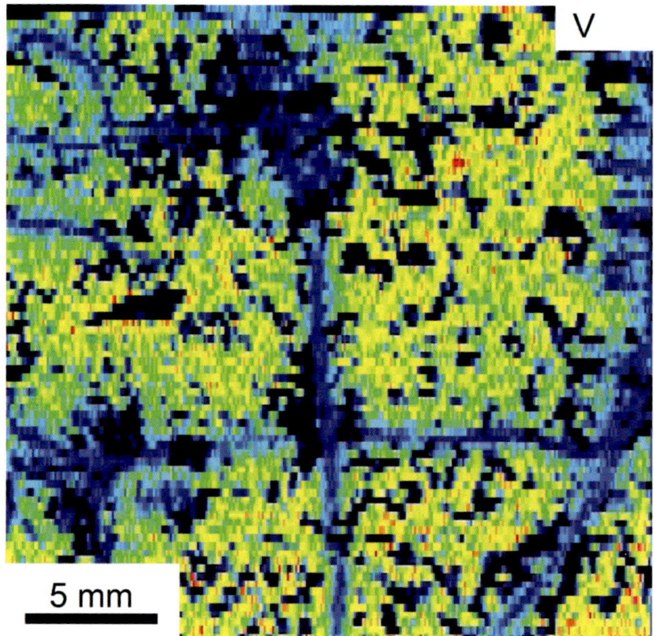

Plate 8.7   Gel–soil contact artefacts in DGT imaging applications in unsaturated soil. Vanadium distribution around lupin roots (blue, elongated features). The black areas, where no V was taken up by the DGT gel, are caused by numerous small entrapped air bubbles. Vanadium was found to be a very good indicator of soil-gel contact discontinuities in this study, as the soil background V concentration yielded a good signal in LA-ICPMS analysis, while the blank V concentration on the resin gel was negligible. Unpublished data of Andreas Kreuzeder, Jakob Santner, Vanessa Scharsching and Walter W. Wenzel.

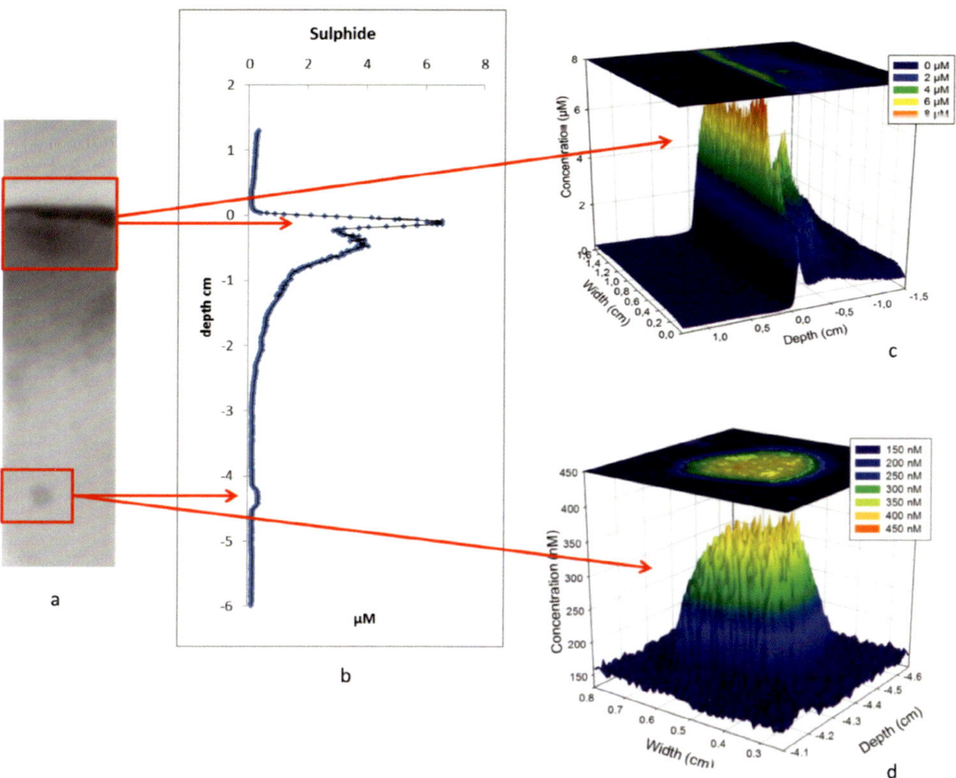

Plate 8.8   Two-dimensional density plot superimposed on a three-dimensional graphic. Combination of visual formats to display sulphide release in estuarine sediment: (a) greyscaled scanned DGT-CID image of silver iodide (AgI) gel. (b) 1-DGT representation of sulphide release. (c, d) Combined two- and three-dimensional plots of the sediment–water interface and microniche zones. Reprinted from Gao et al. [118], with permission from Elsevier.

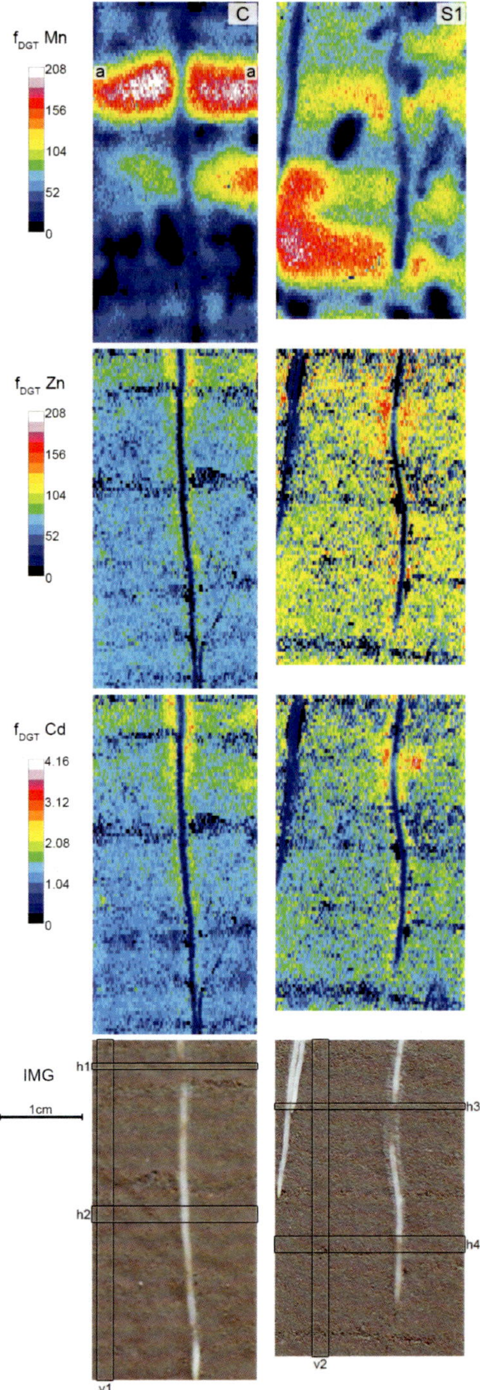

Plate 8.10   Distribution of Mn, Cd and Zn around a *Salix x smithiana* root. C and S1 indicate soil treatments without (C) and with (S1) elemental sulphur addition. For this image, the DGT resin gel was applied for 20 h on soil saturated to 80 per cent maximum water holding capacity. At the start of the experiment, the soil was filled into the growth container and carefully compacted layerwise. DGT-labile Mn is clearly elevated in two of these soil layers, indicating that the bulk soil density is higher than in the layers above and below. As a result, the water saturation of the soil pores is increased, leading to anaerobia and a localised reductive dissolution of Mn. Zinc, which is not redox-sensitive, does not show elevated concentrations in these more strongly compacted layers. Reproduced from Hoefer et al. [88].

0.5M, and increases with ionic strength down to 10^{-3}M, where it is close to 50 percent (a detailed explanation for this increase is provided in Section 5.3.5). Additionally, metal accumulations in systems with excess of ethylenediamine indicate increasing accumulations of Cu and Co as ionic strength decreases, consistent with a partially labile character of these positively charged complexes.

The electrostatic effects on metal complexes with humic substances are of special relevance from an environmental point of view. DGT accumulations in the presence of partially labile humic or fulvic complexes are expected to decrease with decreasing ionic strength, due to the negative charge of these substances at environmentally relevant pH values. However, these effects need to be studied in more detail since other phenomena can modify these predictions. For instance, humate enrichment in the diffusive gel (with respect to the bulk solution) due to Donnan partitioning or other interactions could also be relevant for metal speciation [41, 65], though other work [66] found a minimal impact in most cases.

5.3.5 Kinetic Effects in the Resin Layer. Ligand-Assisted Dissociation

The application of equation 5.36 to complexes of Ni, Cd and Co with NTA at pH 7.5 allowed fitting of k_d-values to achieve consistency with the experimental accumulations [48]. The resulting k_d-values were compared with those predicted by Eigen's mechanism:

$$k_d = \frac{K^{os}k_{-w}}{K} \tag{5.37}$$

where the outer sphere stability constant K^{os} is estimated from the electrostatic interaction energy between two point charges of values $+2$ and -3 corresponding to the free metal and ligand (NTA^{3-}), respectively, at the standard closest approach in the outer sphere complex of 0.5 nm [67], and k_{-w} is the rate constant for the release of a water molecule from the first hydration shell of the metal. DGT data indicate a faster NiNTA and CoNTA dissociation rate in comparison to the Eigen prediction, while a reasonable value was found for CdNTA. As NiNTA dissociation takes place almost entirely in the resin domain due to its partially labile character, a ligand-assisted mechanism (also called adjunctive mechanism) parallel to the Eigen one might be responsible for this accelerated NiNTA dissociation in the presence of the resin. In the other limit, Cd dissociation takes place mostly at the gel-resin interface and, thus, penetration into the resin layer is not physically relevant due to the labile character of this complex. Accordingly, the kinetic constant in this labile case is closer to Eigen expectations.

The experimentally observed lability of NiNTA ($\xi = 0.7$) is significantly higher than the theoretical prediction based on the Eigen mechanism, which is $\xi = 0.006$ at $I = 0.5$M (as computed with equation 5.36).

These results suggest the coexistence of two dissociation mechanisms, namely: Eigen's and ligand-assisted, the latter being faster than the former, but only accessible within the resin domain. These processes can be included in the theoretical model without apparent changes in equation 5.36, just by using the Eigen value of the dissociation constant in the computation of m (equation 5.10) and the ligand-assisted value in the computation of λ_{ML} (equation 5.9). Results of this model were in good agreement with the experimental data of the M-NTA complexes [48].

In the ligand-assisted pathway, dissociation of complexes follows a second-order mechanism that can be written as an exchange reaction between the charged resin sites and the charged complexes to produce the occupied sites and the free ligand. A new question is, then, to what extent does the ligand assisted dissociation constant vary with the ionic strength. The dependence predicted by the classical theory of chemical kinetics of charged species is consistent with the experimental measurements of accumulation in back resin layers. As indicated in Section 5.2.4, the back percentage of Ni in the system NiNTA increases as ionic strength decreases. This effect cannot be justified by the Donnan partitioning, but follows if a decrease of the kinetic dissociation constant takes place as ionic strength decreases. The increase of the percentage of Ni accumulation in a back resin disc is not accounted for in the Eigen mechanism either, because it predicts a kinetic dissociation constant independent of ionic strength. Clearly, further work is required to understand fully the processes taking place in the DGT devices, and the dependence of the kinetic constants on ionic strength (especially considering that the resin sites are not charged points, but Chelex beads of hundredths of microns in size).

5.3.6 Influence of the Settling of the Beads in the Resin Disc

The resin beads used in DGT often tend to settle to one side of the resin during casting of the binding layers. This feature might have an influence on the metal accumulation, since a significant fraction of the dissociation reaction layer of partially labile complexes lies within the resin domain, as explained in previous sections. For this reason, the relevance of an inhomogeneous distribution of the binding resin on metal accumulation was recently assessed [32] by numerical simulation of DGT devices with binding groups occupying only one half of the resin layer, as a reasonable simplified model for the distribution inside the actual resin discs. Results indicate a reduction in the metal accumulation in the inhomogeneous device with respect to the homogeneous one for partially labile complexes, while fully labile or inert complexes remain unaffected.

Maximum discrepancies between non-homogeneous devices and equivalent ideal discs (i.e., with a completely even distribution of resin beads) are reported in Table 5.1. These values are calculated with concentrations of resin sites high enough to avoid saturation effects and kinetic constants for the metal binding to the resin high enough to achieve perfect sink conditions.

A simple approximate analytical expression can be used to reproduce the accumulation in R/2 devices:

$$J = \frac{D_M c_M^*}{\delta^g} + \frac{D_{ML} c_{ML}^*}{\delta^g} \xi^{(r/2)} + k_d c_{ML}^* \left(1 - \xi^{(r/2)}\right) \, \min\left[m, \delta^r/2\right] \tag{5.38}$$

where $\xi^{(r/2)}$ stands for the lability degree calculated when $\delta^r/2$ is used as δ^r in equation 5.23 and $\min\left[m, \delta^r/2\right]$ indicates that the last factor of the third right term is m when $m < \delta^r/2$

Table 5.1 *Effect of the settling of Chelex beads. Accumulation (n_M) in a DGT device with Chelex homogeneously dispersed in the resin disc (R) and accumulation in a DGT device with resin beads settled to one half of the volume (R/2) of the resin disc. Parameters: $c_{T,M} = 0.01$ mol m^{-3}, $D_M = 7.07 \times 10^{-10}$ m^2 s^{-1}, $D_{ML} = 4.95 \times 10^{-10}$ m^2 s^{-1}, $\delta^r = 0.4$ mm, $\delta^g = 0.852$ mm and $t = 24$ h*

K'	k'_a(s^{-1})	k_d(s^{-1})	ξ (R)	n_M (nmoles) (R)	n_M (nmoles) (R/2)	Difference (%)
10	5×10^{-3}	5×10^{-4}	0.34	54.8	54.0	1.5
10^2	5×10^{-2}	5×10^{-4}	0.29	43.0	37.4	13.0
10^3	5×10^{-1}	5×10^{-4}	0.26	39.7	27.8	30.0
10^4	5	5×10^{-4}	0.25	38.4	24.3	36.7
10^7	5×10^3	5×10^{-4}	0.25	38.3	22.5	41.3
10^9	5×10^5	5×10^{-4}	0.25	38.3	22.5	41.3

or $\delta^r/2$ in the opposite case. Equation 5.38 can be obtained assuming a thickness of the resin disc equal to the layer that contains Chelex plus a "reaction layer" term that takes into account the metal produced by dissociation of the complex in the layer without Chelex ($0 < x < \delta^r/2$).

5.4 Conclusions

DGT provides an integrated measurement of the availability of metals (and other analytes) in natural media. Integration includes time, species and processes present. The standard DGT measurement is the DGT concentration, c_{DGT}, usually interpreted as the labile concentration of the system. This chapter has considered the question of which species contribute to c_{DGT} and to what extent, thus providing information relevant to unravelling the functioning of natural systems. The main findings are summarized as a series of statements.

- The lability degree quantifies the percentage of the contribution of a complex to the accumulation with respect to the maximum contribution that would arise if the complex was fully labile.
- A measurement procedure for the lability degree of a complex (see equation 5.6) as well as expressions for its theoretical calculation in terms of the physicochemical parameters (equations 5.23 and 5.24) have been provided.
- The lability degree is not an intrinsic characteristic of a complex, but it depends on the characteristics of the DGT device, amongst which the thickness of the resin disc is especially relevant.
- c_{DGT} can be seen as the effective free metal concentration in a virtual system with only metal (no ligands present), that would result in the measured accumulation (see equation 5.20).

- When the mobilities of the complexes are similar to that of the free metal, c_{DGT} measures the labile fraction of the total metal present in the system, justifying it being referred to as the labile concentration.
- The lability degree determined in single ligand systems can be used to estimate the behavior of complex mixtures in DGT with reasonable accuracy (see equations 5.17 and 5.22).
- Patterns of the DGT accumulations have been discussed along this chapter. Linear accumulations with time indicate "steady-state" conditions. They arise with labile or partially labile complexes even when the resin fails to act as a perfect sink for the metal. Loss of linearity indicates that bulk equilibrium or saturation effects are approached.
- Metals complexed to stronger ligands than Chelex can still be accumulated in DGT devices. The ratio of the "effective affinities" is a key parameter that determines how thermodynamic factors affect accumulation by DGT (see equation 5.28).
- Faster dissociation of complexes than predicted with the standard Eigen theory has been observed for the NiNTA complex, most likely due to a ligand-assisted (or exchange) mechanism. This is another factor, in addition to penetration of complexes in the resin domain, which determines why most metal complexes behave as labile in DGT.
- The settling of Chelex to approximately one half of the resin disc volume in the standard DGT device does not affect measurement of labile complexes, but it reduces the accumulation of partially labile complexes (see Table 5.1).
- In systems with charged partially labile complexes, electrostatic effects between the charged resin sites and the complexes can modify the accumulations. The accumulation of negatively charged complexes decreases as the ionic strength decreases, while the opposite effect is found for positively charged complexes. Labile complexes are almost free from electrostatic effects since they fully dissociate without penetration into the resin domain. Approximate analytical expressions (see equation 5.36) predict the accumulation in the presence of electrostatic effects.
- Electrostatic Donnan factors can be estimated through expression 5.34 or, experimentally, through the accumulation of cations, like Na^+, with negligible specific binding to the resin sites. The resulting partition factors can be used in other systems whenever pH and ionic strength are fixed.
- The total accumulation, as well as the mass distribution, in DGT devices with a stack of two resin discs, allows the estimation of the different physicochemical parameters of the system. Kinetic dissociation constants of partially labile complexes can be determined with equation 5.33. No kinetic information on fully labile complexes can be obtained because accumulations are only dependent on the transport. Likewise, the kinetic association constant of the metal with the resin sites when the resin does not act as a perfect sink can be determined with equation 5.25 or 5.27.
- Dissociation of charged complexes within the resin layer has been suggested to occur via a ligand-assisted mechanism. This mechanism prescribes an ionic strength dependence of the dissociation rate constants, in contrast with the Eigen model.

Acknowledgments

We thank all our co-workers along these years of research projects related with DGT. Martin Jiménez-Piedrahita kindly provided Figures 5.1, 5.4, 5.5 and 5.6 and Table 5.1. This contribution was financially supported by the Spanish Ministry of Education and Science (Projects CTM2012-39183 and CTM2013-48967).

References

1. O. A. Garmo, O. Royset, E. Steinnes and T. P. Flaten, Performance study of diffusive gradients in thin films for 55 elements, *Anal. Chem.* 75 (2003), 3573–3580.
2. V. E. dos Anjos, G. Abate and M. T. Grassi, Comparison of the speciation of trace metals in freshwater employing voltammetry, diffusive gradients in thin films (DGT) and a chemical equilibrium model, *Quimica Nova* 33 (2010), 1307–1312.
3. W. W. Bennett, P. R. Teasdale, D. T. Welsh et al., Inorganic arsenic and iron(II) distributions in sediment porewaters investigated by a combined DGT-colourimetric DET technique, *Environ. Chem.* 9 (2012), 31–40.
4. H. P. van Leeuwen, Steady-state DGT fluxes of nanoparticulate metal complexes, *Environ. Chem.* 8 (2011), 525–528.
5. E. Navarro, F. Piccapietra, B. Wagner et al., Toxicity of silver nanoparticles to Chlamydomonas reinhardtii, *Environ. Sci. Technol.* 42 (2008), 8959–8964.
6. J. G. Panther, P. R. Teasdale, W. W. Bennett, D. T. Welsh and H. J. Zhao, Titanium dioxide-based DGT technique for in situ measurement of dissolved reactive phosphorus in fresh and marine waters, *Environ. Sci. Technol.* 44 (2010), 9419–9424.
7. Y. L. Zhang, S. Mason, A. McNeill and M. J. McLaughlin, Optimization of the diffusive gradients in thin films (DGT) method for simultaneous assay of potassium and plant-available phosphorus in soils, *Talanta* 113 (2013), 123–129.
8. J. G. Panther, R. R. Stewart, P. R. Teasdale et al., Titanium dioxide-based DGT for measuring dissolved As(V), V(V), Sb(V), Mo(VI) and W(VI) in water, *Talanta* 105 (2013), 80–86.
9. C. E. Chen, H. Zhang, G. G. Ying and K. C. Jones, Evidence and recommendations to support the use of a novel passive water sampler to quantify antibiotics in wastewaters, *Environ. Sci. Technol.* 47 (2013), 13587–13593.
10. A. Caillat, P. Ciffroy, M. Grote, S. Rigaud and J. M. Garnier, Bioavailability of copper in contaminated sediments assessed by a DGT approach and the uptake of copper by the aquatic plant Myriophyllum aquaticum, *Environ. Toxicol. Chem.* 33 (2014), 278–285.
11. D. M. Costello, G. A. Burton, C. R. Hammerschmidt and W. K. Taulbee, Evaluating the performance of diffusive gradients in thin films for predicting Ni sediment toxicity, *Environ. Sci. Technol.* 46 (2012), 10239–10246.
12. F. Degryse and E. Smolders, Cadmium and nickel uptake by tomato and spinach seedlings: Plant or transport control?, *Environ. Chem.* 9 (2012), 48–54.
13. W. Davison, G. Fones, M. Harper, P. Teasdale and H. Zhang, Dialysis, DET and DGT: In situ diffusional techniques for studying water, sediments and soils, in *In situ monitoring of aquatic systems: Chemical analysis and speciation*, ed. J. Buffle and G. Horvai (Chichester: Wiley, 2000), pp. 495–569.
14. W. Davison and H. Zhang, Progress in understanding the use of diffusive gradients in thin films (DGT) – Back to basics, *Environ. Chem.* 9 (2012), 1–13.

15. M. P. Harper, W. Davison and W. Tych, Temporal, spatial, and resolution constraints for in situ sampling devices using diffusional equilibration: Dialysis and DET, *Environ. Sci. Technol.* 31 (1997), 3110–3119.

16. H. Ernstberger, W. Davison, H. Zhang, A. Tye and S. Young, Measurement and dynamic modeling of trace metal mobilization in soils using DGT and DIFS, *Environ. Sci. Technol.* 36 (2002), 349–354.

17. M. H. Tusseau-Vuillemin, R. Gilbin and M. Taillefert, A dynamic numerical model to characterize labile metal complexes collected with diffusion gradient in thin films devices, *Environ. Sci. Technol.* 37 (2003), 1645–1652.

18. F. Degryse, E. Smolders and R. Merckx, Labile Cd complexes increase Cd availability to plants, *Environ. Sci. Technol.* 40 (2006), 830–836.

19. N. J. Lehto, W. Davison, H. Zhang and W. Tych, An evaluation of DGT performance using a dynamic numerical model, *Environ. Sci. Technol.* 40 (2006), 6368–6376.

20. J. L. Levy, H. Zhang, W. Davison, J. Galceran and J. Puy, Kinetic signatures of metals in the presence of Suwannee river fulvic acid, *Environ. Sci. Technol.* 46 (2012), 3335–3342.

21. A. Kreuzeder, J. Santner, H. Zhang, T. Prohaska and W. W. Wenzel, Uncertainty evaluation of the diffusive gradients in thin films technique, *Environ. Sci. Technol.* 49 (2015), 1594–1602.

22. J. Santner, A. Kreuzeder, A. Schnepf and W. W. Wenzel, Numerical evaluation of lateral diffusion inside diffusive gradients in thin films samplers, *Environ. Sci. Technol.* 49 (2015), 6109–6116.

23. W. Davison, Defining the electroanalytically measured species in a natural water sample, *J. Electroanal. Chem.* 87 (1978), 395–404.

24. S. Mongin, R. Uribe, C. Rey-Castro et al., Limits of the linear accumulation regime of DGT sensors, *Environ. Sci. Technol.* 47 (2013), 10438–10445.

25. J. Puy, R. Uribe, S. Mongin et al., Lability criteria in diffusive gradients in thin films, *J. Phys. Chem. A* 116 (2012), 6564–6573.

26. J. Galceran and J. Puy, Interpretation of diffusion gradients in thin films (DGT) measurements: A systematic approach, *Environ. Chem.* 12 (2015), 112–122.

27. J. Galceran, J. Puy, J. Salvador, J. Cecília and H. P. van Leeuwen, Voltammetric lability of metal complexes at spherical microelectrodes with various radii, *J. Electroanal. Chem.* 505 (2001), 85–94.

28. P. L. R. van der Veeken, J. P. Pinheiro and H. P. van Leeuwen, Metal speciation by DGT/DET in colloidal complex systems, *Environ. Sci. Technol.* 42 (2008), 8835–8840.

29. K. Zielinska, R. M. Town, K. Yasadi and H. P. van Leeuwen, Partitioning of humic acids between aqueous solution and hydrogel. 2. Impact of physicochemical conditions, *Langmuir* 31 (2015), 283–291.

30. M. R. Shafaei-Arvajeh, N. Lehto, O. A. Garmo and H. Zhang, Kinetic studies of Ni organic complexes using diffusive gradients in thin films (DGT) with double binding layers and a dynamic numerical model, *Environ. Sci. Technol.* 47 (2013), 463–470.

31. S. Mongin, R. Uribe, J. Puy et al., Key role of the resin layer thickness in the lability of complexes measured by DGT, *Environ. Sci. Technol.* 45 (2011), 4869–4875.

32. M. Jimenez-Piedrahita, A. Altier, J. Cecilia et al., Influence of the settling of the resin beads on diffusion gradients in thin films measurements, *Anal. Chim. Acta* 885 (2015), 148–155.

33. R. Uribe, J. Puy, J. Cecilia and J. Galceran, Kinetic mixture effects in diffusion gradients in thin films (DGT), *Phys. Chem. Chem. Phys.* 15 (2013), 11349–11355.

34. H. P. van Leeuwen and R. M. Town, Outer-sphere and inner-sphere ligand protonation in metal complexation kinetics: The lability of EDTA complexes, *Environ. Sci. Technol.* 43 (2009), 88–93.

35. H. P. van Leeuwen and R. M. Town, Protonation effects on dynamic flux properties of aqueous metal complexes, *Collect. Czech. Chem. Commun.* 74 (2009), 1543–1557.

36. R. Uribe, S. Mongin, J. Puy et al., Contribution of partially labile complexes to the DGT metal flux, *Environ. Sci. Technol.* 45 (2011), 5317–5322.

37. D. Alemani, J. Buffle, Z. Zhang, J. Galceran and B. Chopard, Metal flux and dynamic speciation at (bio)interfaces. Part III: MHEDYN, a general code for metal flux computation; application to simple and fulvic complexants, *Environ. Sci. Technol.* 42 (2008), 2021–2027.

38. J. Galceran, J. Puy, J. Salvador et al., Lability and mobility effects on mixtures of ligands under steady-state conditions, *Phys. Chem. Chem. Phys.* 5 (2003), 5091–5100.

39. H. Zhang and W. Davison, Performance characteristics of diffusion gradients in thin films for the in situ measurement of trace metals in aqueous solution, *Anal. Chem.* 67 (1995), 3391–3400.

40. K. W. Warnken, W. Davison, H. Zhang, J. Galceran and J. Puy, In situ measurements of metal complex exchange kinetics in freshwater, *Environ. Sci. Technol.* 41 (2007), 3179–3185.

41. P. L. R. van der Veeken and H. P. van Leeuwen, Gel-water partitioning of soil humics in diffusive gradient in thin film (DGT) analysis of their metal complexes, *Environ. Chem.* 9 (2012), 24–30.

42. L. Six, E. Smolders and R. Merckx, The performance of DGT versus conventional soil phosphorus tests in tropical soils-maize and rice responses to P application, *Plant Soil* 366 (2013), 49–66.

43. S. C. Apte, G. E. Batley, K. C. Bowles et al., A comparison of copper speciation measurements with the toxic responses of three sensitive freshwater organisms, *Environ. Chem.* 2 (2005), 320–330.

44. Z. S. Zhang, J. Buffle, R. M. Town, J. Puy and H. P. van Leeuwen, Metal flux in ligand mixtures. 2. Flux enhancement due to kinetic interplay: Comparison of the reaction layer approximation with a rigorous approach, *J. Phys. Chem. A* 113 (2009), 6572–6580.

45. J. P. Pinheiro, J. Salvador, E. Companys, J. Galceran and J. Puy, Experimental verification of the metal flux enhancement in a mixture of two metal complexes: The Cd/NTA/glycine and Cd/NTA/citric acid systems, *Phys. Chem. Chem. Phys.* 12 (2010), 1131–1138.

46. J. Salvador, J. L. Garcés, E. Companys et al., Ligand mixture effects in metal complex lability, *J. Phys. Chem. A* 111 (2007), 4304–4311.

47. J. L. Levy, H. Zhang, W. Davison, J. Puy and J. Galceran, Assessment of trace metal binding kinetics in the resin phase of diffusive gradients in thin films, *Anal. Chim. Acta* 717 (2012), 143–150.

48. J. Puy, J. Galceran, S. Cruz-Gonzalez et al., Metal accumulation in DGT: Impact of ionic strength and kinetics of dissociation of complexes in the resin domain, *Anal. Chem.* 86 (2014), 7740–7748.

49. R. M. Town, P. Chakraborty and H. P. van Leeuwen, Dynamic DGT speciation analysis and applicability to natural heterogeneous complexes, *Environ. Chem.* 6 (2009), 170–177.

50. S. Tankere-Muller, W. Davison and H. Zhang, Effect of competitive cation binding on the measurement of Mn in marine waters and sediments by diffusive gradients in thin films, *Anal. Chim. Acta* 716 (2012), 138–144.

51. J. P. Gustafsson, Visual MINTEQ version 3.0. www.lwr.kth.se/English/Oursoftware/vminteq/index.htm. 2010

52. M. Pesavento, R. Biesuz, M. Gallorini and A. Profumo, Sorption mechanism of trace amounts of divalent metal-ions on a chelating resin containing iminodiacetate groups, *Anal. Chem.* 65 (1993), 2522–2527.

53. M. Pesavento, R. Biesuz and G. Alberti, Characterization of the sorption equilibria of nickel(II) on two complexing resins by the Gibbs-Donnan model, *ANN. CHIM. ROME.* 89 (1999), 137–146.

54. M. Pesavento, R. Biesuz, F. Baffi and C. Gnecco, Determination of metal ions concentration and speciation in seawater by titration with an iminodiacetic resin, *Anal. Chim. Acta* 401 (1999), 265–276.

55. W. Davison and H. Zhang, In-situ speciation measurements of trace components in natural waters using thin-film gels, *Nature* 367 (1994), 546–548.

56. H. P. van Leeuwen and J. Galceran, Biointerfaces and mass transfer, In *Physicochemical kinetics and transport at chemical-biological surfaces*, ed. H. P. van Leeuwen and W. Koester (Chichester: Wiley, 2004), pp. 113–146.

57. K. W. Warnken, W. Davison and H. Zhang, Interpretation of in situ speciation measurements of inorganic and organically complexed trace metals in freshwater by DGT, *Environ. Sci. Technol.* 42 (2008), 6903–6909.

58. J. Levy, H. Zhang, W. Davison and R. Groben, Using diffusive gradients in thin films to probe the kinetics of metal interaction with algal exudates, *Environ. Chem.* 8 (2011), 517–524.

59. E. Uher, M. H. Tusseau-Vuillemin and C. Gourlay-France, DGT measurement in low flow conditions: Diffusive boundary layer and lability considerations, *Environ. Sci.: Processes Impacts* 15 (2013), 1351–1358.

60. N. Fatin-Rouge, A. Milon, J. Buffle, R. R. Goulet and A. Tessier, Diffusion and partitioning of solutes in agarose hydrogels: The relative influence of electrostatic and specific interactions, *J. Phys. Chem. B* 107 (2003), 12126–12137.

61. H. Ohshima, Electrical double layer, In *Electrical phenomena at interfaces. Fundamentals, measurements and applications*, eds. H. Ohshima and K. Furusawa (New York: Marcel Dekker, 1998), pp. 1–18.

62. H. P. van Leeuwen, R. M. Town and J. Buffle, Chemodynamics of soft nanoparticulate metal complexes in aqueous media: Basic theory for spherical particles with homogeneous spatial distributions of sites and charges, *Langmuir* 27 (2011), 4514–4519.

63. L. P. Yezek, P. L. R. van der Veeken and H. P. van Leeuwen, Donnan effects in metal speciation analysis by DET/DGT, *Environ. Sci. Technol.* 42 (2008), 9250–9254.

64. J. F. L. Duval, J. P. Pinheiro and H. P. van Leeuwen, Metal speciation dynamics in monodisperse soft colloidal ligand suspensions, *J. Phys. Chem. A* 112 (2008), 7137–7151.

65. K. Yasadi, J. P. Pinheiro, K. Zielinska, R. M. Town and H. P. van Leeuwen, Partitioning of humic acids between aqueous solution and hydrogel. 3. Microelectrodic dynamic speciation analysis of free and bound humic metal complexes in the gel phase, *Langmuir* 31 (2015), 1737–1745.

66. W. Davison, C. Lin, Y. Gao and H. Zhang, Effect of gel interactions with dissolved organic matter on DGT measurements of trace metals, *Aquatic Geochemistry* 21 (2015), 281–293.

67. F. M. M. Morel and J. G. Hering, Complexation, In *Principles and applications of aquatic chemistry* (New York: Wiley, 1993), pp. 319–420.

6

Applications in Natural Waters

HELÉNE ÖSTERLUND, ANDERS WIDERLUND AND JOHAN INGRI

6.1 Introduction

This chapter is aimed at providing a short checklist for experts and non-experts, during planning of a DGT sampling programme in running freshwater, in lakes, brackish, and marine waters. Although DGT techniques are simple in theory, their practical implementation usually becomes more complicated. Facing a field campaign, a number of decisions must be taken in order to be able to obtain relevant field data. DGT field data are spread over many different publications, and it is difficult to get an overview of how well DGT performs during routine conditions. This checklist is therefore accompanied by field data, illustrating the robustness and precision of the DGT system.

Detailed interpretation of DGT measurements from the field is associated with a range of uncertainties and questions that need to be addressed before deployment. Parameters that must be considered have been classified into three major groups: basic parameters, field parameters and evaluation parameters.

The checklist has its origin in how to calculate the concentration from DGT measurements. Once the DGT device is retrieved and the accumulated amount, M, of each analyte of interest is determined, the DGT labile concentrations, C_{DGT}, can be estimated using equation 6.1 ([1] and Chapter 1) – the "DGT equation":

$$C_{\mathrm{DGT}} = \left(\frac{M \Delta g}{D t A} \right) \tag{6.1}$$

where Δg is the diffusion layer thickness, D is the diffusion coefficient, t is the deployment time and A is the area of the opening in the DGT device. Chapter 2 provides a detailed assessment of the limitations of this equation and the variants that should be used depending on its application.

6.2 Basic Parameters

6.2.1 Binding Layer

The binding layer determines the analytes that can be measured. For measurement of trace metals, the original Chelex binding gel is recommended.

The DGT technique can be applied to any element or compound for which a selective binding agent can be found and no adverse interactions between analyte and diffusive layer are present. The DGT technique was first characterised for Cd, Cu, Fe, Mn, Ni and Zn [1]. Later Garmo et al. [2] evaluated the Chelex DGT and concluded that it was also useful for the measurements of Pb, Co, Al, Ga, Y and the rare earth elements (REE). For shorter deployment times, of up to half a day, the Chelex DGT can also be used to measure major elements [3].

Fe oxide is the most commonly used binding agent for the measurement of oxyanions. It has been evaluated and applied for P, V, As, Mo, Se, Sb and W [4–6] in freshwater. Alternative binding agents for oxyanions are titanium dioxide available in proprietary form as Metsorb [7–9] and Zr-oxide [10].

There is also a range of binding agents characterised and applied for specific analytes, such as thiols for Hg [11] and methyl-Hg [12], TEVA for Tc [13] and ammonium molybdo-phosphate for Cs [14]. A detailed compilation of analytes and binding layers is provided in Chapter 4.

6.2.2 Diffusive Layer

Where possible, it is recommended to use the normal open pore DGT with APA gel.

The pore size of the normal DGT APA gel is still not known accurately, but particles with diameters >5 nm are unlikely to contribute effectively to the DGT measurement [15]. This is not a definitive cut-off, but means that hydrated metal ions and complexes <5 nm diffuse relatively freely in the gel and larger complexes are retarded. Large complexes will therefore only contribute marginally to the total accumulated analytical mass. Analytes bound to ligands that are small enough to diffuse through the diffusive gel will be included if the time of diffusion prior to binding is long enough for the complex to dissociate (see Chapter 5 for a detailed appreciation of how DGT measures solution complexes). In practise, organic ligands present in natural waters are often smaller than 5 nm in size, but the diffusion coefficients of their complexes are lower than those of the free ions (see Chapter 3 for details on diffusion coefficients). Consequently, if the diffusion coefficient of the free ion is used in the calculation of c_{DGT}, the concentration of this complexed metal will be underestimated. This is illustrated for example by Warnken et al. [16]. Measurements of trace metals, applying DGT in combination with discrete sampling and subsequent analysis of the dissolved fraction, were conducted in thirty-four streams. The results showed that Cd and Zn determined by DGT corresponded well to the dissolved concentrations in the discrete samples. On the contrary, Cu concentrations measured by DGT were often lower than the corresponding values of the dissolved Cu concentration. With the aid of speciation modelling using WHAM, this apparent discrepancy was quantitatively accounted for by the lower diffusion coefficient of Cu–fulvic acid complexes [17]. The importance of D is discussed further in Section 6.4.4.

Full quantification of c_{DGT} in terms of complexed and uncomplexed metal can be achieved by using several DGT devices containing more than one type of diffusive gel characterised by different pore sizes [18, 19]. These papers showed the capability of DGT for characterising more fully solution concentrations in natural systems, but they relied on a quite elaborate set of measurements.

Other workers have adopted a simpler approach [20]. DGT devices with a diffusive gel having a more restricted pore size, known as a restricted gel, RG (see Chapter 3 and [18]) have been used. Again, the pore size is not known exactly, but the diffusion coefficients of complexes and free ions are lower in the RG compared to the APA gel [17]. Some laboratory studies have indicated that particles much larger than 5 nm have access to both RG and APA gels [21], which would imply that the pore size is larger than previously thought and that no differences between the RG and APA DGT results should be expected. However, the retardation of metal complexes of fulvic and humic acid was more pronounced in the RG and increased with molecular weight, even though the nominal size range considered of 2–3.8 nm was modest [17]. These operational results suggest that size fractionation can be expected if DGT deployments use both types of diffusive gel.

When executing calculations of DGT-labile concentrations from RG DGT deployments the difference in diffusion coefficients between the protective membrane filter and RG must be considered as well as for the diffusive boundary layer (DBL) (which is discussed below) (see Chapter 2).

The restricted DGT gel has been little studied in the field. Figure 6.1 shows an example when APA and RG DGT were deployed in situ [20]. The APA DGT and RG DGT concentrations closely follow each other both in depth profiles and time series. This clearly underlines the robustness of the DGT method. The labile complexes are probably smaller than the pore size of the RG diffusive gel since the APA and RG DGT samplers gave similar results for Cu and Ni measurements. Although the results from the Gotland Deep indicate limited benefit of using APA and RG DGT probes in combination, there are other studies in both freshwater [18, 19] and seawater [22] settings where larger differences have been obtained. However, if only one DGT version is to be used, the APA variant is to be preferred, mainly because the data collected will be directly comparable with the majority of other studies, thus facilitating the interpretation.

In some cases use of the APA gel may not be appropriate. For example, Hg and some organic compounds are known to bind to the APA gel, which has led to an agarose gel being used instead [23, 24]. Trace metals, particularly Cu and Pb, bind slightly to the APA gel, but this does not usually present a problem [25]. It would only be an issue if metal concentrations were very low and deployment times were less than a day, which is unlikely to arise (see Chapter 3).

6.2.3 Gel Thickness

It is recommended to use the standard diffusive gel thickness (0.08 cm).

The standard diffusion gel thickness is 0.08 cm. When partnered with a 0.4-mm binding gel and a 0.14-mm membrane filter, it is a good fit in a standard DGT holder. It is sufficiently thick to minimise the effects of the diffusive boundary layer (DBL) on the measurement (see Chapter 2) while still providing good sensitivity. The choice of gel thickness will

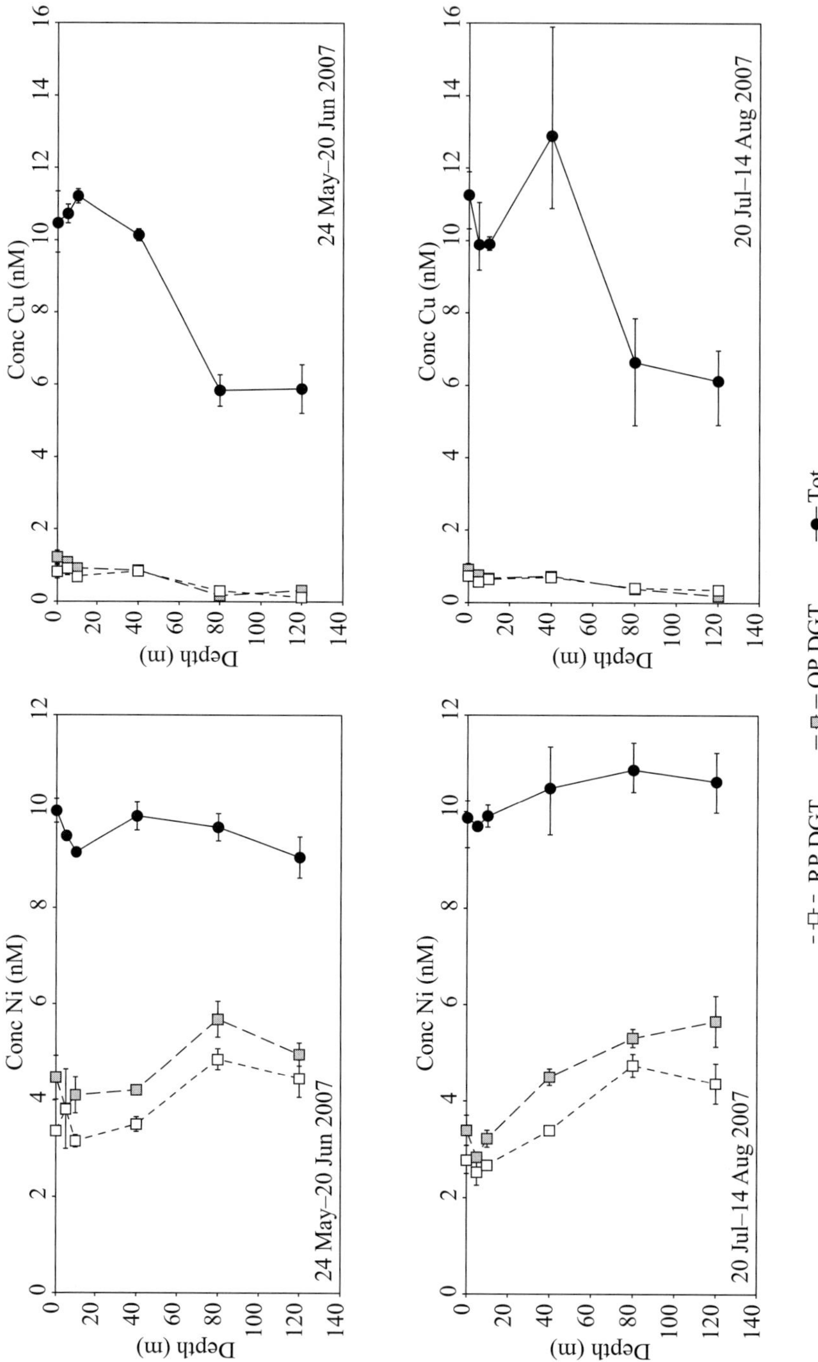

Figure 6.1 Vertical profiles of total and DGT labile concentrations of Ni (left) and Cu (right) from two sampling occasions in the Gotland Deep of the Baltic Sea, 2007. The profiles from the open pore (OP) and restricted pore (RP) DGT measurements follow each other, illustrating the robustness of the DGT technique. The uncertainty bars illustrate maximum and minimum concentrations for the DGT [20]. (Reprinted from Marine Chemistry, H. Österlund, J. Gelting, F. Nordblad, D. C. Baxter and J. Ingri, Copper and nickel in ultrafiltered brackish water: Labile or non-labile?, 34–43, 2012, with permission from Elsevier.)

affect sensitivity and the detection limit, since DGT devices with thinner diffusion gels will accumulate a greater mass of an analyte than simultaneously deployed devices with a thicker gel.

Simultaneously deployment of devices with thinner and thicker diffusive gels is most commonly used for providing an in situ estimation of the thickness of the DBL [26]. Chapter 2 provides a detailed listing of published estimates of DBL thickness, obtained both in situ and in the laboratory. Such estimates should use analytes which are fully labile. If there is only partial lability the diffusive gel thickness, as well as the binding layer thickness, will to a certain extent determine which species of an element will contribute to the total accumulated mass. This is because metal–ligand complexes will have longer time for dissociation in the thicker gels. A set of DGT devices with various gel thicknesses can therefore also be used for investigations of dissociation characteristics of metal–ligand complexes [27, 28] (see Chapter 3 for further details).

6.3 Field Parameters

6.3.1 Deployment of DGT Samplers

It is recommended to use at least two samplers at each sampling location.

Figure 6.2 shows different set-ups for DGT sampling in natural waters, with the examples of deployment of devices in the Baltic Sea and in a freshwater stream. To avoid a pronounced DBL (c.f. Section 6.4) the DGT holder must not be shielded or exposed to stagnant water [29]. In streams and rivers, this is not a problem. In lakes and open marine waters the set-up shown in Figures 6.2a and b, respectively, can be used to minimize the DBL. Many different holder designs have been used and devices have also been simply tied to a line. The prime requirement of holders is that they are insufficiently bulky to impeded water flow in the general vicinity and they must not directly impede water flow at the surface of the device. It is recommended to deploy a minimum of two DGT samplers at each sampling location, or as a minimum at some locations. Analysing two samplers or more gives a good indication of the overall precision in the analyte concentrations and highlights potential problem data points. Figure 6.2c shows how three DGT devices can be mounted in an acrylic transparent holder, which can be easily manufactured, washed and reused in subsequent sampling campaigns.

6.3.2 Temperature

In situ temperature loggers are recommended at each sampling point.

The diffusion coefficient of the analyte within the diffusion layer is temperature dependent and therefore it is important that the deployment temperature is known. If tabulated values of D are unavailable at this temperature a known value at a standard temperature, such as 25°C, can be readily corrected to the deployment temperature using equation 2.25 in Chapter 2. Adding temperature loggers at each sampling point is a good way of measuring

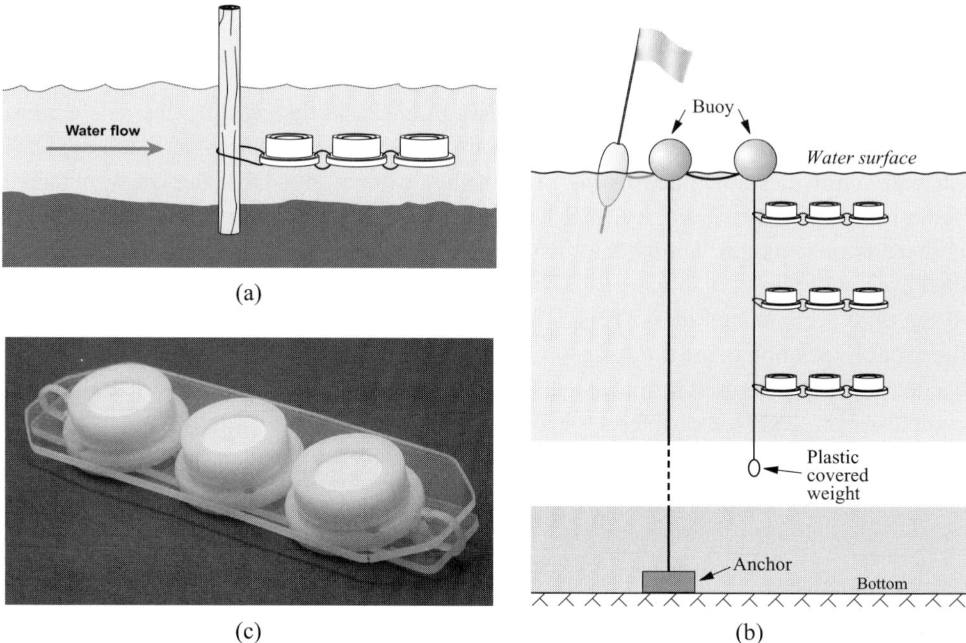

Figure 6.2   Schematic illustrations of arrangements when measuring in streams and lakes (a) and open sea (b). The pole in (a) helps to assure that the devices are placed at a well-defined depth. The picture in (c) shows how DGT devices can be mounted in acrylic glass holders for easy deployment and retrieval. The top and bottom plates that keep the DGT devices in place are locked with cable ties.

temperature to ensure the calculation is performed accurately, especially if the temperature is likely to change appreciablely during deployment. If the system where the DGT sampling is carried out can be expected to have even temperature during night and day time and not vary significantly during the deployment, the temperature can be measured upon set-out and retrieval of the devices. This can be expected in, for example, oceans. However, in systems where the temperature can be expected to vary during the time of the day or due to seasonal changes, continuous logging of the temperature must be preferred. For every degree from the true average temperature, the inaccuracy increases by around 3 per cent. The example in Figure 6.3 illustrates the temperature variation measured in a stream in Northern Sweden from August 9 to 30. The average temperature using all measured values corresponds to 12.5°C, but if the average temperature is calculated from first and last values the average temperature is 15.6°C. This temperature difference corresponds to an almost 10 per cent higher value of the diffusion coefficient and would lead to underestimation of the DGT-labile concentrations to the same extent (see equation 6.1). This error is in the same range as the overall precision of DGT measurements, estimated by Kreuzeder et al. [30].

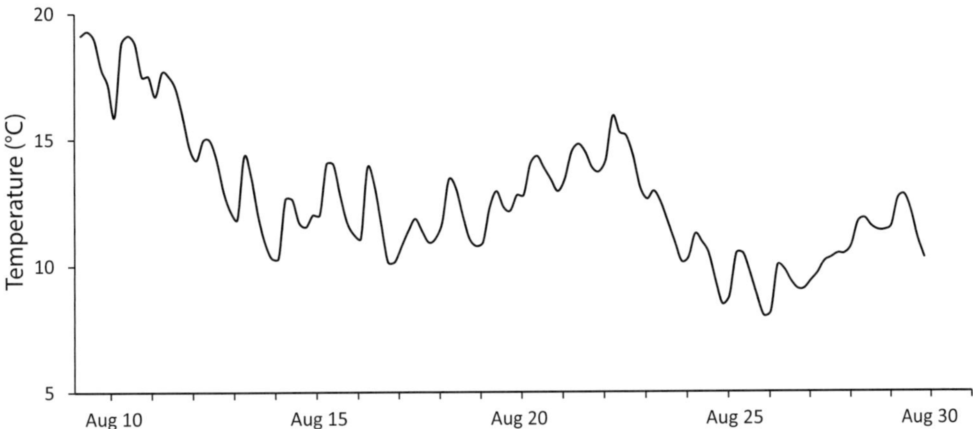

Figure 6.3   Example of variation of temperature at a sampling site in Northern Sweden measured by a temperature logger during three-weeks deployment. Depending on whether the mean temperature is determined from deployment and retrieval values (15.6°C) or from the temperature logger (12.5°C), the difference in calculated DGT labile concentration is almost 10 per cent.

Temperature loggers also show if equipment or measurements have been disturbed by, for example, removal and replacement of the set-up. Furthermore, for streams and rivers, temperature is a useful parameter that provides insight into hydrological events in the catchment.

6.3.3 Deployment Time

Deployment time determines the detection limit and the approach to saturation of the binding layer. This important parameter must be carefully considered early in each sampling project.

The optimum deployment time depends on in-situ analyte concentrations. Longer deployment times will improve the detection limit since greater amounts of analyte will be accumulated. Deployment time is often a compromise because of the difference in accumulation rates between analytes at high and low concentrations. When a Chelex DGT device is fresh, all binding sites of the adsorbent are occupied by easily exchangeable Na ions. As soon as the device is placed in water and accumulation begins, the Na ions are exchanged by ions with higher affinity for the Chelex resin. Ca and Mg are found in relatively high concentrations and will soon exchange with Na. However, these ions are also adsorbed comparably weak to the binding sites and more strongly binding trace metals can still be accumulated by replacements of Ca and Mg. The following selectivity series is reported by the manufacturer of Chelex for chloride and nitrate solutions and can serve as a guide: $Cu^{+2} \gg Pb^{+2} > Fe^{+3} > Cr^{+3} > Ni^{+2} > Zn^{+2} > Co^{+2} > Cd^{+2} > Fe^{+2} > Mn^{+2} > Ba^{+2} > Ca^{+2} \ggg Na^{+}$ [31]. The total capacity of the Chelex binding layer is high but finite. Zhang and Davison [1] determined the maximum capacity of the normal Chelex DGT to be

Table 6.1 *Minimum, mean and maximum APA DGT concentrations ($\mu g\ L^{-1}$) of Cd, Ni, Zn and Cu measured in the Landsort Deep of the Baltic Sea at 5 m depth in 2004 and 2005. (Reprinted from [20] with permission from Elsevier.)*

Analyte	Year	Min	Mean	Max
Ni	2004	3.3	4.2	5.2
	2005	4.2	4.5	5.2
Cu	2004	0.81	1.4	1.8
	2005	0.94	1.2	1.7

0.65 mg Cd, approximately 6 µmol, although some later versions of the device have used higher amounts of Chelex, attaining a capacity of up to 13 µmol [32]. In ocean water the capacity for Ca is reached in a few hours [32], effectively ruling out measurement by DGT, but measurements can be made using short deployment times of a few days or less in waters with lower Ca concentrations, including fresh and brackish waters [33]. Trace metals will simultaneously accumulate and effectively lower the capacity for Ca, but their concentrations are usually so low that this effect is negligible for short deployment times. Their low concentrations and strong binding means that trace metals such as Cu and Pb are likely to accumulate linearly with respect to time, for deployments in excess of a year in ocean waters, while for the higher concentrations in coastal waters capacity may be exceeded a month [1]. In the exceptional case encountered in some sediment porewaters, where there are very high concentrations of Fe (200 µM), it can displace Mn. A consequence is that measurements of Mn at 20 µM by DGT were accurate for deployments up to 6 h, but Mn was substantially underestimated for a longer deployment of 21 h [32]. Mn has been measured successfully by DGT in freshwaters where the typical two-week deployment time and low concentrations of analytes ensures that the capacity of the device is not approached by Mn or other ions which bind more effectively [16]. However, even for short time deployments of a few hours, the amount of Mn accumulated from a solution with high Ca and Mg to emulate seawater, but a lower pH (5.5) is only 25 per cent of the amount expected [32]. Such near-linear accumulation with time at a less than theoretical slope is characteristic of slow binding to the resin, as explained in Chapter 5. It may be that when the resin is saturated with Ca and Mg, which it will be after 2 h, that binding of Mn is slow, as has been shown directly for Mn in the absence of Ca and Mg, but at even lower pH [34].

Maximum, minimum and mean concentrations of DGT measurements using an APA diffusive gel of Cu and Ni in Landsort Deep in 2004 and 2005 are presented in Table 6.1. Consistency of the results between the two years affirms the robustness and reproducibility of the DGT technique. The differences in deployment times between the two years, i.e. 11–29 and 57–86 days in 2004 and 2005, respectively, do not appear to have affected the accumulation efficiency.

While capacity considerations encourage short deployment times, issues arise if they are too short. For simple solutions, the steady-state equation 6.1 is applicable after a few hours. However, where the analyte is present as higher molecular weight complexes in solution, such as with humic substances, the transient time before steady state is established, will be increased due to the slower diffusion of these complexes, and hence deployment times should be longer (see Chapters 2 and 5). The transient time may be further increased by adsorption of ions, such as Cu and Pb, to the diffusive gel, which may be further enhanced by the presence of humic substances (Chapter 3). To overcome these possible effects deployment times in natural waters should be at least a few days.

6.3.4 Fouling

Remember to inspect each sampler for fouling upon retrieval.

The formation of biofilms on the surface of DGT devices can potentially interfere with DGT measurements and reduce the amount of analyte trapped in the binding gel. Biofilms composed of algae, bacteria, fungi, and products from cell metabolism are commonly referred to as biofouling [35]. Adhesion of suspended particulate matter is another type of fouling [36]. These various types of fouling may result in physicochemical interactions between fouling and analytes, which affect the thickness of the DGT diffusion boundary layer and/or the diffusion coefficient of an analyte. In this physicochemical interaction, the fouling can be seen as an additional inert diffusion layer on the surface of the protective DGT filter membrane. Another principally different type of interaction is when the fouling actively interacts with an analyte, for example, through biosorption, complexation and precipitation of insoluble compounds [35].

A special case of this interactive type of fouling has been observed in a stream receiving mining-affected water discharging from a sulphide ore concentration plant [37]. In this case, fouling by what appeared to be pure Fe oxides was observed on the protective DGT filter membrane. Measurements of DGT–P using DGT devices with Fe oxide as binding agent [4] were performed at two stations along the stream, located before and after a lake. The DGT devices were deployed for six to fourteen days in the stream, which had filtered (<0.45 μm) Fe concentrations in the range 0.2–0.3 mg/L and a neutral pH. Seasonal variations in stream water concentrations of soluble reactive P were measured using DGT together with the conventional molybdenum blue colorimetric method, permitting comparison of the two methods under field conditions during a period of six months (Figure 6.4). The DGT and molybdenum blue methods showed similar seasonal variations, with a concentration maximum appearing during late June–early July. However, at the lake inlet where Fe oxide fouling was observed, DGT–P concentrations were on average only 19 per cent of the concentrations measured with the molybdenum blue method. At the lake outlet, where no Fe oxide fouling was present, average DGT–P concentrations were 62 per cent of those obtained with the molybdenum method (Figure 6.4).

A direct quantitative comparison of the two methods is difficult since the precise form of P measured with the molybdenum method is unknown and can, for example, include

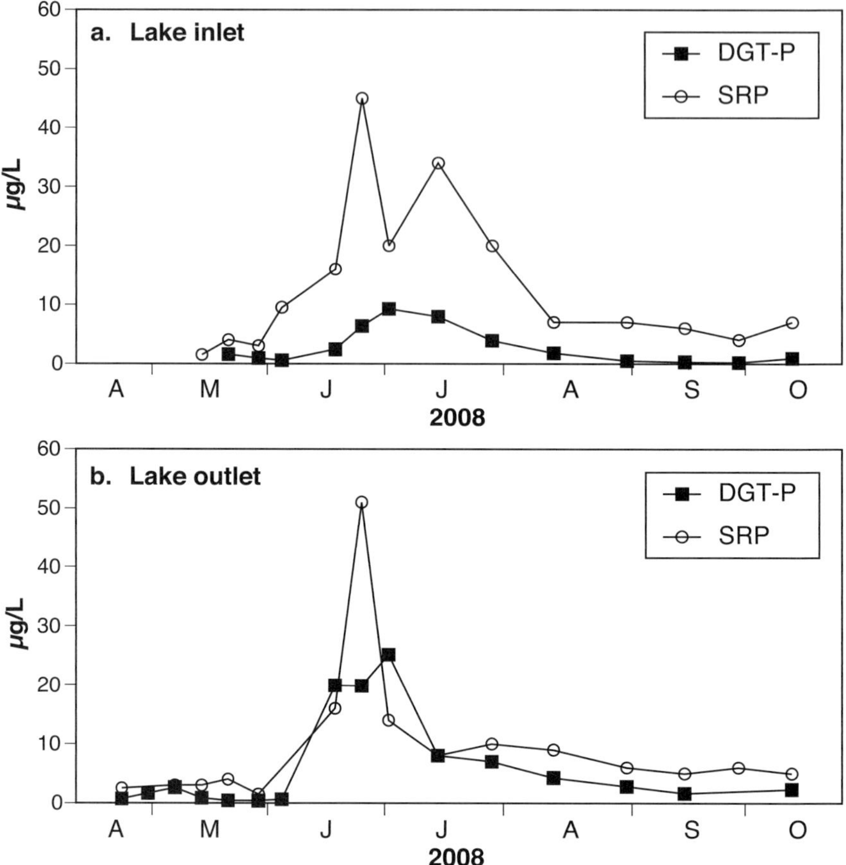

Figure 6.4   Seasonal variations in mining-affected stream water of DGT–P and soluble reactive phosphorus (SRP) concentrations measured with the molybdenum blue colorimetric method. Sampling period: April–October 2008. (a) Sampling station at lake inlet where Fe oxide fouling had formed on the DGT protective membrane filter. (b) Sampling station at lake outlet with no Fe oxide fouling on the protective filter. At the lake inlet, DGT–P concentrations were, on average, only 19 per cent of those measured with the molybdenum blue method.

P bound to colloids. However, this example illustrates that fouling caused by Fe oxide forming on DGT devices is likely to compete with the Fe oxide binding agent used to capture soluble reactive phosphorus. The resulting potential underestimation of DGT–P concentrations was supported by analyses of protective filters, diffusive gels, and binding gels from the same DGT device, showing that 14–40 per cent of P entering the device was trapped on the protective filter. Since the P content found in the diffusive gel was negligible, the Fe oxide binding gel accumulated practically all of the remaining 60–86 per cent of the P that entered the DGT.

The impact of biofouling is both difficult to predict and, in standard DGT deployments used for monitoring purposes, difficult to quantify. Although some combined field and laboratory studies indicate a considerable impact on metal concentrations [35, 36], there are also studies showing no apparent effects of biofouling [14, 38, 39]. Moderate to severe biofouling is generally readily visible [36], and data obtained from DGT devices displaying any signs of biofouling should be interpreted with caution. As can be expected, biofouling is most likely to represent a problem during long-term (>10 days) DGT deployments in nutrient-rich waters [35, 36]. It has also been suggested that biofouling may be more intense during the warmer and more productive summer season, with minimal impact of biofouling during winter [41]. However, there are examples of long-term deployments (2–4 weeks), where no visible biofouling was observed [40]. In this study, the absence of biofouling was supported by the generally good agreement between Mn, Zn and Cd concentrations measured by DGT and ultra-filtration (<1 kDa).

To prevent the formation of biofouling, it is sensible to keep DGT deployment times as short as possible, preferably less than seven to ten days [35, 36]. However, in a study where radiocesium was measured in a lake by DGT, biofilm growth was considered not to be significant for deployment times up to one month [14]. In addition to short deployment times, treatments of DGT filter membranes and diffusive gels with antibiotics or metal-iodides (Cu and Ag) have been used to prevent biofouling [36]. Experiments with different types of DGT filter membranes indicate that polycarbonate membranes may improve the quantification of Cd, Cu, Mn, Pb and V compared to the more commonly used polyethersulfone membranes [35].

6.3.5 Ionic Strength and pH

Caution should be exercised at extremes of pH and ionic strength.

During the early development of DGT, investigations indicated that DGT could be used in waters with ionic strengths as low as 10 nM [1]. However, later work showed that precision and accuracy of DGT measurements tend to deteriorate when the ionic strength is less than 1 mM [40–44]. Warnken et al. [44] demonstrated that the problem could be controlled by extensive and thorough washing of the diffusive gels after polymerization. This procedure reduced the amount of unreacted polymerization products in the gels, which can confer a negative charge. With a negatively charged gel there is increased accumulation of cations (see Chapter 3) and any variation in gel charge lowers precision [15]. This effect is most marked in waters with ionic strengths below 1 mM because there are insufficient ions to adequately screen the charge [15]. Exhaustive washing of hydrated gels resulted in decreased accumulation rates of cations at ionic strengths below 1 mM, due to a slight positive charge associated with a pH <7. At ionic strengths ≥1 mM, the slight charge is insufficient to affect DGT measurements of cations. When additional measurements of diffusion coefficients were performed at lower ionic strength, it was concluded that DGT can be used with preserved precision down to ionic strengths of 0.1 mM [17, 44]. In systems which contain charged, partially labile species there can be other effects associated with ionic strength [45], as considered in Chapter 5.

For most analytes, reliable DGT measurements can be performed over a wide pH range. Limitations are associated with the characteristics of the binding agent and the diffusive gel.

The Chelex resin consists of a styrene divinylbenzene copolymer chemically bonded iminodiacetic acid (IDA). IDA forms tridentate chelates with polyvalent cations and has a high affinity for metal ions [31]. At lower pH values IDA becomes protonated, and the chelating efficiency is impaired. The pH at which decreased adsorption appears varies between analytes and depends on the selectivity of IDA. For example, Cd measurements can be conducted down to pH 4–5 [1, 29], while Cu uptake in synthetic solutions resulted in accurate results at least down to pH 2 [29] (see Chapter 4). In practice, measurements can be made at lower pH by calibrating for the effect of pH on the DGT measurement. This was successfully done for determination of labile Mn in acid mine drainage [46]. Below pH 4.0, the adsorption of Mn decreased linearly with pH to 55 per cent at pH 3.0. Taking this correction into account revealed that >85 per cent of the dissolved Mn measured in discrete samples was DGT-labile. Evaluation of ferrihydrite DGT for anionic determination has shown that accurate measurements of phosphate can be undertaken in the pH range 2–10 [4], and for arsenic and molybdate between pH 3–8 [9, 48].

Swelling of the DGT diffusive gel occurs at pH values <1 and ≥11 [40]. However, measurements conducted at pH 12.9 showed that even though the accumulation rate is decreased, there is still a linear uptake over time [29]. Therefore, DGT could probably be used at high pH after appropriate calibration by determination of an effective diffusion coefficient at the given pH.

6.4 Evaluation Parameters

6.4.1 Blank Values

It is important to set aside blank devices from each sampling occasion.

Any scatter in the amount of analyte measured in blank DGT devices, which have not been deployed, will contribute to inaccuracy and imprecision of DGT measurements. As blank levels can change from batch to batch, blanks from different batches of samplers should not be mixed.

Measurement of some trace metals including Zn, Cu and Pb are very prone to contamination. All handling with the DGT devices should therefore be carried out under clean bench/clean room conditions. However, the mounting of the devices in the field during deployment and retrieval has also been shown to be a highly significant contributory factor [48]. Table 6.2 lists typical blank values of unexposed DGT devices for selected elements. With additional caution lower values can be achieved, but it is easy to exceed these values due to contamination problems.

Table 6.2 *Typical blank values for unexposed DGT devices*

Analyte	Binding agent	ng/gel	Ref
Cd	Chelex	0.035	[49]
Cu	Chelex	0.95	[49]
Pb	Chelex	0.095	[49]
Zn	Chelex	14	[49]
As	Ferrihydrite	2.0	[6]
Mo	Ferrihydrite	2.1	[6]
Sb	Ferrihydrite	0.16	[6]
P	Ferrihydrite	8.0	Unpublished

6.4.2 Precision of DGT Measurements

Precision can be readily estimated from replicate measurements.

Repeatability better than 5 per cent has been achieved for Cu and Cd measurements in laboratory tests [44], but inferior precision must be expected in field studies. When the precision of DGT measurement was estimated from several duplicates [14, 35–39] deployed in the Baltic Sea, the relative standard deviations (RSD) were determined to be 16 per cent for Cu and 9 per cent for Ni [20]. The relative difference between accumulated masses for each duplicate was in this case regarded as an independent measurement, and the relative differences of all duplicates were pooled to estimate the RSD following a procedure described by Minkkinen [50]. A selection of samples from this sampling campaign is shown in Figure 6.1. From measurements including Al, Cd, Cu, Fe, Mn, Ni, Pb and Zn in freshwater streams, where deployment times were 24 h or 14 days, RSD values ranging between 6 and 12 per cent have been reported [16]. The lowest values were reported for Cd and highest for Cu and Pb. Another approach to investigate precision of DGT measurements was presented by Kreuseder et al. [30]. They investigated the combined uncertainty of DGT measurements under well-controlled experimental conditions for determination of As, Cd, Cu and P applying the Kragten method. It was concluded that under the prevailing conditions, the relative uncertainty was around 10 per cent at a confidence level of 95 per cent. The major contributors to the overall uncertainty were found to be the elution factor for analyte recovery, the diffusion coefficient and the diffusion layer thickness.

6.4.3 Diffusive Boundary Layer (DBL) and Effective Sampling Area

The DBL can be calculated if at least two samplers, with different gel thickness, are used at a sampling site. Knowledge of the DBL provides reassurance on data fidelity.

The DBL thickness depends on the flow velocity of the ambient water [26] and can be determined from deployments of multiple DGT devices with different diffusive gel thicknesses [27] (see Chapter 2). In laboratory studies it has been shown that for DGT devices deployed in solutions with no agitation, the DBL thickness can be up to 1.5 mm, which is around 1.5 times the diffusive layer thickness in standard DGT devices [26]. If this DBL was not taken into consideration, the DGT-labile concentration would be underestimated (less than half). In flow rates above 2 cm s^{-1} the DBL thickness is around 0.2 mm and does not significantly affect the overall precision [25]. Consequently, with appropriate experimental setup, laboratory measurements using DGT should certainly be accurate to within 10 per cent and possibly 5 per cent.

Chapter 2 presents a table that shows that there is a wide range in estimates of the DBL thickness obtained from in situ deployments. Many are less than 0.4 mm, especially for situations where flow is known to be high. However, for some field studies, despite flow rates above 2 cm s^{-1}, considerably higher values of the DBL thicknesses, up to the same level as for stagnant water, were measured [9, 48]. The study presented by Turner et al. [9] mounted the devices on large, flat, plastic plates, which are likely to have impeded flow, but it illustrated the potential problems arising from variable flow rates. Metsorb DGT was deployed for consecutive periods of a week in two rivers for four to five months for determination of uranium. DBL thicknesses were measured using multiple DGT devices with different diffusive layer thicknesses on four occasions in each river and applied to the whole dataset. The DBL thickness was only poorly related to the flow rate of the river water, suggesting that other factors such as biofouling might be at play. For one of the rivers the DBL thickness ranged from 0.062 to 0.086 cm, so using a mean value of 0.073 cm for calculation of the DGT concentration would at worst introduce an error of less than 8 per cent. The other river had a much more variable DBL thickness of 0.037–0.141 cm, so that using the mean thickness of 0.078 cm could for the extreme examples introduce errors of up to 36 per cent. If the DBL thickness had been disregarded completely the concentration would have been underestimated by typically 30 per cent, which is lower than claimed by the authors who inappropriately used the effective sampling area in this calculation. For long-term monitoring in waters with fluctuating flow rates it is desirable to measure the DBL thickness for each DGT deployment, but the above example shows that for one of the studied rivers a mean value of the DBL thickness would have sufficed. If DBL is not considered, modest variations in DGT labile concentration must be interpreted with caution [48]. Inaccuracies associated with the DBL are unlikely to be a concern for many applications, for example if the purpose is to assess whether a threshold concentration is exceeded [48].

6.4.4 Diffusion Coefficients

Selection of appropriate diffusion coefficients must be carefully evaluated for each field campaign.

Chapter 3 provides values of diffusion coefficients at 20°C or 25°C. In an information sheet distributed by the supplier DGT Research Ltd, diffusion coefficients for the most commonly analyzed metals as well as phosphate are listed for temperatures ranging between 0–35°C [51]. Good agreements are generally observed when comparing results by different research groups, and disagreements can often be attributed to discrepancies in the experimental procedures. This, in part, is because the diffusion coefficients are not only element specific, but also differ between species of the same element. The relative uncertainty of D determined from diffusion cell measurements is generally quite low, around 2 per cent. For example Scally et al. [17] reported 2.5 per cent and 1.2 per cent for Cu and Ni, respectively, and Kreuzeder et al. [30] reported values between 2.0 and 2.7 per cent for As, Cd, Cu and P. However, these figures are calculated from one single observation and repeated measurements would probably expand these values. From four independent measurements of the Cd diffusion coefficient, 4.0 per cent RSD has been obtained [52] which correspond to 6.4 per cent uncertainty (at 95 per cent confidence level). The impact of the uncertainty from D determination on the overall precision of DGT measurements is therefore not considered to be dominant, but neither can it be neglected [30].

If humic ligands are complexed with the analytes of interest the diffusion coefficient of these complexes will be significantly lower, which was already briefly discussed in Section 6.2.2. Fulvic acids have been shown to have values of D corresponding to around 20 per cent of the value of free metal ions [17]. Simply using the diffusion coefficients of free metal ions for the calculation of DGT concentrations would therefore underestimate concentrations for elements that are commonly complexed with organic ligands. Note, however, that this DGT concentration, which effectively weights the concentration of species according to their diffusion coefficients, is appropriate for comparison with bio-uptake of the element, if this step is transport limited, as treated in detail in Chapter 9.

The DGT-measured concentration can be related to other more conventional measurements using two main approaches. If the total dissolved concentration of a metal has been measured independently, along with other parameters that enable speciation calculations to be performed, the proportion of metal bound to humic substances can be estimated using an equilibrium speciation model [16, 53, 54]. For the WHAM model this is termed the colloidal bound fraction or carry out ultra-filtration of complementing discrete samples [38]. Figure 6.5a plots the dissolved concentration versus the DGT labile concentration of copper from measurements carried out in thirty-eight freshwater streams located in Northern England [16]. For most of the measurements c_{DGT} is underestimated compared to the dissolved copper concentration. In Figure 6.5b the dissolved concentrations have been converted to the maximum dynamic concentrations and a much better fit is then observed. The maximum dynamic concentration is a theoretical value of the DGT labile concentration, which considers the diffusion coefficients of species and assumes they are fully labile (see Chapter 5 for further details). It is calculated by adding the free metal concentration to the concentration of metal complexed by humic substances multiplied by 0.2. The latter value compensates for the lower diffusion coefficient of fulvic acids, assuming that they are the dominant complexing ligand. The concentrations of the species were derived from equilibrium speciation modelling with WHAM [16]. The pattern generally seen when using this approach is that Cd and Zn are not complexed by humic substances, Cu is often

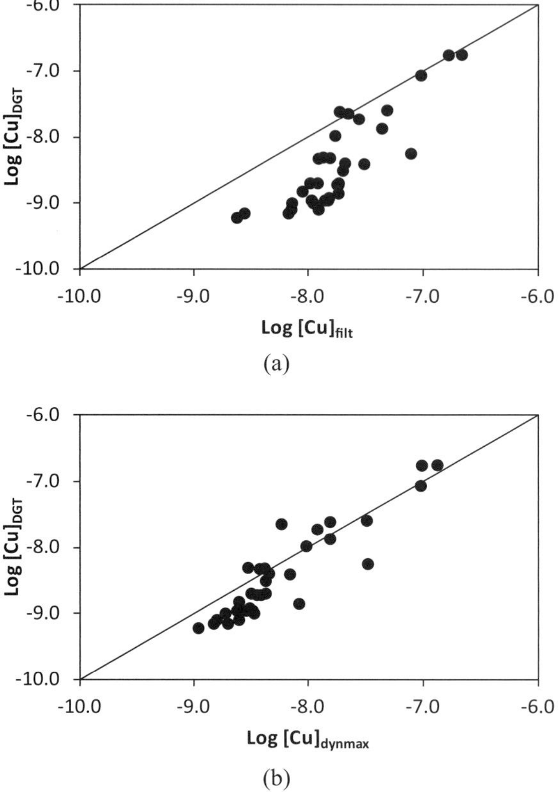

Figure 6.5   Filtered concentrations (mol L^{-1}) of Cu are plotted against DGT labile concentrations from measurements conducted in thirty-eight freshwater streams in Northern England (a). In general, the DGT labile Cu is lower than the corresponding filtered concentration. In (b), the filtered concentrations have been recalculated to represent the maximum dynamic concentrations which are based on complexation of Cu by fulvic acids predicted using the WHAM speciation code and the complexes' lower diffusion coefficient. The closer fit to the 1:1 line implies that Cu-fulvic acid complexes to a large extent are labile [16]. Reprinted with permission from Environmental Science and Technology, 43, K. W. Warnken, A. J. Lawlor, S. Lofts et al., In situ speciation measurements of trace metals in headwater streams, 7230–7236. Copyright (2009) American Chemical Society.

complexed, but still fully labile, as in the example in Figure 6.5, and Al, Mn and Pb are present in colloidal forms [55].

Speciation codes such as WHAM were developed for geochemical modelling of freshwaters. For seawater and brackish water another approach has been used. Forsberg et al. [38] described how the complexed fraction of trace metals can be evaluated using ultra-filtration in combination with DGT measurements. The 1 kDa pore size of the ultra-filter was in the same size range as the normal open pore diffusive gel. Figure 6.6 show the results from sampling in the Landsort Deep, Baltic Sea, at 5 m depth. Concentrations of Mn, Zn and Cd measured by DGT agreed with the concentrations measured in 1 kDa ultrafiltered

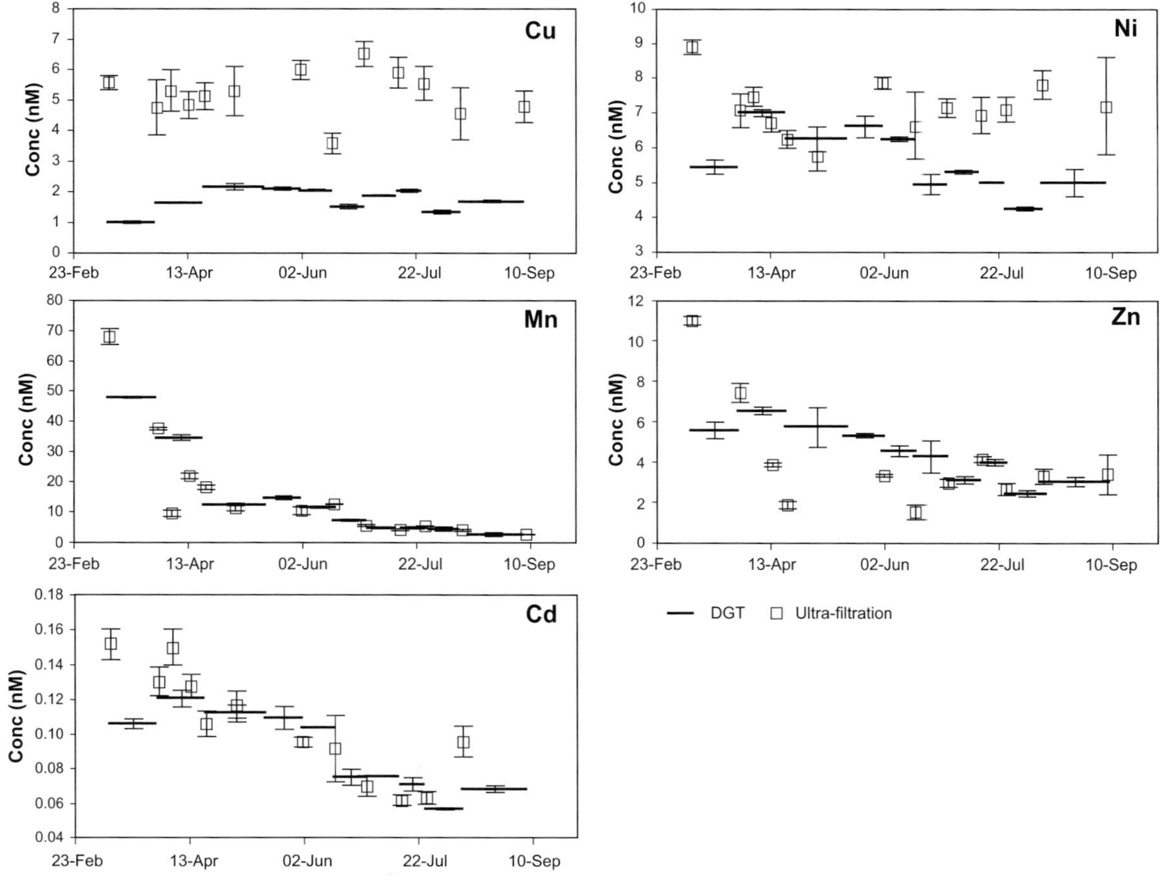

Figure 6.6   Concentrations of Mn, Zn, Cu, Ni and Cd in the Landsort Deep of the Baltic Sea, measured by DGT and ultrafiltration [38]. Reprinted with permission from Environmental Science and Technology, 40, J Forsberg, R. Dahlqvist, J. Gelting-Nyström et al., Trace metal speciation in brackish water using diffusive gradients in thin films and ultrafiltration: comparison of techniques, 3901–3905. Copyright (2009) American Chemical Society.

samples, indicating that no significant concentrations of metal-ligand complexes with these metals were present. However, for Cu and to some extent also Ni, the DGT-labile concentrations were lower compared to their ultrafiltered counterparts. The differences between the concentrations obtained by the two fractionation techniques (ultrafiltered – DGT-labile) were used to recalculate the DGT-labile concentration assuming that organic complexes (smaller than 1 kDa) had lower diffusion coefficients than those of the free metal ions, corresponding to the values estimated by Scally et al. [17]. After recalculation it could be concluded that on average 70 per cent of Ni was present as free metal ions and 30 per cent bound to small organic complexes. For Cu the same calculations indicated 20 per cent free Cu and 80 per cent complexed to organic ligands.

6.5 Concluding Remarks and Future Development

DGT has been thoroughly tested together with other speciation methods [53, 54], including equilibrium speciation modelling Warnken et al. [16], in the field during routine sampling.

Increasingly, DGT is being seen both as a routine tool in monitoring and metal speciation studies, and a new tool that adds new dynamically relevant information not available from traditional techniques. The increasing numbers of DGT publications indicate that DGT, together with ultra-filtration, and other methods, provides new information not obtained by each method alone. When planning environmental projects the possible use of DGT should be considered, even though such measurements are not the prime focus of the work. Unexpected new information will often come to light, extending the scientific value of the project. Two examples are given below.

It has been assumed that aquatic moss and DGT both measure the bioavailable fraction in water [56]. This was tested in a study by Öhlander et al. [57]. DGT, ultra-filtration, membrane filtration and aquatic moss were simultaneously applied in a contaminated freshwater stream. Ultrafiltered concentrations were lower than DGT labile concentrations, which in turn were lower than 0.2 µm filtered concentrations, indicating the presence of labile colloids discriminated by ultra-filtration. The lack of correlation between metal concentrations in aquatic moss and DGT prompted further work, which showed that iron-rich particles contaminated the moss, even though the moss was thoroughly cleaned.

The study of oceanic paleocirculation is to a large extent based upon analysis of Nd isotopes in ferromanganese crusts, and other sedimentary records. Isotopic signatures found in these records are believed to reflect the composition of the ocean water in which they have formed. However, if the isotopic composition in the solute phase is different than that of bulk water it would have an impact on the interpretation of the records. The concentrations and isotopic composition of Nd in fresh, brackish and seawater have been determined with DGT [58]. At each sampled site, the isotopic composition of Nd in the water is the same, within errors, as that obtained from the DGT measurements (Figure 6.7a), supporting published interpretations. The importance of the colloidal fraction is clearly illustrated (Figure 6.7b), underlining that DGT can provide multiple information that would be very difficult to obtain with other methods.

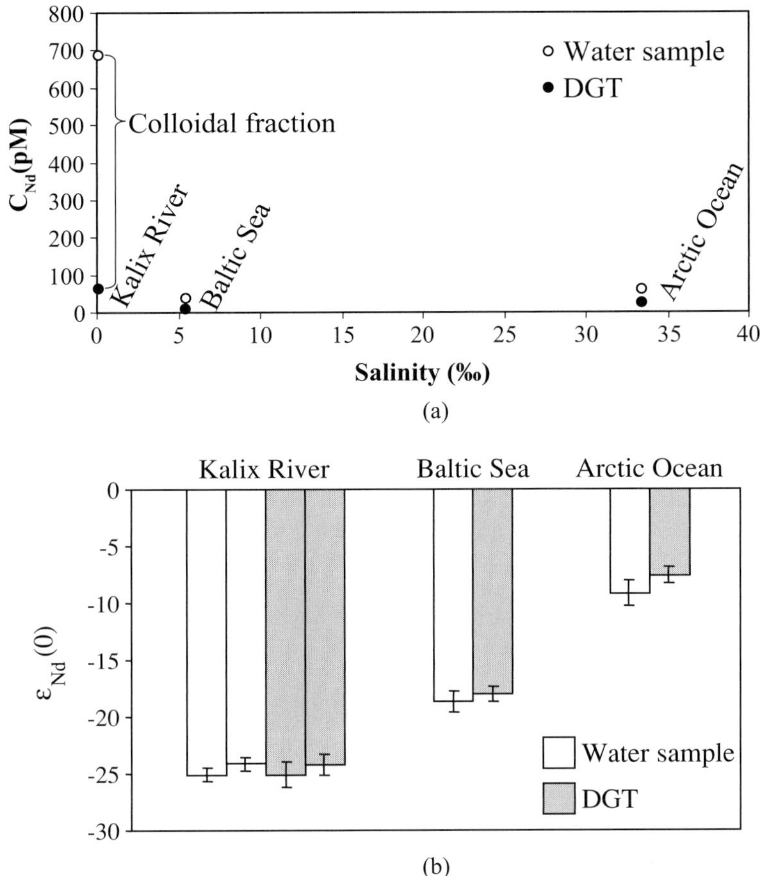

Figure 6.7   The concentration (a) and isotopic composition (b) of Nd in fresh, brackish and seawater determined with DGT [58]. $\varepsilon_{Nd}(0)$ denotes the deviation of ^{143}Nd/^{144}Nd ratio in parts per 10,000 from the bulk-earth value. At each sampled site, the isotopic compositions of Nd in the water show the same values, within errors, as those obtained with DGT. The importance of the colloidal fraction is clearly illustrated, especially for the freshwater sample, underlining that DGT can provide additional information that would be difficult to obtain with other methods. Reprinted from Earth and Planetary Science Letters, 133(1–2), R. Dahlqvist, P. S. Andersson and J. Ingri, The concentration and isotopic composition of diffusible Nd in fresh and marine waters, 9–16, (2005), with permission from Elsevier.

Future applications of DGT are likely to include studies of the isotopic composition of dissolved species of elements, as several studies have indicated the potential for DGT in this area. Diffusion alone can cause detectable changes in Fe and Zn isotopes [59], but Malinovsky et al. [60] showed there was no fractionation of Zn isotopes due to the diffusion process in the DGT sampler, within the reported precision of MC–ICP–MS measurements, provided quantitative elution was obtained. Similarly, Desaulty et al. [61] concluded that DGT can be used for Zn and Pb isotopes, opening up the way to the use of DGT in studies of the aquatic geochemistry of these elements. The DGT technique has also been used, in combination with laser ablation multicollector ICP–MS, to study two-dimensional sulphur isotope variations of dissolved porewater sulphide in freshwater and marine sediments [62].

References

1. H. Zhang and W. Davison, Performance characteristics of diffusion gradients in thin films for the in situ measurement of trace metals in aqueous solutions, *Anal. Chem.* 67 (1995), 3391–3400.
2. Ø. A. Garmo, O. Røyset, E. Stiennes and T. P. Flaten, Performance study of diffusive gradients in thin films for 55 elements, *Anal. Chem.* 75 (2003), 3573–3580.
3. Ø. A. Garmo, N. J. Lehto, H. Zhang et al., Dynamic aspects of DGT as demonstrated by experiments with lanthanide complexes of a multidentate ligand, *Environ. Sci. Technol.* 40 (2006), 4754–4760.
4. H. Zhang, W. Davison., R. Gadi and T. Kobayashi, In situ measurement of dissolved phosphorus in natural waters using DGT, *Anal. Chim. Acta.* 370 (1998), 29–38.
5. J. Luo, H. Zhang, J. Santner and W. Davison, Performance characteristics of diffusive gradients in thin films equipped with a binding gel layer containing precipitated ferrihydrite for measuring arsenic(V), selenium(VI), vanadium(V), and antimony(V), *Anal. Chem.* 82 (2010), 8903–8909.
6. H. Österlund, S, Chlot, M. Faarinen et al., Simultaneous measurements of As, Mo, Sb, V and W, *Anal. Chim. Acta* 682 (2010), 59–65.
7. J. G. Panther, R. R. Stewart, P. R. Teasdale et al., Titanium dioxide-based DGT for measuring dissolved As(V), V(V), Sb(V), Mo(VI) and W(VI) in water, *Talanta* 105 (2013), 80–86.
8. H. L. Price, P. R. Teasdale and D. F. Jolley, An evaluation of ferrihydrite- and Metsorb™-DGT techniques for measuring oxyanion species (As, Se, V, P): Effective capacity, competition and diffusion coefficients, *Anal. Chim. Acta* 803 (2013), 56–65.
9. G. S. C. Turner, G. A. Mills, M. J. Bowes et al., Evaluation of DGT as a long-term water quality monitoring tool in natural waters, Uranium as a case study, *Environ. Sci.: Processes Impacts* 16 (2014), 393–403.
10. Q. Su, J. Chen, H. Zhang et al., Improved diffusive gradients in thin films (DGT) measurement of total dissolved inorganic arsenic in waters and soils using a hydrous zirconium oxide binding layer, *Anal. Chem.* 86 (2014), 3060–3067.
11. H. Dočekalová and P. Diviš, Application of diffusive gradient in thin films technique (DGT) to measurement of mercury in aquatic systems, *Talanta* 65 (2005), 1174–1178.
12. O. Clarisse and H. Hintelmann, Measurements of dissolved methylmercury in natural waters using diffusive gradients in thin film (DGT), *J. Environ. Monit.* 8 (2006), 1242–1247.
13. M. A. French, H. Zhang, J. M. Pates, S. E. Bryan and R. C. Wilson, Development and performance of the diffusive gradients in thin-films technique for the measurement of technetium-99 in seawater, *Anal. Chem.* 77 (2005), 135–139.
14. C. Murdock, M. Kelly, L.-Y. Chang, W. Davison and H. Zhang, DGT as an in situ tool for measuring radiocesium in natural waters, *Environ. Sci. Technol.* 35 (2001), 4530–4535.
15. H. Zhang and W. Davison, Progress in understanding the use of diffusive gradients in thin films (DGT) back to basics, *Environ. Chem.* 9 (2012), 1–13.

16. K. W. Warnken, A. J. Lawlor, S. Lofts et al., In situ speciation measurements of trace metals in headwater streams, *Environ. Sci. Technol.* 43 (2009), 7230–7236.
17. S. Scally, W. Davison and H. Zhang, Diffusion coefficients of metals and metal complexes in hydrogels used in diffusive gradients in thin films, *Anal. Chim. Acta* 558 (2006), 222–229.
18. H. Zhang and W. Davison, Direct in situ measurements of labile inorganic and organically bound metal species in synthetic solutions and natural waters using diffusive gradients in thin films, *Anal. Chem.* 72 (2000), 4447–4457.
19. K. W. Warnken, W. Davison and H. Zhang, Interpretation of in situ speciation measurements of inorganic and organically complexed trace metals in freshwater by DGT, *Environ. Sci. Technol.* 42 (2008), 6903–6909.
20. H. Österlund, J. Gelting, F. Nordblad, D. C. Baxter and J. Ingri, Copper and nickel in ultrafiltered brackish water: Labile or non-labile?, *Mar. Chem.* 132–133 (2012), 34–43.
21. P. L. R. van der Veeken, J. P. Pinheiro and H. P. van Leeuwen, Metal speciation by DGT/DET in colloidal complex systems, *Environ. Sci. Technol.* 42 (2008), 8835–8840.
22. W. Baeyens, A. R. Bowie, K. Buesseler et al. Size-fractionated labile trace elements in the Northwest Pacific and Southern Oceans, *Mar. Chem.* 126 (2011), 108–113.
23. P. Pelcova, H. Docekalova and A. Kleckerova, Development of the diffusive gradient in thin films technique for the measurement of labile mercury species in waters, *Anal. Chim. Acta* 819 (2014), 42–48.
24. J.-L. Zheng, D.-X. Guan, J. Luo et al., Activated charcoal based diffusive gradients in thin films for in situ monitoring of bisphenols in waters, *Anal. Chem.* 87 (2015), 801–807.
25. O. A. Garmo, W. Davison and H. Zhang, Effects of binding of metals to the hydrogel and filter membrane on the accuracy of the diffusive gradients in thin films technique, *Anal. Chem.* 80 (2008), 9220–9225.
26. K. W. Warnken, H. Zhang and W. Davison, Accuracy of the diffusive gradients in thin-films technique: Diffusive boundary layer and effective sampling area considerations. *Anal. Chem.* 78 (2006), 3780–3787.
27. K. W. Warnken, W. Davison, H. Zhang, J. Galceran and J. Puy, In situ measurements of metal complex exchange kinetics in freshwater, *Environ. Sci. Technol.* 41 (2007), 3179–3185.
28. M. R. S. Arvajeh, N. Lehto, O. A. Garmo and H. Zhang, Kinetic studies of Ni organic complexes using diffusive gradients in thin films (DGT) with double binding layers and a dynamic numerical model, *Environ. Sci. Technol.* 47 (2013), 463–470.
29. J. Gimpel, H. Zhang, W. Hutchinson and W. Davison, Effect of solution composition, flow and deployment time on the measurement of trace metals by the diffusive gradient in thin films technique, *Anal. Chim. Acta.* 448 (2001), 93–103.
30. A. Kreuzeder, J. Santner, H. Zhang, T. Prohaska and W. W. Wenzel, Uncertainty evaluation of the diffusive gradients in thin films technique, *Environ. Sci. Technol.* 49 (2015), 1594–1602.
31. Bio-Rad Laboratories, Chelex 100 and Chelex 20 chelating ion exchange resin instruction manual (2005).
32. S. Tankéré-Muller, W. Davison and H. Zhang, Effect of competitive cation binding on the measurement of Mn in marine waters and sediments by diffusive gradients in thin films, *Anal. Chim. Acta* 716 (2012), 138–144.
33. R. Dahlqvist, H. Zhang, J. Ingri and W. Davison. Performance of the diffusive gradients in thin films technique for measuring Ca and Mg in freshwater, *Anal. Chim. Acta* 460 (2002), 247–256.

34. J. L. Levy, H. Zhang, W. Davison, J. Puy and J. Galceran, Assessment of trace metal binding kinetics in the resin phase of diffusive gradients in thin films, *Anal. Chim. Acta* 717 (2012), 143–150.

35. E. Uher, H. Zhang, S. Santos, M.-H. Tusseau-Vuillemin and C. Gourlay-Francé, Impact of biofouling on diffusive gradient in thin film measurements in water, *Anal. Chem.* 84 (2012), 3111–3118.

36. C. Pitchette, H. Zhang, W. Davison and S. Sauvé, Preventing biofilm development on DGT devices using metals and antibiotics, *Talanta* 72 (2007), 716–722.

37. S. Chlot, A. Widerlund and B. Öhlander, Interaction between nitrogen and phosphorus cycles in mining-affected aquatic systems – Experiences from field and laboratory measurements, *Environ. Sci. Pollut. Res.* 20 (2013), 5722–5736.

38. J. Forsberg, R. Dahlqvist, J. Gelting-Nyström and J. Ingri, Trace metal speciation in brackish water using diffusive gradients in thin films and ultrafiltration: Comparison of techniques, *Environ. Sci. Technol.* 40 (2006), 3901–3905.

39. M. Wallner-Kersanach, C. F. F de Andrade, H. Zhang, M. R. Milani and L. F. H. Niencheski, In situ measurement of trace metals in estuarine waters of Patos Lagoon using diffusive gradients in thin films (DGT), *J. Braz. Chem. Soc.* 20 (2009), 333–340.

40. H. Zhang and W. Davison, Diffusional characteristics of hydrogels used in DGT and DET techniques, *Anal. Chim. Acta.* 398 (1999), 329–340.

41. M. R. Sangi, M. J. Halstead and K. A. Hunter, Use of the diffusion gradient thin film method to measure trace metals in fresh waters at low ionic strength, *Anal. Chim. Acta.* 456 (2002), 241–251.

42. M. C. A.-D. L. Torre, P.-Y. Beaulieu and A. Tessier, In situ measurement of trace metals in lakewater using the dialysis and DGT techniques, *Anal. Chim. Acta* 418 (2000), 53–68.

43. A. J. Peters, H. Zhang and W. Davison, Performance of the diffusive gradients in thin films technique for measurement of trace metals in low ionic strength freshwaters, *Anal. Chim. Acta* 478 (2003), 237–244.

44. K. W. Warnken, H. Zhang and W. Davison, Trace metal measurements in low ionic strength synthetic solutions by diffusive gradients in thin films, *Anal. Chem.* 77 (2005), 5440–5446.

45. J. Puy, J. Galceran, S. Cruz-Gonzalez et al., Metal accumulation in DGT: Impact of ionic strength and kinetics of dissociation of complexes in the resin domain, *Anal. Chem.* 86 (2014), 7740–7748.

46. J. Søndergaard, In situ measurements of labile Al and Mn in acid mine drainage using diffusive gradients in thin films, *Anal. Chem.* 79 (2007), 6419–6423.

47. S. Mason, R. Hamon, A. Nolan, H. Zhang and W. Davison, Performance of a mixed binding layer for measuring anions and cations in a single assay using the diffusive gradient in thin films technique. *Anal. Chem.* 77 (2005), 6339–6346.

48. R. Buzier, A. Charriau, D. Corona et al., DGT-labile As, Cd, Cu and Ni monitoring in freshwater: Toward a framework for interpretation of in situ deployment, *Environ. Poll.* 192 (2014), 52–58.

49. L. Sigg, F. Black, J. Buffle et al., Comparison of analytical techniques for dynamic trace metal speciation in natural freshwaters, *Environ. Sci. Technol.* 40 (2006), 1934–1941.

50. P. Minkkinen, Monitoring precision of routine analyses by using duplicate determinations, *Anal. Chim. Acta* 191 (1986), 369–376.

51. DGT Research Ltd, DGT – for measurements in waters, soils and sediments, (n.d.) Available at www.dgtresearch.com.

52. H. Österlund, M. Faarinen, J. Ingri and D. C. Baxter, Contribution of organic arsenic species to total arsenic measurements using ferrihydrite-backed diffusive gradients in thin films (DGT), *Environ. Chem.* 9 (2012), 55–62.

53. J. Gimpel, H. Zhang, W. Davison and A. Edwards, In situ trace metal speciation in lake surface waters using DGT, *Environ. Sci. Technol.* 37 (2003), 138–146.

54. E. R. Unsworth, K. W. Warnken, H. Zhang et al., Model predictions of metal speciation in freshwaters compared to measurements by in situ techniques, *Environ. Sci. Technol.* 40 (2006), 1942–1949.

55. H. Zhang and W. Davison, Use of diffusive gradients in thin-films for studies of chemical speciation and bioavailability, *Environ. Chem.* 12 (2015), 85–101.

56. P. Diviš, J. Machát, R. Szkandera and H. Dočekalová, In situ measurement of bioavailable metal concentrations at the downstream on the Morava River using transplanted aquatic mosses and DGT technique, *Int. J. Environ. Res.* 6 (2012), 87–94.

57. B. Öhlander, J. Forsberg, H. Österlund et al., Fractionation of trace metals in a contaminated freshwater stream using membrane filtration, ultrafiltration, DGT and transplanted aquatic moss, *Geochem.: Exploration, Environ., Anal.* 12 (2012), 303–312.

58. R. Dahlqvist, P. S. Andersson and J. Ingri, The concentration and isotopic composition of diffusible Nd in fresh and marine waters, *Earth Planet. Sci. Lett.* 233 (2005), 9–16.

59. I. Rodushkin, A. Stenberg, H. Andrén, D. Malinovsky and D. C. Baxter, Isotopic fractionation during diffusion of transition metal ions in solution, *Anal. Chem.* 76 (2004), 2148–2151.

60. D. Malinovsky, R. Dahlqvist, D. C. Baxter, J. Ingri and I. Rodushkin, Performance of diffusive gradients in thin films for measurement of the isotopic composition of soluble Zn, *Anal. Chim. Acta* 537 (2005), 401–405.

61. A.-M. Desaulty, C. Bodard, T. Laurioux et al., Using DGT passive samplers and MC-ICP-MS to determine Pb and Zn isotopic signature of natural water, *Proced.: Earth Planet. Sci.* 13 (2015), 76–79.

62. A. Widerlund, G. M. Nowell, W. Davison and D. G. Pearson, High-resolution measurements of sulphur isotope variations in sediment pore-waters by laser ablation multicollector inductively coupled plasma mass spectrometry, *Chem. Geol.* 291 (2012), 278–285.

7

Principles and Application in Soils and Sediments

NIKLAS J. LEHTO

7.1 Introduction

Soon after the first work reporting on the use of diffusive gradients in thin-films (DGT) to measure trace metal ion concentrations in seawater [1], a gel holding device was adapted from its slightly older diffusive equilibria in thin-films (DET) counterpart [2] and used to accommodate a planar DGT probe. The adapted probe was used to investigate solute distributions around the sediment-water interface of a freshwater sediment at high-resolution [3]. The use of DET and DGT for high resolution measurements of solute concentrations and fluxes in sediments and soils is described in the appropriate context in Chapter 8, however the work by Zhang et al. [3] remains of great relevance to this chapter. It was the first to recognize the differences between DGT measurements in purely aqueous media and those carried out in a porous environment, such as a soil or sediment. Subsequent work on soils and sediments has helped to develop and then validate the theories that underpin the use of DGT in these dynamic porous media and thus allow insight into processes that govern the cycling of trace elements and nutrient in soils and sediments and their availability to terrestrial and aquatic biota.

7.2 Principles of DGT Use in Soils and Sediments

The principles that underpin DGT measurements in solution are assumed to be understood as they are considered in Chapters 2–6. The requirements that apply to interpreting DGT measurements described therein, such as ionic strength, gel pore size, solute gel interactions and competition effects between different solute species (including H^+ ions) should also be given due consideration when carrying out DGT measurements in soils and sediments. The primary focus of this chapter will be to consider how the presence of solid phases adjacent to the DGT device influences the interpretation of the results. While soils and sediments have important morphological and biogeochemical differences, the principles discussed here apply to both, unless stated otherwise.

When deployed in a soil or sediment, solutes in the hydrogel layer in a DET probe rapidly equilibrate with the adjacent solution or porewater and the concentration of solute in the hydrogel is measured at the end of the deployment [2]. A DGT probe deployed in a soil or sediment introduces a sink for free ions in the porewater, which creates a diffusive flux of those ions from the porewater into the probe. This flux is regulated by a thin film of porous hydrogel and a membrane filter, known as the 'diffusive layer.' As the free ions are progressively removed, their concentration in the adjacent porewater decreases and brings about a state of disequilibrium between the free ions in the porewater, their complexes and ions sorbed onto soil solid phases. The dissociation of labile complexes within the diffusion and resin layers, and in the porewater, in response to the removal of the free ions results in a reduction in their respective concentrations and, consequently, a diffusive flux of the complexes toward the resin layer. Furthermore, ions and complexes bound to the soil solid phase binding sites can desorb to approach (or buffer) the equilibrium between the free ion and solid phase-bound species. It follows that the total diffusive flux of solute through the diffusive layer is determined by the diffusion coefficients of the free ion, and its various complexes that are labile within the timeframe of that particular deployment, and the diffusive gradients of the free ion and its complexes (i.e., the difference between the concentration at the interface and at the sink, divided by the thickness of the diffusion layer). Therefore, the total amount of a given ion that is bound by the resin layer during a DGT deployment is determined by (1) the diffusive gradients in the diffusion layer, (2) the diffusive fluxes of labile solute from the porewater (including the free ion) and (3) the ability of the solid phase to resupply solute if the concentration in the porewater becomes depleted.

A full understanding of the various components that contribute to the total amount of solute bound by the DGT resin in any environmental system requires considerable information about the species present and their dynamic behavior during the deployment. Often, quite a lot of this type of information is unavailable, due to limitations in the types and numbers of measurements that can be carried out in any experimental campaign. To overcome these limitations it is necessary to make a few broad assumptions. In general, a majority of dissolved inorganic complexes of trace elements can be considered to be fully labile within the duration of most DGT deployments and have similar diffusion coefficients to the hydrated cation (Hg being a notable exception). On the other hand, complexes with large dissolved organic ligands often have smaller diffusion coefficients and can be less labile. Where organic complexes are likely to make a significant contribution to the amount of solute bound by the resin, this should be duly acknowledged in the interpretation of the results. If slow diffusing, partially labile organic complexes are unlikely to be important, it is reasonable to assume that the solute sampled by the DGT resin is a combination of the diffusive fluxes of the free ion and its inorganic complexes. Ultimately, the solute sampled by a DGT device is representative of the fraction that is 'DGT-labile' within the particular experimental conditions and should always be acknowledged as such.

By analyzing the total amount of an ion bound by the sink (usually called the 'binding layer'), it is possible to estimate an operationally defined time-averaged concentration of the DGT-labile fraction at the probe–soil interface, c_{DGT}, using equation 7.1 derived from Fick's first law of diffusion:

$$c_{DGT} = \frac{M \delta^{mdl}}{D A_p t} \tag{7.1}$$

where M is the total amount of the ion bound by the binding layer during a certain deployment time (t). D is the diffusion coefficient of the species of interest in the material diffusion layer, whose thickness is δ^{mdl} (where organic complexes are thought to be insignificant, the diffusion coefficient of the hydrated ion is often used) the defined sampling area is A_p. Equation 7.1 represents the simplest derivation applicable to DGT measurements, versions of this equation that are more specialized can be found in Chapter 2. For deployments where the deployment media has sufficient capacity to supply solute to the DGT device interface with a flux that is the same or greater than the diffusive flux into the DGT device, and in the absence of partially labile or inert complexes or slowly diffusing labile complexes, c_{DGT} can be used to approximate the total concentration of a given ion and its complexes in solution.

In soils and sediments, the interpretation of c_{DGT} is complicated by the presence of solid particles. The rest of this chapter will look at how the understanding of DGT deployments in these media has evolved and informed modeling approaches to interpret these measurements. Subsequent references to c_{DGT} are made as referred to in the cited literature, unless stated otherwise; however, the complexities associated with interpreting it, as described above, should be duly recognized. The term 'solute' is used acknowledging that the total uptake of an ion by DGT is likely to be affected by a number of its DGT-labile species in a given environmental system.

Zhang et al. [3] used a numerical model to simulate diffusion of a single ion into a DGT device in one dimension. They showed how the concentration of the dissolved ion at the device interface might change during a deployment in both a well-mixed case, such as a well-stirred solution, and in a quiescent system, where the only transport mechanism from the environment normal to the device interface would be diffusion. In the latter case, the dissolved ion concentration would be expected to be depleted at and away from the device interface due to the constant demand of solute by the DGT device, resulting in decreasing fluxes into the device as the deployment progressed. Under these circumstances, c_{DGT} would not provide an accurate estimate of bulk solution concentration of that ion.

The findings of this modeling and complementary DET data were used to interpret in-situ measurements of trace metal fluxes to a DGT device in freshwater sediment. This analysis revealed three types of response by the sediment to the DGT. For Zn and Cd, fluxes were sustained by the sediment to such an extent that c_{DGT} was similar to the DET-measured bulk porewater concentrations. c_{DGT} for Cu, Ni and Fe were somewhat lower, while c_{DGT} for Mn was considerably lower than the DET-determined bulk water concentration. These results suggested that the combined diffusive and desorptive fluxes of Zn and Cd from the sediment in response to their uptake could match the initial flux into the DGT, thus allowing their solution concentrations at the device interface to be sustained at a level similar to those in the bulk solution, as conceptualized by (a) in Figure 7.1. Conversely for Mn, they suggested that the flux from the solid phase was so low that diffusion from the sediment porewater was the only process supplying the metal to the DGT (Figure 7.1c), while the fluxes of Cu, Ni and Fe from the sediment porewater were partially sustained by a combination of desorption and diffusion (Figure 7.1b).

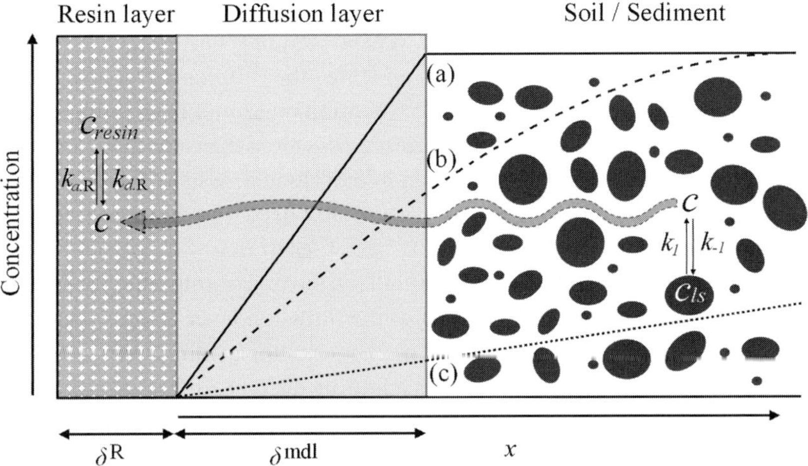

Figure 7.1 Concentrations of an ionic species in a DGT device and adjacent porewater during deployment for (a) the sustained case, (b) the intermediate case and (c) the diffusive only case. The wavy grey line illustrates the diffusive pathway of the mobile species. Adapted with permission from Zhang et al. [4]. Copyright (1998) American Chemical Society.

It is useful to consider the ratio, R_c, of c_{DGT} to the bulk porewater solute concentration (c^{soln}) (equation 7.2).

$$R_c = c_{DGT}/c^{soln} \tag{7.2}$$

For deployments of a set duration, using probes with a common diffusive layer thickness, R_c is indicative of a sediment's capability to supply solute to the DGT. Over the length of a deployment in a soil or a sediment, R_c will rapidly reach its maximum value as solute diffuses into the diffusive layer and a linear gradient between the binding gel and the device interface is established. As the deployment progresses, R_c will progressively decrease. The extent of this decrease over a given deployment time is determined by the sediment's ability to resupply solute to the device interface. For a 24-h deployment of DGT devices, with a commonly used diffusion layer thickness (0.093 cm), the three modes of solute supply to the DGT identified by Zhang et al. [3] in sediments, and subsequently in soils [4], can be represented by R_c and described as follows:

1. *Sustained case*: continuous supply from the solid- and solution phases at a rate almost equal to the flux into the DGT (accounting for measurement errors, $R_c > 0.8$).
2. *Partially sustained case*: some supply from the solid phase, but insufficient to sustain R_c ($0.2 \leq R_c \leq 0.8$) (also known as the '*intermediate case*')
3. *Diffusive case*: diffusion is the major process supplying solute to the DGT ($R_c < 0.2$).[1]

---

[1] The *diffusive case* is often cited when $R_c < 0.1$ [4, 5]; however, when lateral supply in a three-dimensional environment is corrected for [6], R_c as high as 0.2 can be observed without supply from the solid phase.

Theoretical one-dimensional solute profiles through the DGT probe and the adjacent soil or sediment for each of these three cases are represented in Figure 7.1.

When considering R_c, it is important to recognize the differences in how the two constituent values that define it are obtained. c^{soln} is often measured from a filtered porewater sample that has been acidified and will therefore provide a measure that closely approximates the total dissolved solute concentration in that sample. Where dissolved species that are partially labile or inert to DGT during a particular deployment are present in solution in significant amounts, their contribution to c^{soln} can result in low R_c values.

As indicated by Fick's first law (viz. equation 7.1), the flux of solute along a concentration gradient is inversely related to the thickness of the diffusion layer. Zhang et al. [3] showed that it is possible to test a soil's or sediment's ability to resupply solute to the device interface by doing a series of deployments with probes featuring different diffusion layer thicknesses. They demonstrated that the flux of Zn from a freshwater sediment to the binding gel had a linear relationship with the diffusion layer thickness, while the flux of Ni had a linear relationship only between diffusion layer thicknesses of 0.21 and 0.13 cm, but deviated once it was decreased further. They argued that, within this linear range, the supply from the sediment-sustained Zn and Ni concentrations at the DGT probe interface and that the calculated c_{DGT} could be used to provide estimates of porewater concentrations for these metals. The same approach was used in the first reported application of the DGT technique to investigate solute supply in soils [4], where it was shown that Cu, Ni, Zn and Cd fluxes from sewage sludge amended soils were sustained over a 16-h deployment when the diffusion layer thickness was >0.13 cm, thus allowing the porewater concentrations to be estimated.

In the absence of a representative measurement of c^{soln} (such as may be the case for a highly stratified sediment, or other highly dynamic and/or heterogeneous environments), or other supporting measurements (as above), it is inappropriate to assume that the solute flux to a DGT probe (or subsection of a probe) in a given deployment is *sustained*, *partially sustained* or *diffusive*, as discussed above. In these cases results are often reported either simply as the average flux of solute to a given area of binding gel (A) over the length of the deployment ($M/(t \times A)$), or as the time-averaged solute concentration at the device interface, c_{DGT} (equation 7.1). Both are specific to the duration of the deployment.

7.3 DGT Deployments in Soils and Sediments

The natural spatial (vertical and horizontal) and temporal heterogeneity in soils and sediments presents numerous challenges for a representative characterization of these environments. These challenges apply equally to DGT users. The procedures for sample preparation and subsequent deployment of the various types of DGT probes available for characterizing these environments, under laboratory conditions and in situ, are described in detail in Chapter 10. Subsequent sections in this chapter will focus on how these measurements can be interpreted.

Detailed interpretation of DGT measurements of solute resupply dynamics in a soil or sediment often requires the use of numerical models, along with complementary techniques, including: DET (or other means of determining solute concentrations in porewater at an appropriate resolution to DGT), DGT probes with different diffusion layer thicknesses or sequential extractions used to determine labile solid phase concentrations. Section 7.4 will describe the numerical modeling used to interpret DGT deployments in soils and sediments, and illustrate how it can be applied to gain information about solute dynamics in a porous reactive environment. The application of DGT measurements and associated numerical modeling in soils is discussed in Section 7.5; corresponding work in sediments is discussed in Section 7.6.

7.4 Modeling Diffusion of Solute to DGT in Soils and Sediments

Equation 7.1 describes the total flux into a DGT binding layer over the deployment as a hypothetical constant porewater solute concentration at the device interface (c_{DGT}). It is essentially the same equation as used for deployments in well-stirred, simple solutions, where there is a steady-state flux, such that c_{DGT} is independent of deployment time. However, for soils and sediments, such steady-state conditions are rarely reached because there is no convective supply of solutes and there are interactions between solutes and solids. The temporal and spatial variations in solute concentration within the perturbed soil or sediment prevent the use of an analytical expression that represents the flux into the DGT device accurately. Instead, a time-dependent model that accounts for the supply of solute by diffusion and release from a finite solid phase reservoir is needed to describe the flux of solute to a DGT deployed in a soil.

7.4.1 DGT-induced Fluxes in Soils and Sediments (DIFS)

Harper et al. [5] developed a numerical model using the Mathworks' package MATLAB, which was soon followed by a publicly available software tool for simulating DGT devices in soils and sediments [7]. Known as DGT-induced fluxes in sediments (DIFS), it simulates the diffusion of a dissolved ion from a soil or a sediment porewater into a DGT probe over the course of its deployment, while accounting for the equilibrium between dissolved and solid phase-bound ions through a pair of forward and reverse rate expressions. By considering these reaction and transport processes, the model simulates the temporal and spatial variation in ion concentrations in the solution and solid phases and then uses this to emulate the DGT-measured flux. Although the DIFS model was originally designed for use in sediments, most of its use has been to simulate DGT deployments in soils because of the availability of data on homogenized soil samples. The remainder of this section considers DIFS mainly as a model for soils, however the same principles apply to sediments.

The model consists of three compartments: the binding gel, the dissolved phase and the sorbed phase, and two different subdomains: the diffusive layer and the soil. The diffusion medium in the soil, the soil porewater, is defined in terms of the soil porosity (φ). The model assumes that all the soil pores are filled with water. The porosity of the soil can be determined separately, or calculated from the soil bulk density (ρ^{b}) using the density of solid phase material (ρ^{s}) (equation 7.3).

$$\varphi = 1 - \frac{\rho^{b}}{\rho^{s}} \tag{7.3}$$

Bielders et al. [8] considered that 2.65 g cm^{-3} was a representative value of ρ^s for most soil solid phases. Although this value has been used predominantly, it should be noted that ρ^s can vary between different soil texture classes [9].

Dissolved ion transport in the model occurs by molecular diffusion alone and is described using Fick's Law of Diffusion, where the diffusion coefficient of the ion (D) (cm^2 s^{-1}) is specific to the experimental conditions (temperature and viscosity of the diffusion medium) and x represents distance perpendicular to the DGT surface (equation 7.4)

$$\frac{\partial c^{soln}}{\partial t} = D\frac{\partial^2 c^{soln}}{\partial x^2} \tag{7.4}$$

The model assumes single uniform porosities within the material diffusive layer (usually 0.95) and within the soil (specific to the soil). Diffusion coefficients of numerous trace metals in different types of hydrogels have been measured (see Chapter 3 and Scally et al. [10]). For the commonly used polyacrylamide gels cross-linked with an agarose derivative, the rates of diffusion are slightly less than those in water.

The rate of diffusion of solute in the soil porewater is determined by the ion and the soil being considered. The soil limits the diffusion rate through the effectively increased length, or tortuosity (θ_T), of the diffusive pathway caused by the presence of solid particles. In the DIFS model, this modified diffusion coefficient is expressed as an effective diffusion coefficient, D^S, and is specified separately. Boudreau [11] reports a number of proposed relationships for sediments that relate tortuosity to the more easily determined porosity (φ), which can be used to calculate D^S from the ion diffusion coefficient in water (D) for a given temperature. The most commonly used relationship is shown in equation 7.5:

$$D^S = \frac{D}{\theta_T} = \frac{D}{1 - \ln{(\varphi)^2}} \tag{7.5}$$

In the DIFS model, the solute concentration at the binding gel-diffusive gel interface is defined by a constant zero-concentration (Dirichlet) boundary condition, with the opposite (far end of the soil subdomain) boundary represented by an insulating (Neumann) boundary condition. The model considers the diffusion of an ion from the soil porewater, through the diffusive layer to the resin layer–diffusive layer interface where it is instantly removed. The diffusive flux to this boundary is then integrated over the length of the deployment to provide the mass of the ion bound by the binding gel. Under simulated soil conditions where depletion extends to the far end of the soil subdomain, the model automatically increases the size of the subdomain and runs the simulation again. This iteration continues until the soil subdomain is large enough to contain within it the entire depletion zone.

The diffusion of ions in the soil (equations 7.4 and 7.5) is coupled with the interaction between the mobile ions in the solution phase and ions bound to a single immobile solid phase in the soil. This interaction is simulated by a pair of first-order reactions that are regulated by rate constants, k_1 and k_{-1} (both have units of s^{-1}). Herein these reactions are described as sorption and desorption, but in practice may involve a range of processes acting together to effect the partitioning of solute between the two phases. The model assumes that there is no interaction between the ion and the diffusive layer and does not consider mobile binding phases, such as dissolved ligands. That is, for metals it applies to simple inorganic species that effectively instantaneously equilibrate and have the same diffusion coefficient. The rates of change of dissolved and sorbed concentrations in the soil are described by equations 7.6a and 7.6b:

$$\frac{\partial c^{\text{soln}}}{\partial t} = -k_1 c^{\text{soln}} + k_{-1} P_{\text{c}} c^{\text{ls}} \tag{7.6a}$$

$$\frac{\partial c^{\text{ls}}}{\partial t} = \frac{k_1}{P_{\text{c}}} c^{\text{soln}} - k_{-1} c^{\text{ls}} \tag{7.6b}$$

where c^{ls} (mol g^{-1}) is the concentration of sorbed ion that can desorb within the considered simulation time and is thus considered labile. Particle concentration (P_{c}) is used to relate the sorbed concentration to a porewater concentration and is expressed as mass of particles (M_{p}) in a given volume of soil (V_{s}) (g cm^{-3}). It can be calculated from the soil porosity (equation 7.7):

$$P_{\text{c}} = \frac{M_{\text{p}}}{V_{\text{s}}} = \frac{\rho^{\text{s}}(1 - \varphi)}{\varphi} \tag{7.7}$$

The model assumes a linear adsorption isotherm where K_{dl} (cm^3 g^{-1}) is the partition coefficient between ions in the dissolved phase (mol cm^{-3}) and labile ions sorbed to the solid phase. When the ions in the dissolved phase and the solid phase are at equilibrium, there is no interaction between the two. In this case, the ratio of k_1 to k_{-1} is expressed by equation 7.8.

$$K_{\text{dl}} = \frac{c^{\text{ls}}}{c^{\text{soln}}} = \frac{1}{P_{\text{c}}} \frac{k_1}{k_{-1}} \tag{7.8}$$

The model also uses an index, t_{c}, to describe the response time (s) of labile sorbed ions to a disruption in equilibria, which can be expressed in terms of k_1 and k_{-1} (equation 7.9).

$$t_{\text{c}} = \frac{1}{k_1 + k_{-1}} = \frac{1}{k_{-1}(1 + K_{\text{dl}} P_{\text{c}})} \approx \frac{1}{k_{-1} K_{\text{dl}} P_{\text{c}}} \tag{7.9}$$

t_{c} can be considered as the time taken by the components of K_{dl} to reach 63 percent of their equilibrium values after the solution concentration has been depleted to zero [12]. It is important to note that, while the rate constants k_1 and k_{-1} define the rates of a simple sorption/desorption reaction between dissolved ions and ions sorbed onto solid phase binding sites in the DIFS model, in practice they represent an aggregation of processes that can affect the amount of DGT-labile solute available to be bound by the DGT resin.

7.4.2 DIFS Modeling Software

Harper et al. reconfigured the DIFS model from Mathworks' MATLAB into a model that works in a single dimension (1D) normal to the DGT probe interface [7]. This was compiled within C++ into a software package (forthwith "1D-DIFS") that could be easily used on an average PC. As well as modeling diffusive flux in soils, the model featured a subroutine to calculate one of R_c, K_{dl} or t_c if the other two parameters are known. Another important feature of the model is its ability to account for very large depletion distances in the soil (defined here as the perpendicular distance from the center of the device interface where the ion concentration is less than 95 percent of its initial value). When the simulated porewater concentration at the far end of the soil/sediment subdomain from the device interface falls below 99 percent of its initial value, the model automatically increases the domain size and continues to do so iteratively until depletion is no longer observed [7].

The limitations of the one-dimensional approach for simulating longer time period DGT deployments were recognized at the initial stages of modeling DGT deployments. A one-dimensional model describes diffusive flux in a single dimension that is normal to the device interface and implies a planar surface of infinite area; it does not account for lateral diffusion effects at the edges of the DGT probe. Where the extent of ion depletion into the adjacent environment is negligible, the error arising from the one-dimensional approach is minimal; however, when the solute depletion extends further away from the device interface, the one-dimensional model underestimates diffusive supply to a DGT probe due to the increasing contribution that lateral diffusion makes to the diffusive flux into the device.

Sochaczewski et al. [6] adapted the DIFS model to consider diffusion into a DGT piston-type soil probe in two dimensions (2D). The model (forthwith, "2D-DIFS") was configured to solve equations governing the ion transport and reactions within the model domain using the finite element method, thus optimizing the model solution by focusing on the areas with the steepest gradients in ion concentrations (binding gel–diffusive gel interface and diffusive gel-soil interface). 2D-DIFS was packaged within a graphical user interface where the user could specify certain physical characteristics of the soil DGT probe and use mathematical relationships between porosity (φ) and particle concentration (P_c) to determine one from the other (equation 7.7) and, by extension: D^S from D (equation 7.5). The user can also enter experimentally determined values for P_c and D^S, in which case the value for φ is not required. The 2D-DIFS model can be run at a variety of different model resolutions with a view to managing the requirements placed on computer performance. The model also has an operating mode dedicated to estimating K_{dl} and t_c from time-series R_c measurements in a given soil (see Section 7.5.1). Corrective equations have been provided to correct the two-dimensional results for a three-dimensional environment for situations where this is likely to be necessary [6]. Unlike its predecessor, the two-dimensional model only uses one size of soil/sediment subdomain for each simulation (40 mm × 40 mm) [6].

7.4.3 Interpreting Soil and Sediment Dynamics

The DIFS model can be used to gain insight into the respective roles of K_{dl} and t_c in influencing the concentration of solute in both the solution and solid phases away from the device interface. This helps us further understand the soil conditions under which the flux into the DGT device is sustained or partially sustained by supply from the solid phase, or when there is only supply by diffusion.

The initial use of DGT was for measuring solution concentrations. However, as discussed in Section 7.2, the concentration in the porewater of a soil or sediment is only obtained when the combination of diffusive flux from the porewater and resupply from the solid phase is capable of sustaining the concentration at the DGT device interface throughout the deployment (when $R_c \geq 0.8$). In the DIFS model the maximum resupply flux from the solid phase at any given time is given by the product of the concentration of labile ions on the solid phase at that time (c_t^{ls}; at $t = 0$, $c^{ls} = c^{soln} \times K_{dl}$), the amount of solid phase present (P_c) and the desorption rate constant (k_{-1}). A modeling study was carried out to investigate how K_{dl}, P_c and k_{-1} combine to determine the circumstances where the interfacial concentration is sustained, or, in other words: what does a sustained case tell a DGT user about the soil or sediment? For the model runs described below an R_c value ≥ 0.9 is considered to be representative of the sustained case.

Considering a standard DGT probe ($\delta^{mdl} = 0.093$ cm) deployed for 24 h in a soil or sediment with a relatively low P_c of 0.5 g cm^3 and a rapid rate of resupply ($k_{-1} \geq 0.81 \times 10^{-4}$ s^{-1}) of a dissolved ion ($D = 5 \times 10^{-6}$ cm^2 s^{-1}), the K_{dl} of the sediment would have to be at least 2500 cm^3 g^{-1} for c_{DGT} to approximate c^{soln} (Figure 7.2). For k_{-1} less than 10^{-4} s^{-1}, the rate of resupply is insensitive to k_{-1}, implying that the amount of labile ions close to the device interface determines the rate of supply. When there is more solid phase present, but similar resupply kinetics, the capacity of the solid phase becomes less important. This is shown for soils or sediments with a high rate constant, where a P_c of 1, 1.5

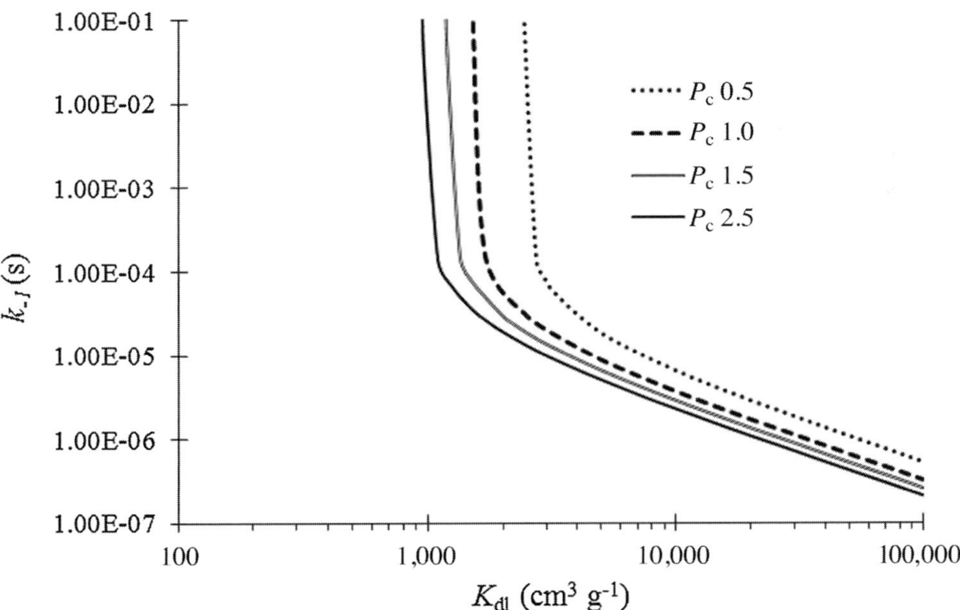

Figure 7.2 Solid phase capacity (K_{dl}) and desorption rate constant (k_{-1}) required to sustain the concentration of a dissolved ion at the DGT probe interface over 24 h when particle concentration (P_c) is varied.

or 2.5 g cm^{-3}, K_{dl} would need to be greater than 1,500, 1,200 or 1,000 cm^3 s^{-1}, respectively, to sustain the dissolved ion concentration at the DGT probe interface. Above a K_{dl} of 5,000 cm^3 g^{-1} higher particle concentrations (*viz* sorbed solute) allow for smaller desorption rate constants. Consequently a k_{-1} higher than 1.3×10^{-6} s^{-1}, 1.1×10^{-6} s^{-1} or 8.6×10^{-7} s^{-1} is sufficient to sustain solution concentrations at the probe interface when P_c is 1, 1.5 or 2 g cm^{-3}, respectively.

Under conditions where a soil or sediment cannot sustain the dissolved ion concentrations at the device interface, K_d and t_c have a key role in determining the extent to which solute concentrations are depleted away from the DGT device interface. Considering a soil or sediment with a P_c of 1 g cm^{-3} and a fast response time (t_c 1 s; k_{-1}: 1.64×10^{-2} s^{-1}), when K_d is reduced to 60 cm^3 g^{-1}, the R_c value at the end of the 24-h deployment has decreased to 0.54, indicating a *partially sustained* case. The depletion distance of dissolved ions in the solution phase was 2.1 mm; the corresponding distance for labile ions bound to the solid phase was 1.5 mm. For soils or sediments with the same P_c, but exceptionally low K_d values of 30, 5 and 1 cm^3 g^{-1}, the depletion distances in porewater dissolved ion concentrations (and c^{ls}) increase to 2.8 mm (2.2 mm), 5.8 mm (5.2 mm) and 9.8 mm (9.2 mm), respectively.

If K_d is kept constant at 60 cm^3 g^{-1}, but the response time of the solid phase-bound ions is increased to 1,000 s (k_{-1}: 1.64×10^{-5} s^{-1}), there is a minimal change in R_c (0.52), indicating that the slower reaction kinetics have not greatly influenced the DGT uptake within this time. However, the depletion distance of dissolved ions in the porewater is increased to 2.7 mm, while the depletion distance of sorbed ions has remained at 1.5 mm. The slower rate of desorption reduces the amount of solid phase-bound ions that can desorb in response to a localized reduction in dissolved ion concentrations within the 24 h and dissolved ion concentrations are depleted to a greater extent.

The above simulations of the *sustained case* were carried out using 1D-DIFS, while variation in c and c^{ls} in the *partial case* were modeled using 2D-DIFS on a personal computer using the Windows XP operating system. In the latter simulation, depletion distances are likely to be slightly overestimated [6], however the relative effects of K_d and t_c on c and c^{ls} are still relevant. In general terms, the ability of a soil to supply solute to the DGT interface is more sensitive to changes in K_{dl} than t_c, except when K_{dl} is high, when t_c becomes the critical term. In the *sustained* and *partial* cases, porosity has a lesser effect on the diffusive flux [5]. It should be noted that, for any given value of K_{dl}, there will be value of t_c below which the value of R_c will not be affected. Below this threshold value, the amount of metal available to the DGT is regulated by the amount of labile metal on the solid phase (i.e., c^{ls}, via K_{dl}) and the rate of diffusional transport [13, 14].

In the absence of solid phase supply, the depletion distance from the DGT probe interface will be informed by the diffusive characteristics of the soil. Three-dimensional modeling of a DGT probe deployed in soil/sediment shows that porewater can be depleted laterally up to 20 mm away and vertically 15 mm from the center of the device interface when the *diffusive case* applies (Figure 7.3). As discussed above, the depletion distances are likely to be much lower in the *partial* or *sustained* cases.

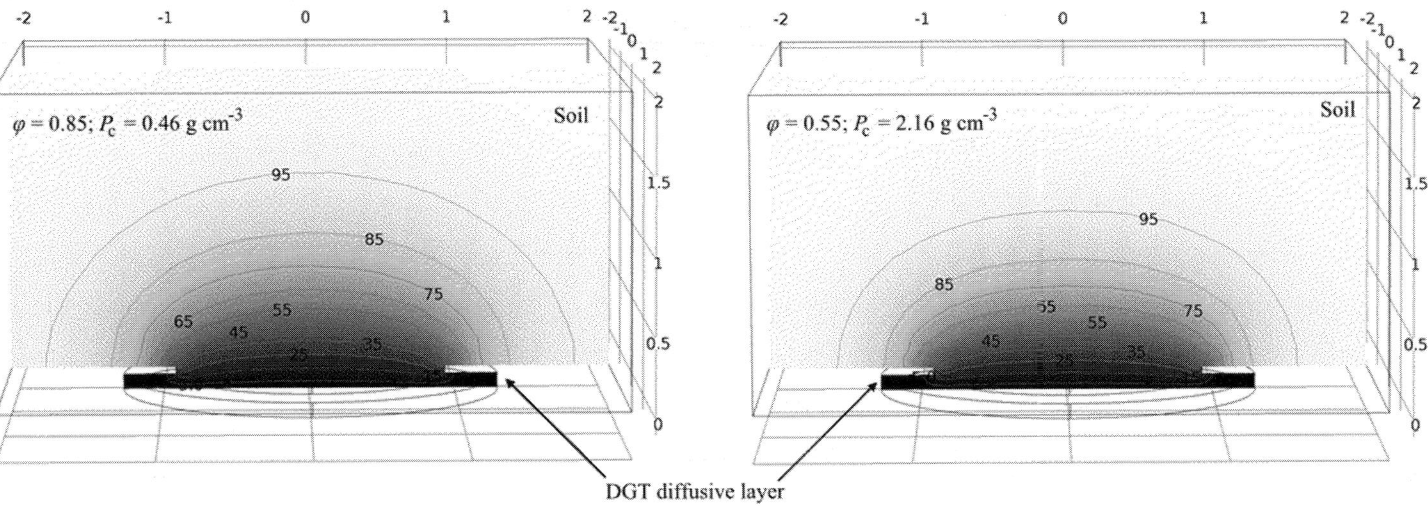

Figure 7.3 Two-dimensional cross-sections (at $y = 0$) of dissolved ion concentrations in the diffusive layer and in the adjacent soil around a DGT soil probe after 3D-DIFS simulations of a 24-h deployment. Results at two different porosities, modeled under the *diffusive case* ($K_{dl} = 0.001$ cm³ g⁻¹; $t_c = 1 \times 10^6$ s; $D = 5 \times 10^{-6}$ cm² s⁻¹), are shown. The greyscale intensity shows the final relative dissolved ion concentrations (dark = low concentration). The contour lines show concentration isopleths, with the value indicating the concentration along the line as a percentage of the initial value. The centre of the probe interface is at the origin, the distances on the x, y and z-axes are in centimeter.

7.5　Analysis of Solute Resupply Dynamics in Soils

7.5.1　Measuring K_{dl} and t_c Using DGT and DIFS

When DGT devices are deployed over variable periods of time, the depletion of solute concentration and the consequent effects on R_c can be used to estimate the resupply characteristics of the soil solid phase [5]. Ernstberger et al. [15] carried out a series of twenty-one DGT deployments in homogenized slurries of a single soil over times ranging from 4 h to 19.5 days. The results of these deployments were used to obtain time series of R_c values for Ni, Cu, Zn and Cd. These deployments were then modeled using 1D-DIFS where K_{dl} and t_c were varied until a good visual fit between the modeled and measured R_c was obtained. The best model fits to R_c values from the longer deployments (>100 h) could be obtained by modifying K_{dl}, whereas the time-dependent dynamic of R_c in the shorter deployments (<100 h) was more sensitive to t_c [15].

In a subsequent study, Ernstberger et al. [13] expanded on this work by using time-series DGT deployments to analyze Ni, Zn and Cd resupply dynamics in five soils with different textures, pHs and soil solution concentrations. The 1D-DIFS model provided good fits to all three metals in the five soils (Figure 7.4).

The resupply of Cd and Zn from the soil solid phases was characterized by rapid desorption kinetics, illustrated by k_{-1} ranging between 17×10^{-6} and $135 \times 10^{-6}\,s^{-1}$, while Ni showed more kinetic limitation ($3.6 \times 10^{-6} \le k_{-1} \le 43 \times 10^{-6}\,s^{-1}$). The DGT/DIFS-determined K_{dl} values for Cd and Zn were found to be closely linked to values determined using isotopic dilution with both increasing with pH from 4.1 to 7.1 (measured in soil slurries at 220 percent of field capacity – determined under 50 mbar suction). For Ni, the K_{dl} values estimated using DGT/DIFS were lower than for the other two metals and showed a poor relationship to soil pH, which was attributed to Ni forming less labile solid phases at pH 7 and higher. These two studies helped to validate the DIFS model and the use of time series of R_c values for determining K_{dl} and t_c.

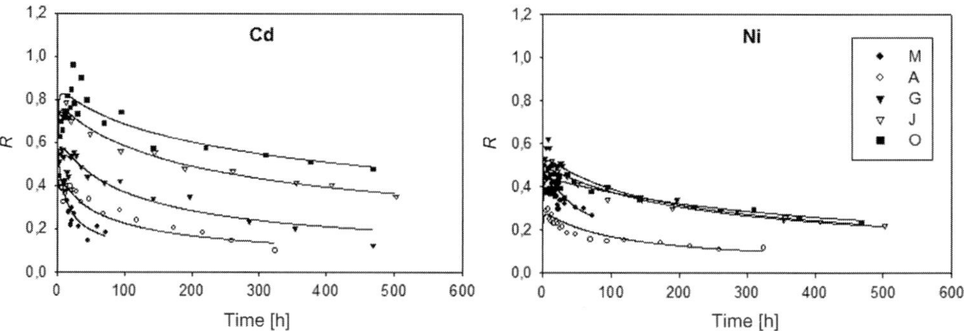

Figure 7.4　Time-series R_c for Cd and Ni measured in five soils by Ernstberger et al. [13]. DIFS model fits for soil and metal-specific K_{dl} and t_c are shown. The soil types were a silty loam (M), brown sand (A), pelo-vertic alluvial gley (G), argillic pelosol (J) and a humic rendzina (O). Reprinted with permission from Ernstberger et al. [13]. Copyright (2005) American Chemical Society.

Lehto et al. [16] revisited some of the data analyzed by Ernstberger et al. [13] using the 2D-DIFS model. They analyzed the uncertainty in determining K_{dl} and t_c using the two-dimensional model and demonstrated how variability in the R_c time series data determined the accuracy with which K_{dl} and t_c could be measured using DIFS (Figure 7.5). They found that the one-dimensional model tended to overestimate the metal desorption rate constants when significant depletion was observed in the porewater solute concentrations. For Ni, the most kinetically limited metal, the k_{-1} values estimated using the one-dimensional model were between three to six times greater than the values estimated using the two-dimensional model. The biggest difference occurred for the soil where Ni had the lowest R_c values across the time series (0.28 at 5.9 h, 0.12 at 324.1 h) and, thus, the greatest depletion of porewater solute away from the device interface. Given that a one-dimensional model does not account for lateral diffusion to a DGT probe, it appears that the one-dimensional model compensates for this additional flux at long time periods by ascribing faster rates of desorption from the solid phase. These simulations disproved Ernstberger et al. [15] suggestion that the one-dimensional approach would be more likely to overestimate K_{dl} than affect t_c (and, by extension, k_{-1}).

The 1D−DIFS model allows for the estimation of K_{dl}, t_c or R_c for a given DGT measurement if the other two have been established previously, either by using DGT or by other means [7]. Values of R_c were obtained from 24-h DGT deployments and combined with either total Zn concentrations [17] or isotopic dilution measurements of Zn and Cd [14] in a combination of field soils and soils spiked with different concentrations of metal salts. The desorption rate constants measured in the field soils were between two and three orders of magnitude smaller than the spiked soils. In several cases the desorption rate constants for Zn and Cd were too large to be determined using this method [14]. Fitz et al. [18] demonstrated, by comparing k_{-1} values for As in rhizosphere soil and the bulk soil, that an As hyperaccumulator, *Pteris vittata*, depleted the labile fraction of As in its rhizosphere. Muhammad et al. [19] used 2D-DIFS modeling along with DGT-determined R_c values and K_{dl} values obtained by extraction to estimate t_c values for Cd and Zn and showed that moderate acidification of a rhizosphere soil may result in faster rates of resupply and thus enhance phytoextraction of these metals. Given the low R_c values that were measured in this study, it is possible that the porewater concentrations of the two metals may have been depleted at the far end of the thin soil layer used in this study. The study appropriately used the 2D-DIFS model only to provide qualitative comparisons of t_c between samples; the values themselves were not reported.

DGT-measured metal fluxes in soils have been used to study metal immobilization caused either by changing the pH through liming [20, 21] or by introducing iron oxide minerals to the soil [21]. Kovaříková et al. [22] interpreted metal fluxes into DGT probes containing diffusive gels with different pore sizes to differentiate between inorganic and organic-complexed metals in sewage sludge-amended soils. The leaching of Cu, Cr and As from treated wood mulch into soil was assessed using DGT fluxes in packed soil columns [23]. Conesa et al. [24] investigated the potential for DGT to analyze metal supply in mine tailings. They measured metal fluxes into DGT with two δ^g values (0.04 and 0.07 cm)

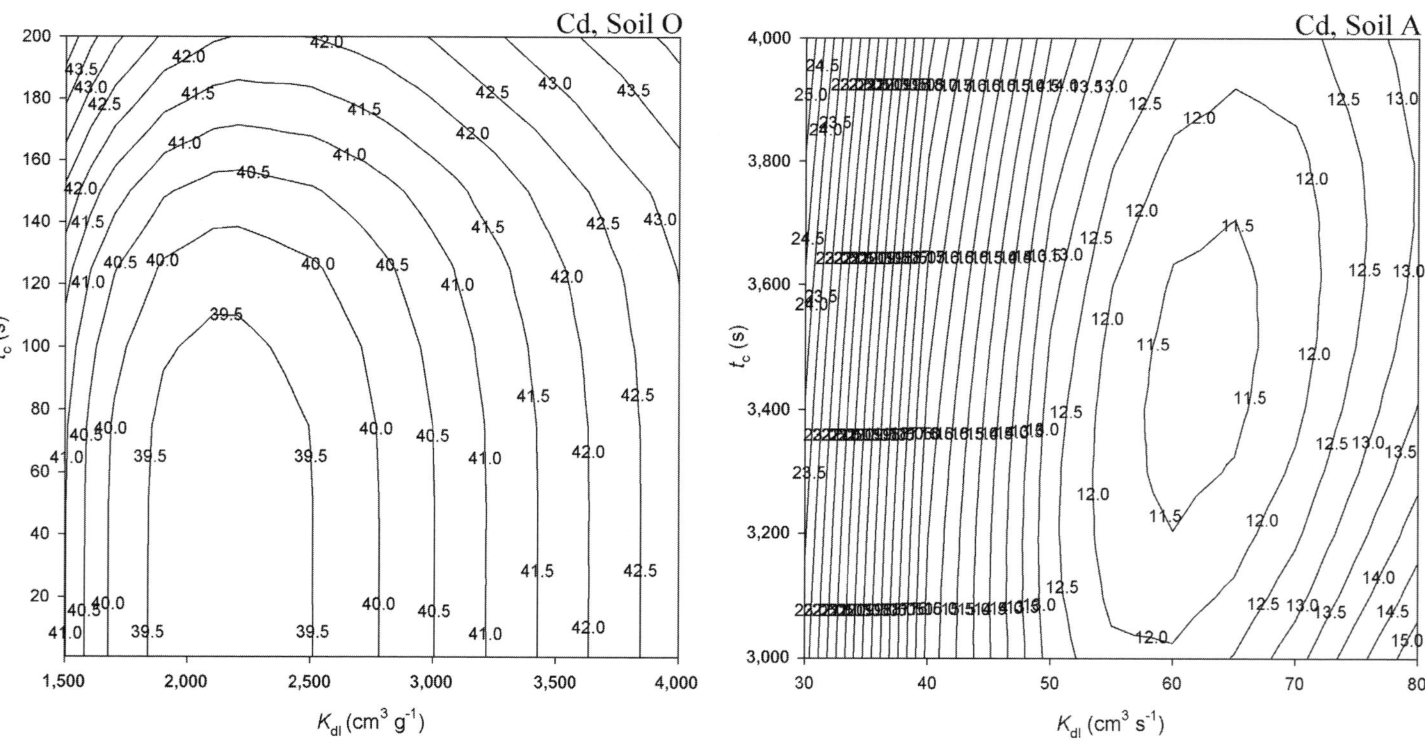

Figure 7.5   Contour plots showing error values for different values of K_{dl} and t_c, determined by 2D-model fits to the R_c time-series measurements from Cd in soils O and A (Figure 7.4). The combinations of K_{dl} and t_c values whose error values are the lowest provide the best model fits to the measured R_c values. Adapted from Lehto et al. [16], Copyright 2008, with permission from Elsevier.

deployed in mine tailings with pHs ranging from 3 to 7.2, under saturated and free draining water regimes. Out of the four metals analyzed (Zn, Cd, Pb and Cu), Zn fluxes were most affected by high H^+ concentrations. Except for Pb, the fluxes of metals into both types of probes were higher at all pHs considered when the tailings were saturated. Hattab et al. [25] observed a change in Cu and Zn fluxes to DGT in Cu-contaminated soils four years after amendment with organic matter and dolomite and phytostabilization with poplar and willow.

7.5.2 Soil Resupply Dynamics of Organic Compounds

While most work with DGT has involved the analysis of inorganic solutes and their resupply fluxes, DGT devices using an XAD18 binding layer and an agarose diffusive gel [26] have been used to analyze the resupply characteristics of trace polar organic contaminants in soils [27]. Chen et al. [27] spiked two soils with similar textures, but different pHs (6.5 and 5.4 in H_2O) and organic matter contents (4.8 and 9.3 percent), with equal amounts of four antibiotics (three sulphonamides and trimethoprim). After two weeks of aging, they carried out an acetonitrile extraction on the soils and time series deployments of DGT. Results were analyzed using the 2D-DIFS model to measure the resupply characteristics (K_{dl} and t_c) as described by Ernstberger et al. [15]. They observed lower solution phase concentrations, but larger solid phase pools for trimethoprim than for the sulphonamides. The DGT determined K_{dl} in the two soils (74 and 50 cm^3 g^{-1}) was higher than that determined by acetonitrile extraction (15 and 16 cm^3 g^{-1}) for the trimethoprim, while the K_{dl} determined by both methods for the sulphonamides were very similar in both soils. Trimethoprim had the fastest response rates, but they differed for the two soils ($t_c = 1091$ and 127 s). Response times for the sulphonamides were longer overall (t_c ranged from 1,637 s to 18,465 s), suggesting that the supply of the latter group of antibiotics was diffusion limited.

7.5.3 DGT-determined Effective Concentration (c_E) as a Predictor of Plant Bioavailability

The use of DGT to assess the bioavailability of trace metals and other elements to a variety of biota and the limitations of the approach has been the subject of a number of excellent reviews [28, 29] and a detailed discussion of the topic can be found in Chapter 9. This section will consider how DGT deployments in soils can be used to interpret how certain soil characteristics govern element availability for plant uptake.

Several studies have shown good correlations between concentrations of trace elements in plant tissues and c_{DGT} [28, 29]. For the diffusional limited cases of plant uptake and DGT, there is diffusional supply from the soil solution and additional supply from release of solute bound to the solid phase. As the time-averaged concentration at the DGT interface, c_{DGT}, is less than c^{soln}, which can be confusing as the former has the additional supply from the solid phase. To overcome this conceptual difficulty Zhang et al. [30] introduced the concept of effective concentration, c_E. In each soil they measured the mass of metal taken up by the DGT (δ^{mdl}: 0.093 cm) over a 24 h DGT deployment, M^{DGT}, and used this to calculate c_{DGT} for the deployment (equation 7.1). They then used the DIFS model to simulate that deployment in a *diffusion only* scenario by specifying the diffusion characteristics of the soil accurately, but assuming poor soil resupply characteristics (i.e., very low K_{dl}, very high t_c). The model-estimated mass of metal that would be provided by diffusion alone to the DGT device, M^{diff}, was then used to estimate a theoretical time-averaged concentration at the DGT interface, c^{diff}, using equation 7.1, which was subsequently used to calculate the ratio, R_{diff}, (equation 7.10).

$$R_{diff} = c^{diff}/c^{soln} \tag{7.10}$$

When using the DIFS model to calculate M^{diff}, it should be noted that additional solute supply by lateral diffusion in three dimensions makes an important contribution to the mass of solute taken up by a DGT probe. Therefore, M^{diff} obtained from 2D-DIFS should be corrected using the correction functions reported by Sochaczewski et al. [6] to make quantitative estimates of c_E. R_{diff} and c_{DGT} were used to calculate the effective metal concentration experienced by the DGT device, c_E, (equation 7.11):

$$c_E = c_{DGT}/R_{diff} \tag{7.11}$$

c_E is often described as a hypothetical constant dissolved concentration at the device interface that would be needed to supply the total uptake by a DGT probe during its deployment, without supply from the solid phase [28, 31]. Hence, it should always be greater than c^{soln}. The contribution of different dissolved species to determining c_{DGT} is also likely to apply to c^{diff}, and by extension, c_E, and should be duly recognized when interpreting the results.

Zhang et al. [30] found that c_E provided the best correlation with plant tissue Cu concentrations ($r^2 = 0.95$) when compared to Cu extracted using 0.05 M EDTA ($r^2 = 0.55$), Cu^{2+} ion activity ($r^2 = 0.67$) or the dissolved Cu concentration in the extracted soil solution ($r^2 = 0.85$). They suggested that the strong relationship between c_E and tissue Cu concentrations was because the plant uptake was inducing a localized depletion of Cu in the soil porewater adjacent to the root-soil interface and that the solid phase resupply was a key process in making metal available to the roots of the plant in these soils. Because c_E is dependent on deployment time and δ^{mdl}, the *de facto* standard procedure for determining it is the one reported by Zhang et al. [30]. Subsequent work has shown that c_E can achieve similar, if not better, correlations with plant trace metal concentrations than various extraction methods or soil solution concentrations for predicting uptake of a range of trace elements by different plants [17, 31–40].

Equations 7.1, 7.10 and 7.11 can be used to express c_E as $(M^{DGT}/M^{diff}) \times c^{soln}$. It follows that the $M^{DGT}:M^{diff}$ ratio can be considered as a type of dynamic distribution coefficient (K_{dl}^*) that is specific to a certain deployment time and δ^{mdl}. It is important to note that K_{dl}^* is determined by a dynamic sampling of c^{ls} through a well-defined perturbation, where the extent of disequilibrium between the c^{soln} and c^{ls} is determined by (1) the diffusive limitation of solute supply from soil and (2) the rate at which the solid phase is able to resupply solute to the porewater. The importance of k_{-1} in regulating the latter can be systematically analyzed using DGT and the DIFS model (see Section 7.4.3). Moreover, when deploying a DGT probe in a soil that has a moisture content near the soil's maximum water holding capacity, the soil being sampled is usually subject to a moisture content that specific to the soil being considered and can be considered representative of field conditions: the soil:porewater (w/v) ratio is usually more than 1:1 (see Chapter 10). This is different to the numerous extraction methods whose results are often reported alongside c_E. These methods sample a combination of c^{soln} and c^{ls} metal concentrations and mostly achieve this by manipulating the c^{soln}/c^{ls} equilibrium in two ways. Firstly, they often introduce varying amounts of either an ion (e.g., H^+, Ca^{2+}, NH_4^+, HCO_3^-) that competes with the target analyte(s) for the solid phase binding sites, or a chelant to bind the analyte(s) of interest (e.g., diethylenetriaminepentaacetic acid, DTPA, or ethylenediaminetetraacetic acid, EDTA). Secondly, they change the soil solid:liquid ratio. Soil:extractant (w/v) ratios ranging from 1:2 to 1:100 are commonly used, irrespective of the soil being analyzed. These changes in the physico-chemical conditions of the soil mean that there is often an associated change in the ionic strength and/or pH of the solution. The fundamental differences in how DGT and various extraction methods sample the pools of trace elements are key to understanding why c_E often provides a different measure of labile trace elements in soils.

Six et al. [41] carried out an isotope dilution study comparing the performance of DGT, anion binding resins and a number of extraction methods – including a number of standard soil P tests – for measuring the proportion of solid phase bound P that is available to maize (*Zea mays* L.). Their study considered two soils with different P-binding characteristics, which were subjected to two rates of P amendment. They found that DGT was the only method that consistently sampled the same pool of labile P in the soils as the maize plant, while many of the extraction methods significantly overestimated the size of the bioavailable pool. These findings were subsequently confirmed by Mason et al. [42] who showed that DGT accurately represented the pool of labile P accessed by wheat in twelve out of fourteen different agricultural soils and performed better than a simple resin extraction or Colwell-P, a commonly used extraction method to quantify soil available P [43]. These are the only studies that have looked at how the representation of 'labile pools' of P, as determined by DGT and other methods, correspond to the pool of P sampled by a plant. They help to understand why DGT appears to perform well as a technique for relating the availability of soil P to plant yield [44–47], although this may be restricted to cases when the supply of P from the soil to the root is diffusion limited [47].

7.6 Solute Resupply Dynamics in Sediments

A majority of DET and DGT deployments in marine and freshwater sediments have been carried out with a view to investigating this highly heterogeneous environment by determining the spatial distribution of metal and nutrients to allow the deduction of element cycling processes and fluxes into and out of sediments. These types of analyses are discussed in an appropriate context in Chapter 8. This section focuses on selected examples where combinations of different methods have established information about the capability of the sediments to resupply solute in both heterogeneous and homogenized systems.

7.6.1 Solute Resupply in Heterogeneous Systems

Zhang et al. [48] used a specially designed twin sediment probe to analyze metal supply in an undisturbed box core of sediment collected from a Scottish-sea loch. Planar DGT probes with different thicknesses of diffusive gels ($\delta^g = 0.04$ cm and $\delta^g = 0.12$ cm) were on opposite sides of the same probe. The smallest differences between average interfacial concentrations (c_{DGT}) were identified by the two probes for Fe and As ($c_{i,(\delta g = 0.12)} : c_{i,(\delta g = 0.04)}$ ratio of 0.86 and 1.19, respectively) suggesting that the sediment was able to sustain the concentrations of these elements at the interfaces of the two probes. The other metals (Mn, Co, Ni) showed for the most part higher c^i throughout the depth profile in the probe with $\delta^g = 0.12$ cm, indicating that the flux of these metals was only partially sustained by the sediment across the depth profile. A parallel study in the same sediment reported porewater Fe concentrations determined by core slicing [49]. They found that, while the depth profile of dissolved Fe concentration was broadly similar to that of the DGT-determined profiles (allowing for the reduced resolution of the core slicing method and variation within the box core), the porewater Fe concentrations were approximately twice as high as the c_{DGT} determined by the two DGT probes. This suggests that Fe resupply across the depth profile in that sediment was only partially sustained. The discrepancy between these two conclusions highlights the difficulties associated with determining solute resupply characteristics in such highly heterogeneous environments.

Other studies have shown R_c values close to 1 for Pb and Zn in river sediments [50] and Hg in river and marine sediments [51]. Deployments in river mouth sediments by Pradit et al. [52] indicated high Mn supply at one location; however, this should be interpreted with additional caution given that DGT measurements of Mn may not be fully representative of supply under conditions of high Fe supply [53]. Monbet et al. [54] used a combination of DET and DGT to measure variation in R_c with depth for P in lake sediments. Combined with K_{dl} obtained from sequential extraction measurements, they estimated t_c for both lake sediments and found very slow resupply kinetics (k_{-1} varied between 2.31×10^{-9} and 9.26×10^{-6} s^{-1}). They also observed R_c values above 1 in both sediments, which they attributed to working near the method detection limit for P with both DET and DGT. To date, this is the only attempt at an investigation of sediment resupply dynamics at high resolution. Given that the P-concentrations in the sediment were very low, the error associated with assuming linear relationship between c^{soln} and c^{ls}, as modeled by K_{dl} for P, was likely to be relatively low. At higher concentrations, P-sorption is best modeled using a non-linear sorption isotherm, such as a Langmuir-type expression [55, 56], which is not accounted for in the standard DIFS model. As pointed out in Chapter 8, comparisons between measurements at a small scale in heterogeneous systems should be interpreted with caution, unless measurements are made at exactly the same location, which effectively means within the same probe. Back-to-back measurements come close to this goal, but do not achieve it. This limitation can be partially overcome through deployment of multiple replicate probes within a given area: the additional information about local heterogeneity can help assess the associated uncertainty when comparing analytes from different probes (e.g., [57]).

Solute uptake by DGT can also be determined by localized processes, operating independently of the DGT device, that either produce solute for subsequent uptake (e.g., microbial methylation of mercury to produce monomethylmercury, MMHg) or liberate it from an otherwise inaccessible form (e.g., trace metals sequestered by Fe and Mn oxides that can reductively dissolve). Clarisse et al. [57] hypothesized that microbial methylation of mercury could sustain local concentrations of MMHg at the interfaces of DGTs deployed in mudflat sediments. However, without representative c^{soln} measurements, they could not show this conclusively. Trace metal release arising from transient redox conditions has been best demonstrated in homogenized sediments and is discussed in Section 7.6.2.

7.6.2 Solute Resupply in Homogenized Sediments

The steep redox gradients often observed in sediments prevents them from being homogenized and analyzed "in bulk" in the same way as well-aerated soils. Mixing and sieving a sediment can cause a dramatic disruption to solute equilibria and, once rehydrated, the redox gradients are often rapidly reestablished rendering the results from methods that sample relatively small volumes, such as DGT, difficult to reproduce. The few reported examples where this has been carried out are described further.

Naylor et al. [58] used DGT piston probes with a range of δ^{mdl} in combination with regular sampling of the porewater over time to investigate bimodal metal (Fe, Mn and Ni) dynamics in a homogenized metal sulphide slurry. Firstly, they observed an increase in Ni and Fe concentrations in the porewater, with the highest concentrations for Ni and Fe observed at 95 min. The authors hypothesized that this was most probably due to the release of metals arising from the oxidation of sulphide solid phases. This was followed by a rapid decline in concentrations of all metals, probably caused by their subsequent removal by newly formed iron oxide solid phases; after 625 min, the c^{soln} of all three metals had decreased by over 60 percent from their highest measured values. Sixteen-hour DGT measurements in these slurries showed an increasing c_{DGT} with increasing δ^{mdl} for all probes except the thickest δ^{mdl} (0.0133–0.151 cm), where the c_{DGT} for Fe and Mn remained fairly constant. Normally this would suggest that the slurry was able to resupply these metals at the same rate as the demand by the DGT probes with the two thickest δ^{mdl}; however, when the c_{DGT} are compared with the c^{soln} measurements, the c_{DGT} values are considerably higher. Visual inspection of the figures in that work indicates that c_{DGT} for Ni in all of the probes (with δ^{mdl} ranging from 0.027 to 0.151 cm) are higher than c^{soln} for the metal: using the highest measured c^{soln} at 95 min, the approximate R_c values range between ~44 for the thinnest δ^{mdl} and ~ 155 for the thickest δ^{mdl}. The apparent R_c values for Mn range between ~ 1.1 and ~ 5, while for Fe they are between ~ 0.6 and ~ 2.5. Possible reasons for R_c values exceeding unity include sampling error and/or a remobilization process that inflates the value of c_{DGT}. Some porewater solute could be lost through precipitation of iron oxide and manganese oxide minerals in the sampling process. Alternatively, high R_c values may suggest that there is a supply process that is not measured using the porewater sampling at discrete time intervals, but one that is registered by the DGT probe. For example, a large amount of metal could be released between 0 and 95 min, which results in greatly elevated fluxes into the DGT probe, which, once integrated over time, result in the high R_c values.

These findings highlight the difficulty associated with the use of DGT – which provides a time-integrated measurement of solute supply – to interpret transient phenomena that can release and/or remove metal from solution during the measurement period. In this artificial experimental system of homogenized slurry, the time-dependent release and removal of metal was identified by monitoring c^{soln}. A similar type of monitoring regime is unlikely to be practical in other types of deployment where these processes can occur in discrete areas of sediment. However, the simultaneous deployment of a real-time monitoring device, such as pH- or O_2-sensitive planar optodes, with DGT [59] can provide valuable circumstantial evidence of transient processes that result in metal release/sequestration at an appropriate scale.

Roulier et al. [60] used a combined DGT-DIFS approach to measure the solute resupply characteristics of Cu, Cd and Pb in sieved and homogenized sediments. The considerable variability in their R_c values prevented the determination of K_{dl} and t_c for the three metals with any confidence for all but one sediment. The t_c values they determined in that sediment were similar to those previously measured in soils, while the K_{dl} values were lower for the three metals. Part of the uncertainty came from a high degree of variation in c^{soln} values; the emergence of localized redox gradients during the deployment was cited as a possible cause for this. Dabrin et al. [61] used a similar approach to measure K_{dl} and t_c values for Cd and the bioaccumulation of the metal by three macroinvertebrate species in a series of carbonaceous sediments (54 percent $CaCO_3$ content). They measured very low K_{dl} values for Cd in the spiked sediments (all were less than $2.79\ \mathrm{cm}^3\ \mathrm{g}^{-1}$). Complementary sequential extractions suggested that the Cd was bound up in insoluble carbonates. Out of the three species, DGT only predicted Cd bioaccumulation by a mudsnail, *Potamopyrgus antipodarum*, indicating that the DGT-determined labile pool of Cd in the sediment was similar to that accessed by the organism. Amirbahman et al. [62] used time series DGT deployments to determine K_{dl} and t_c for inorganic Hg and MMHg in a series of homogenized sediments. They also measured the bioaccumulation of the mercury species by three benthic macroinvertebrates. An assumption was made that in these sediments in situ production/removal of MMHg was at steady state. The K_{dl} determined for Hg and MMHg were $35\ \mathrm{cm}^3\ \mathrm{g}^{-1}$ and $390\ \mathrm{cm}^3\ \mathrm{g}^{-1}$, respectively; the k_{-1} values were of 3.89×10^{-5} and $>1.84 \times 10^{-4}\ \mathrm{s}^{-1}$ for Hg and MMHg, respectively. Together the data indicated that MMHg in those sediments was more labile than the inorganic Hg. The ratio of %MMHg:total Hg measured by DGT corresponded to the ratio measured in two of the three invertebrates, which suggests that DGT sampled the same pools of labile Hg species as those organisms.

7.7 Alternative Modeling Approaches to Simulate DGT Deployments in Soil and Sediments

A strength of the DIFS model is that it describes the interaction of DGT with a solute associated with the solution and solid phase of a soil or sediment with the minimum possible number of parameters, which may have to be obtained by fitting data. However, when the necessary assumptions concerning the system are not fulfilled, models that are more sophisticated are required.

7.7.1 Resin Binding Site Saturation Model

Degryse et al. [63] used the DIFS model to predict solution Zn concentrations from DGT deployments in soils. They found that a combination of high Zn and Ca concentrations in the soil solution resulted in saturation of the binding sites in the binding gel, thus invalidating the assumption of a zero-concentration at the resin layer–diffusive layer interface. They modified the model to account for the finite capacity of the resin to bind metal and the preferential binding of Zn over Ca by the Chelex-100 resin binding sites. This revised model was able to provide a superior estimation of Zn uptake by DGT probes deployed in soils with high Zn porewater concentrations (>0.32 mM) compared to the existing 1D-DIFS model. Similar competition behavior for binding sites on Chelex-100 has subsequently been observed between Fe (strongly binding) and Mn (weakly binding), and Fe and Cd to a lesser extent, in synthetic solutions and a marine sediment [53]. A similar modeling approach to Degryse et al. [63] could theoretically help resolve Mn measurements under conditions of high Fe supply).

7.7.2 Partially Labile Solution Species Model

The generally available DIFS models by Harper et al. [7] and Sochaczewski et al. [6] only consider a single dissolved metal species, and the availability of that species to a DGT is determined purely by soil characteristics. As described briefly earlier, and in more detail in Chapter 5, the presence of dissolved metal-complexes that are partially labile or inert within the timescale of a DGT deployment can also limit the amount of metal accumulated by the DGT binding layer. In many natural waters (including soil and sediment porewaters), dissolved organic matter can include numerous important metal-complexing ligands, whose rates of diffusion are lower than those of the free metal ion [64]. When DGT depletes the free ion concentration in soil porewater, the free ion concentration can be partially buffered by supply from slow-diffusing, partially labile complexes [65], as well as the solid phase. Under these conditions, the DIFS model will wrongly interpret this resupply as coming from the solid phase and consequently overestimate the supply from the solid phase.

Introducing into the DIFS model, an advanced solution speciation model, similar to the Windermere Humic Acid Model (WHAM) [66], while simultaneously considering complexation kinetics, would increase greatly its complexity and introduce a large number of additional parameters that would need to be accurately represented. Lehto et al. [67] developed a relatively simple one-dimensional model based on a combination of DIFS and a validated reactive transport model describing DGT deployments in solutions containing metal as free ions and complexed species [68]. By adjusting K_{dl} and t_c it provided acceptable fits to sets of time-series of R_c measurements for Cd, obtained using DGT probes with two different δ^{mdl}. The K_{dl} was considerably lower than previously measured values in soils, because of preferential partitioning to the dissolved ligand in the model. The model was unable to produce a good fit to Cu measurements, probably due to the strong complexation of this metal by organic matter, which could not be adequately described using a single mobile ligand. While this type of model has obvious limitations, it could be used to gain insight into the role of complexation kinetics in DGT uptake processes that are relevant to diffusion limited uptake by plants.

7.7.3 Multiple Desorption Site DIFS Model

Nowack et al. [69] found that the conventional DIFS model was unable to simulate a time series of R_c values for Cu that had been measured in situ. They successfully modeled the in situ data by introducing another solid phase pool of Cu to describe the observed dynamic: a "slow" pool (t_c 1,100 s; K_{dl} 1,500 cm^3 g^{-1}) and a "fast" pool (t_c 30 s; K_{dl} 20 cm^3 g^{-1}). The K_{dl} value of the slow pool was calculated using an equation accounting for pH and organic matter content, rather than by fitting the data. However, when this model was applied to homogenized soils, it was unable to represent the time series R_c data for Cu. A model considering a single pool with a K_{dl} of 2 cm^3 g^{-1} and t_c of 60 s provided adequate fits to the R_c data for Zn obtained in situ.

Ciffroy et al. [70] have developed a version of the DIFS model, DGT-PROFS, that also considers two solid phase pools of solute and considers dissolved ion transport in one dimension. Unlike previous versions of DIFS, DGT-PROFS determines K_{dl} and t_c values for the two pools using a Monte-Carlo type approach where each model parameter can be assigned a probability distribution function. They validated the model through time series DGT deployments in formulated sediments spiked with Cu or Cd. The model fits obtained using the two-site DGT-PROFS model fitted the time series R_c data better than the automated fitting routine included in the 2D DIFS model and provided a systematic assessment of the uncertainty associated with the K_{dl} and t_c determined for the two types of binding sites.

7.8 Future Prospects

This chapter has shown that the pertubation of solute equilibria in soils and sediments simultaneously with measurement of the mean flux to the DGT sink has provided useful new information on solute dynamics. It has also outlined some areas of uncertainty in the use of DGT deployments to assess solute resupply dynamics. Further investigation of these areas will aid the development of DGT as a tool for environmental monitoring and assessing bioavailability.

The time-integrated nature of DGT means that the relative importance of transient features occurring during the deployment can be overlooked. Redox fluctuations in sediments can result in brief pulses of metal supply to the DGT, which can result in misrepresentation of the overall solute fluxes. In plant rhizospheres, pulses of organic acids and phytosiderophores are released at certain times by roots of some plants [71], which can be expected to result in briefly increased metal release from soils. However, it must be acknowledged that, while DGT may underestimate the importance of these temporal changes, other discrete sampling methods may miss them completely. The use of complementary techniques that operate in real time can help give clues to these temporal dynamics. At the very least, they can advise the DGT user to interpret the measurements with appropriate caution.

References

1. W. Davison and H. Zhang, In situ speciation measurements of trace components in natural waters using thin-film gels, *Nature* 367 (1994), 546–548.
2. W. Davison, G. W. Grime, J. A. W. Morgan and K. Clarke, Distribution of dissolved iron in sediment pore waters at submillimetre resolution, *Nature* 352 (1991), 323–325.
3. H. Zhang, W. Davison, S. Miller and W. Tych, In situ high resolution measurements of fluxes of Ni, Cu, Fe, and Mn and concentrations of Zn and Cd in porewaters by DGT, *Geochim. Cosmochim. Acta* 59 (1995), 4181–4192.
4. H. Zhang, W. Davison, B. Knight and S. McGrath, In situ measurements of solution concentrations and fluxes of trace metals in soils using DGT, *Environ. Sci. Technol.* 3 (1998), 704–710.
5. M. P. Harper, W. Davison, H. Zhang and W. Tych, Kinetics of metal exchange between solids and solutions in sediments and soils interpreted from DGT measured fluxes, *Geochim. Cosmochim. Acta* 62 (1998), 2757–2770.
6. Ł. Sochaczewski, W. Tych, B. Davison and H. Zhang, 2D DGT induced fluxes in sediments and soils (2D DIFS), *Environ. Modell. Softw.* 22 (2007), 14–23.
7. M. P. Harper, W. Davison and W. Tych, DIFS – A modelling and simulation tool for DGT induced trace metal remobilisation in sediments and soils, *Environ. Modell. Softw.* 15 (2000), 55–66.
8. C. L. Bielders, L. W. De Backer and B. Delvaux, Particle density of volcanic soils as measured with a gas pycnometer, *Soil Sci. Soc. Am. J.* 54 (1990), 822–826.
9. R. A. McBride, R. L. Slessor and P. J. Joosse, Estimating the particle density of clay-rich soils with diverse mineralogy, *Soil Sci. Soc. Am. J.* 76 (2012), 569–574.
10. S. Scally, W. Davison and H. Zhang, Diffusion coefficients of metals and metal complexes in hydrogels used in diffusive gradients in thin films, *Anal. Chim. Acta.* 558 (2006), 222–229.
11. B. P. Boudreau. *Diagenetic models and their implementation* (Berlin: Springer Verlag, 1997).
12. B. D. Honeyman and P. H. Santschi, Metals in aquatic systems, *Environ. Sci. Technol.* 22 (1988), 862–871.

13. H. Ernstberger, H. Zhang, A. Tye, S. Young and W. Davison, Desorption kinetics of Cd, Zn, and Ni measured in soils by DGT, *Environ. Sci. Technol.* 39 (2005), 1591–1597.

14. H. Zhang, W. Davison, A. M. Tye, N. M. Crout and S. D. Young, Kinetics of zinc and cadmium release in freshly contaminated soils, *Environ. Toxicol. Chem.* 25 (2006), 664–670.

15. H. Ernstberger, W. Davison, H. Zhang, A. Tye and S. Young, Measurement and dynamic modeling of trace metal mobilization in soils using DGT and DIFS, *Environ. Sci. Technol.* 36 (2002), 349–354.

16. N. J. Lehto, Ł. Sochaczewski, W. Davison, W. Tych and H. Zhang, Quantitative assessment of soil parameter (K_D and T_C) estimation using DGT measurements and the 2D DIFS model, *Chemosphere* 71 (2008), 795–801.

17. H. Zhang, E. Lombi, E. Smolders and S. McGrath, Kinetics of Zn release in soils and prediction of Zn concentration in plants using diffusive gradients in thin films, *Environ. Sci. Technol.* 38 (2004), 3608–3613.

18. W. J. Fitz, W. W. Wenzel, H. Zhang, et al., Rhizosphere characteristics of the arsenic hyperaccumulator Pteris vittata L. and monitoring of phytoremoval efficiency, *Environ. Sci. Technol.* 37 (2003), 5008–5014.

19. I. Muhammad, M. Puschenreiter and W. W. Wenzel, Cadmium and Zn availability as affected by pH manipulation and its assessment by soil extraction, DGT and indicator plants, *Sci. Total Environ.* 416 (2012), 490–500.

20. J. Rachou, W. Hendershot and S. Sauvé, Effects of pH on fluxes of cadmium in soils measured by using diffusive gradients in thin films, *Comm. Soil Sci. Plant Anal.* 35 (2005), 2655–2673.

21. E. Lombi, F.-J. Zhao, G. Zhang, et al., In situ fixation of metals in soils using bauxite residue: chemical assessment, *Environ. Pollut.* 118 (2002), 435–443.

22. V. Kovaříková, H. Dočekalová, B. Dočekal and M. Podborská, Use of the diffusive gradients in thin films technique (DGT) with various diffusive gels for characterization of sewage sludge-contaminated soils, *Anal. Bioanal. Chem.* 389 (2007), 2303–2311.

23. T. G. Mercer and G. M. Greenway, Diffusive gradient in thin films (DGT) for profiling leaching of CCA-treated wood waste mulch into the soil environment, *Int. J. Environ. Anal. Chem.* 94 (2014), 115–126.

24. H. M. Conesa, R. Schulin and B. Nowack, Suitability of using diffusive gradients in thin films (DGT) to study metal bioavailability in mine tailings: possibilities and constraints, *Environ. Sci. Pollut. Res.* 17 (2010), 657–664.

25. N. Hattab, M. Motelica-Heino, X. Bourrat and M. Mench, Mobility and phytoavailability of Cu, Cr, Zn, and As in a contaminated soil at a wood preservation site after 4 years of aided phytostabilization, *Environ. Sci. Pollut. Res. Int.* 21 (2014), 10307–10319.

26. C.-E. Chen, H. Zhang and K. C. Jones, A novel passive water sampler for in situ sampling of antibiotics, *J. Environ. Monit.* 14 (2012), 1523–1530.

27. C.-E. Chen, K. C. Jones, G.-G. Ying and H. Zhang, Desorption kinetics of sulfonamide and trimethoprim antibiotics in soils assessed with diffusive gradients in thin-films, *Environ. Sci. Technol.* 48 (2014), 5530–5536.

28. H. Zhang and W. Davison, Use of diffusive gradients in thin-films for studies of chemical speciation and bioavailability, *Environ. Chem.* 12 (2015), 85–101.

29. F. Degryse, E. Smolders, H. Zhang and W. Davison, Predicting availability of mineral elements to plants with the DGT technique: A review of experimental data and interpretation by modelling, *Environ. Chem.* 6 (2009), 198–218.

30. H. Zhang, F.-J. Zhao, B. Sun, W. Davison and S. McGrath, A new method to measure effective soil solution concentration predicts copper availability to plants, *Environ. Sci. Technol.* 35 (2001), 2602–2607.

31. P. N. Williams, H. Zhang, W. Davison, et al., Evaluation of in situ DGT measurements for predicting the concentration of Cd in Chinese field-cultivated rice: Impact of soil Cd:Zn ratios, *Environ. Sci. Technol.* 46 (2012), 8009–8016.

32. J. Song, F.-J. Zhao, Y.-M. Luo, S. P. McGrath and H. Zhang, Copper uptake by Elsholtzia splendens and Silene vulgaris and assessment of copper phytoavailability in contaminated soils, *Environ. Pollut.* 128 (2004), 307–315.

33. M. Koster, L. Reijnders, N. R. van Oost and W. J. Peijnenburg, Comparison of the method of diffusive gels in thin films with conventional extraction techniques for evaluating zinc accumulation in plants and isopods, *Environ. Pollut.* 133 (2005), 103–116.

34. A. L. Nolan, H. Zhang and M. J. McLaughlin, Prediction of zinc, cadmium, lead, and copper availability to wheat in contaminated soils using chemical speciation, diffusive gradients in thin films, extraction, and isotopic dilution techniques, *J. Environ. Qual.* 34 (2005), 496–507.

35. Å. R. Almås, P. Lombnaes, T. A. Sogn and J. Mulder, Speciation of Cd and Zn in contaminated soils assessed by DGT-DIFS, and WHAM/Model VI in relation to uptake by spinach and ryegrass, *Chemosphere* 62 (2006), 1647–1655.

36. J. Cornu and L. Denaix, Prediction of zinc and cadmium phytoavailability within a contaminated agricultural site using DGT, *Environ. Chem.* 3 (2006), 61–64.

37. T. A. Sogn, S. Eich-Greatorex, O. Røyset, A. Falk Øgaard and Å. R. Almås, Use of diffusive gradients in thin films to predict potentially bioavailable selenium in soil, *Comm. Soil Sci. Plant Anal.* 39 (2008), 587–602.

38. J. Soriano-Disla, T. Speir, I. Gómez, et al., Evaluation of different extraction methods for the assessment of heavy metal bioavailability in various soils, *Water Air Soil Pollut.* 213 (2010), 471–483.

39. S. Tandy, S. Mundus, J. Yngvesson, et al., The use of DGT for prediction of plant available copper, zinc and phosphorus in agricultural soils, *Plant Soil* 346 (2011), 167–180.

40. I. Ahumada, L. Ascar, C. Pedraza, et al., Determination of the bioavailable fraction of Cu and Zn in soils amended with biosolids as determined by diffusive gradients in thin films (DGT), BCR sequential extraction, and Ryegrass plant, *Water Air Soil Pollut.* 219 (2011), 225–237.

41. L. Six, P. Pypers, F. Degryse, E. Smolders and R. Merckx, The performance of DGT versus conventional soil phosphorus tests in tropical soils – An isotope dilution study, *Plant Soil* 359 (2012), 267–279.

42. S. D. Mason, M. J. McLaughlin, C. Johnston and A. McNeill, Soil test measures of available P (Colwell, resin and DGT) compared with plant P uptake using isotope dilution, *Plant Soil* 373 (2013), 711–722.

43. J. Colwell, The estimation of the phosphorus fertilizer requirements of wheat in southern New South Wales by soil analysis, *Aust. J. Exp. Agr.* 3 (1963), 190–197.

44. N. W. Menzies, B. Kusumo and P. W. Moody, Assessment of P availability in heavily fertilized soils using the diffusive gradient in thin films (DGT) technique, *Plant Soil* 269 (2005), 1–9.

45. T. M. McBeath, M. J. McLaughlin, R. D. Armstrong, et al., Predicting the response of wheat (Triticum aestivum L.) to liquid and granular phosphorus fertilisers in Australian soils, *Soil Res.* 45 (2007), 448–458.

46. S. Mason, A. McNeill, M. McLaughlin and H. Zhang, Prediction of wheat response to an application of phosphorus under field conditions using diffusive gradients in thin-films (DGT) and extraction methods, *Plant Soil* 337 (2010), 243–258.

47. L. Six, E. Smolders and R. Merckx, The performance of DGT versus conventional soil phosphorus tests in tropical soils – Maize and rice responses to P application, *Plant Soil* 366 (2013), 49–66.

48. H. Zhang, W. Davison, R. J. Mortimer, et al., Localised remobilization of metals in a marine sediment, *Sci. Total Environ.* 296 (2002), 175–187.

49. M. D. Krom, R. J. G. Mortimer, S. W. Poulton, et al., In-situ determination of dissolved iron production in recent marine sediments, *Aquat. Sci.* 64 (2002), 282–291.

50. M. Leermakers, Y. Gao, C. Gabelle, et al., Determination of high resolution pore water profiles of trace metals in sediments of the Rupel river (Belgium) using DET (diffusive equilibrium in thin films) and DGT (diffusive gradients in thin films) techniques, *Water Air Soil Pollut.* 166 (2005), 265–286.

51. P. Diviš, M. Leermakers, H. Dočekalová and Y. Gao, Mercury depth profiles in river and marine sediments measured by the diffusive gradients in thin films technique with two different specific resins, *Anal. Bioanal. Chem* 382 (2005), 1715–1719.

52. S. Pradit, Y. Gao, A. Faiboon, et al., Application of DET (diffusive equilibrium in thin films) and DGT (diffusive gradients in thin films) techniques in the study of the mobility of sediment-bound metals in the outer section of Songkhla Lake, Southern Thailand, *Environ. Monit. Assess.* 185 (2013), 4207–4220.

53. Tankéré-Muller, W. Davison and H. Zhang, Effect of competitive cation binding on the measurement of Mn in marine waters and sediments by diffusive gradients in thin films, *Anal. Chim. Acta* 716 (2012), 138–144.

54. P. Monbet, I. D. McKelvie and P. J. Worsfold, Combined gel probes for the in situ determination of dissolved reactive phosphorus in porewaters and characterization of sediment reactivity, *Environ. Sci. Technol.* 42 (2008), 5112–5117.

55. T. Hiemstra and W. H. Van Riemsdijk, A surface structural approach to ion adsorption: The charge distribution (CD) model, *J. Colloid Interface Sci.* 179 (1996), 488–508.

56. A. Zhou, H. Tang and D. Wang, Phosphorus adsorption on natural sediments: Modeling and effects of pH and sediment composition, *Water Res.* 39 (2005), 1245–1254.

57. O. Clarisse, B. Dimock, H. Hintelmann and E. P. H. Best, Predicting net mercury methylation in sediments using diffusive gradient in thin films measurements, *Environ. Sci. Technol.* 45 (2011), 1506–1512.

58. C. Naylor, W. Davison, M. Motelica-Heino, L. M. van der Heijdt and G. A. van den Berg, Transient release of Ni, Mn and Fe from mixed metal sulphides under oxidising and reducing conditions, *Environ. Earth Sci.* 65 (2012), 2139–2146.

59. H. Stahl, K. W. Warnken, L. Sochaczewski, et al., A combined sensor for simultaneous high resolution 2-D imaging of oxygen and trace metals fluxes, *Limnol. Oceanogr. Methods* 10 (2012), 389–401.

60. J. L. Roulier, M. H. Tusseau-Vuillemin, M. Coquery, O. Geffard and J. Garric, Measurement of dynamic mobilization of trace metals in sediments using DGT and comparison with bioaccumulation in Chironomus riparius: First results of an experimental study, *Chemosphere* 70 (2008), 925–932.

61. A. Dabrin, C. L. Durand, J. Garric, et al., Coupling geochemical and biological approaches to assess the availability of cadmium in freshwater sediment, *Sci. Total Environ.* 424 (2012), 308–315.

62. A. Amirbahman, D. I. Massey, G. Lotufo, et al., Assessment of mercury bioavailability to benthic macroinvertebrates using diffusive gradients in thin films (DGT), *Environ. Sci. Process. Impacts.* 15 (2013), 2104–2114.

63. F. Degryse, E. Smolders, I. Oliver and H. Zhang, Relating soil solution Zn concentrations to diffusive gradients in thin films measurements in contaminated soils, *Environ. Sci. Technol.* 37 (2003), 3958–3965.

64. K. W. Warnken, W. Davison, H. Zhang, J. Galceran and J. Puy, In situ measurements of metal complex exchange kinetics in freshwater, *Environ. Sci. Technol.* 41 (2007), 3179–3185.

65. N. J. Lehto, W. Davison, H. Zhang and W. Tych, An evaluation of DGT performance using a dynamic numerical model, *Environ. Sci. Technol.* 40 (2006), 6368–6376.

66. E. Tipping, Humic ion binding model VI: An improved description of the interactions of protons and metal ions with humic substances, *Aquat. Geochem.* 4 (1998), 3–48.

67. N. J. Lehto, W. Davison and H. Zhang, The use of ultra-thin diffusive gradients in thin-films (DGT) devices for the analysis of trace metal dynamics in soils and sediments: A measurement and modelling approach, *Environ. Chem.* 9 (2012), 115–423.

68. Ø.-A. Garmo, N. J. Lehto, H. Zhang and W. Davison, Dynamic aspects of DGT as demonstrated by experiments with lanthanide complexes of a multidentate ligand, *Environ. Sci. Technol.* 40 (2006), 4754–4760.

69. B. Nowack, S. Koehler and R. Schulin, Use of diffusive gradients in thin films (DGT) in undisturbed field soils, *Environ. Sci. Technol.* 38 (2004), 1133–1138.

70. P. Ciffroy, Y. Nia and J. M. Garnier, Probabilistic multicompartmental model for interpreting DGT kinetics in sediments, *Environ. Sci. Technol.* 45 (2011), 9558–9565.

71. S. M. Reichman and D. R. Parker, Probing the effects of light and temperature on diurnal rhythms of phytosiderophore release in wheat, *New Phytol.* 174 (2007), 101–108.

8

Measurement at High Spatial Resolution

JAKOB SANTNER AND PAUL N. WILLIAMS

8.1 The Need for Versatile High-Resolution Analysis Tools for Assessing the Chemical Heterogeneity of Sediments and Soils

For the mechanistic and quantitative understanding of global biogeochemical cycling it is very often necessary to have detailed knowledge about the controlling, small-scale processes that drive the interaction of earth's atmosphere, biosphere, lithosphere and water bodies. Sediments and soils are central compartments of the global element turnover, functioning as chemical filters and buffers, as well as reaction and storage compartments, and thereby having a strong impact on the composition of the atmosphere and earth's water bodies.

Sediments are stratified substrates, formed by the deposition of particulate material from the overlying water. The rapid consumption of the O_2 supplied from the water and the associated shift to anaerobic respiration leads to steep redox gradients, and consequently to steep concentration gradients of the involved species, down the sediment profile [1, 2]. In contrast, the composition of soils is dominated by the weathering of the parent material and the decomposition of organic *detritus*, either at the soil surface or in the *solum* itself (e.g. decomposing roots and soil fauna), and the mixing of the decomposition and weathering products [3]. The resulting horizontal heterogeneity of both sediments and soils is further enhanced by processes like bioturbation, solute uptake by and exudate release from plant roots, and microbial activity, leading to a complex, three-dimensional mosaic structure of solid and solute distributions. Microsensors [4–6] and layer-wise sampling [7, 8], but also planar pH measurements [9] and autoradiography [10], have been applied traditionally to resolve the spatial solute distribution in these environments.

Many of these approaches were only capable of providing one-dimensional solute profiles, which often also had a spatial resolution in the centimetre to millimetre range. Consequently, a full appreciation of the solute distribution was usually not possible. The introduction of DET and DGT as tools for measuring one-dimensional and two-dimensional solute distributions in the millimetre and submillimetre ranges was a large step towards a more comprehensive appreciation of soil and sediment heterogeneity. Together with planar optodes, a complementary, fluorescence-based imaging technique, DET and DGT are today the state of the art in chemical imaging of solute distributions in aquatic and terrestrial biogeochemical hotspots. In a recent review, Santner et al. [11] provide a

174

comprehensive overview of two-dimensional chemical imaging applications of these methods. In this chapter, we focus on DET and DGT, providing an overview of sampling and analysis setups, method characteristics and limitations, and give a few examples of applications of high-resolution studies using DET and DGT.

8.2 Methodology

8.2.1 Introduction to High-Resolution Analysis Methods

With a wide range of options now available for high-resolution characterisation of DGT/DET, the following sections will review the most commonly used analytical techniques, compare and contrast their functionality and place this in the historical context in which the breakthroughs were made. An overview of DGT and DET high-resolution methods is given in Table 8.1.

Today, precise measurements of porewater solute chemistries at high resolution (< 1 mm) can be obtained relatively easily and affordably using DGT samplers. However, the capture at fine-scale of in situ element gradients by thin-film gels was a significant technological advance in 1991, when a method to measure the distribution of iron in sediment porewaters at sub-millimetre resolution was presented [12]. After deployment in the sediment of the gels within plastic probes, the Fe(II) in the gels was spatially fixed as the oxyhydroxide by exposure of the probe to sodium hydroxide solution. After careful drying, the iron concentrations in the gels were determined using MeV-proton-induced X-ray emission (PIXE) at the Oxford Scanning Proton Microprobe. Yet, with the relatively high cost, long measurement times [24] and limited access to PIXE, coupled with detection thresholds in the parts per million range, it remained a highly specialised research tool, and crucially it was too insensitive to be used for most trace elements [37]. Furthermore, the method was reliant on there being a minimal time between gel retrieval and fixation in NaOH [22]. This proved not to be a problem within the settings of the laboratory, with uncompromised resolutions of as high as 250 μm if gel to solution transfer could be completed in 1–2 s. However, under field situations this presented a greater challenge, with a delay >20 s ruling out the opportunity to measure at a resolution <1 mm, as assessed by one-dimensional modelling of the broadening of a square maxima of finite concentration [60].

8.2.2 Gel Slicing

8.2.2.1 Gel Slicing – the Origins

The solution to these measurement dilemmas was to follow quickly. Firstly, element quantification using widely available and readily affordable detection systems such as ion chromatography (IC) and graphite furnace atomic absorption spectroscopy (GF-AAS), were made possible, by the adoption of a gel slicing approach [15], and in situ studies were facilitated with the improvement of sampling probes [15]. These new techniques expanded the range of measurable determinants, as well as providing a more sensitive analysis for

Table 8.1 *Overview of high-resolution DGT and DET methods*

Analyte	Sampling method	Resin	Spatial dimension	Analytical techniques	Ref
Ag^+, Ba^{2+}, Ca^{2+}, Cd^{2+}, Co^{2+}, Cu^{2+}, Fe(II), Fe(III), Mn^{2+}, Ni^{2+}, Pb^{2+}, Sr^{2+}, Sn^{2+}, Zn^{2+}	DET		1D	PIXE, AAS, ICPOES, ICPMS, colorimetry (FeII/III)	[12–17]
As, Bi, Cr, Mo, Sb, Se, V	DET		1D	ICPMS	[13]
Br^-, Cl^-, NO_2^-, NO_3^-, SO_4^{2-}	DET		1D	HPLC, IC	[15, 16, 18]
Total alkalinity	DET		1D	Titration, colorimetry	[16, 19]
ΣCO_2	DET		1D	GD-FIA	[16]
NH_4^+	DET		1D	GD-FIA	[16, 18, 20]
U, $^{238}U/^{235}U$	DET		1D	ICPMS	[21]
Fe(II), Mn^{2+}	DET		2D	PIXE, AAS	[22, 23]
Fe(II)	DET		2D	CID	[24, 25]
PO_4^{3-}	DET		2D	CID	[26]
$^{29}N_2$, $^{30}N_2$, NO_3^-	DET		2D	GC-IRMS, colorimetry	[27]
Fe^{2+}, PO_4^{3-}	DET		2D	colorimetry, hyperspectral imaging	[28]
Total alkalinity	DET		2D	CID	[29, 30]
Al^{3+}, Cd^{2+}, Co^{2+}, Cu^{2+}, Fe(II), Mn^{2+}, Ni^{2+}, Pb^{2+}, Ra^{2+}, Sn^{2+}, Sr^{2+}, Tl^{2+}, Zn^{2+}	DGT	Chelex 100	1D	AAS, ICPMS	[31–37]
PO_4^{3-}	DGT	Ferrihydrite, Zr-oxide	1D	colorimetry	[38, 39]
As	DGT	Metsorb	1D	ICPMS	[40]
As(III)	DGT	Chelex 100, 3-mercaptopropyl-silica	1D	ICPMS, AAS	[40, 41]

Hg	DGT	Chelex 100, Spheron Thiol	1D	AAS	[42]
CH_3Hg	DGT	3-mercaptopropyl-silica	1D	GC-ICPMS, CVAFS	[43, 44]
Cr, V	DGT	Chelex 100	1D	ICPMS	[35, 36]
U, $^{238}U/^{235}U$	DGT	Spheron-Oxin	1D	ICPMS	[21, 45]
$\delta^{34}S$ in sulphide	DGT	AgI	1D	LA-HR-MC-ICPMS	[46]
Cd^{2+}, Co^{2+}, Cu^{2+}, Fe(II), Mn^{2+}, Ni^{2+}, Pb^{2+}, Zn^{2+}	DGT		2D	PIXE, LA-ICPMS	[47–49]
As(III)	DGT	SPR-IDA	2D	PIXE	[47]
S^{2-}	DGT	AgI, PVC (semi-quantitative)	2D	CID	[24, 50, 51]
PO_4^{3-}, V, As, Mo, Sb, Se, W, U	DGT	ferrihydrite, Zr-oxide	2D	LA-ICPMS	[52–55]
PO_4^{3-}	DGT	Zr-oxide	2D	Slicing and colorimetry, gel staining, CID	[56, 57]
PO_4^{3-}, S^{2-}	DGT	Zr-oxide, AgI	2D	Slicing and colorimetry, CID	[58]
PO_4^{3-}, As, Cu, Cd, Mn, Zn, Co	DGT	Zr-hydroxide + SPR-IDA	2D	LA-ICPMS	[59]

many elements. With a greater ease of sampler retrieval, gels could be more effectively hand-cut into long pieces of *ca.* 1 cm, for further sectioning by a blade into *ca.* 2-mm wide strips. These were then transferred into pre-weighed micro-tubes for subsequent elution and analysis. However, careful cutting of the gels at speed, coupled with the ongoing analyte diffusion, still proved a problem if resolutions of 1 mm or less were required [15, 61–63].

A significant advance in high-resolution sampling occurred when fidelity issues concerning diffusional relaxation, both during equilibration in the soil/sediment and between removal of the probe from sediment and fixation, were largely overcome by the use of the DGT technique for measurements in sediments [37], which secured the solute mobilisation trends in situ. Furthermore, the inbuilt pre-concentration factor of the DGT samplers of typically more than 1000 [60], extended both the range of measureable analyte concentrations and the detection techniques that could be used. For the first time sub-mm fine scale measurements could be obtained for trace metals at very low concentrations [37]. Gel assemblies that featured in this study were the archetype of the modern-day samplers, with deployments based on a variety of schemes, such as a single polyacrylamide gel (0.4 mm) containing at one face a monolayer of Chelex-100 beads, topped by a membrane filter. Samplers with different diffusive layer thicknesses (0.1–2.1 mm), combined with a 0.8-mm thick resin layer, were also trialled. Retrieved probes only required processing just prior to analysis, allowing sectioning to be performed more conveniently and under a more suitable clean-laboratory environment. Binding gels were dissected from the probe housing yielding a 10 cm × 1 cm strip, which could then be simply divided by first using a ruler to obtain ten 1 cm^2 sized units. Each of these pieces was sectioned further to produce strips of the required thickness to ensure volumes in the target range of 10–20 μL. They were eluted in 2M HNO$_3$ using a volume ca. tenfold that of the gel sample [37].

8.2.2.2 Gel Slicing – Mechanisation

The effectiveness of manual sectioning of gels can be highly dependent upon the skill of the researcher. Maintaining both precision and accuracy in the cutting can also be especially problematic for large sample batches or when small gel pieces are required. To improve slicing, a simple but effective guillotine was developed [23]. Designed to accommodate replaceable Teflon razor blades, a vernier scaled micromanipulator controlled the sample-mounting platform, which enabled the precise incisions to be made. This approach produced two new measurements. The first, 1-mm thin slivers, 1-cm wide and 0.4-mm thick of typically a volume of 4 μL, representing a two- to fivefold decrease in gel volume than in earlier works [37] and unprecedented vertical data clarity. The second, consecutive 3 mm × 3 mm × 0.4 mm blocks within a 4 cm × 10 cm sediment deployed gel, was perhaps to prove even more significant, confirming the importance of two-dimensional micro-structure in solute profiles, whilst highlighting pronounced horizontal as well as vertical variance and the existence of very localised porewater maxima (Figure 8.1) [23]. This modest technical advance, provided at the time access to a new level of multi-elemental, spatially resolved measurements, and proved to be a catalyst for much of the subsequent interest in two-dimensional chemical imaging of porewater solutes.

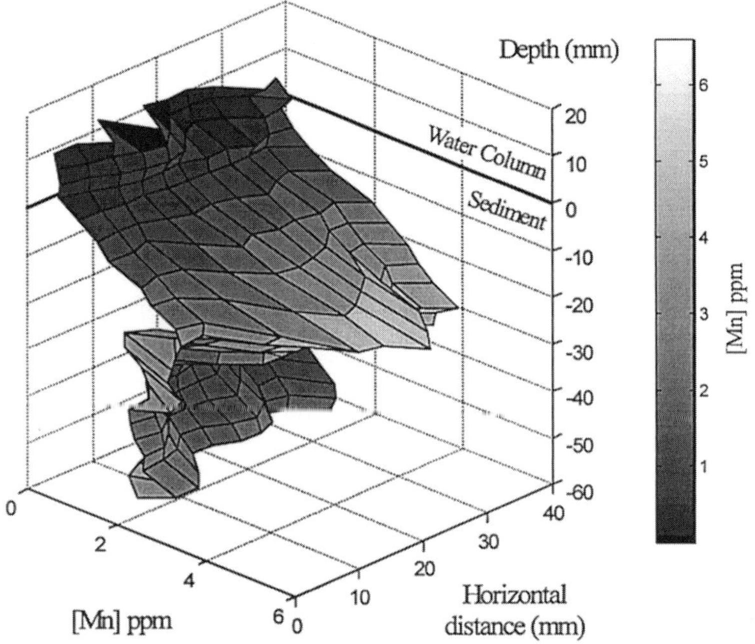

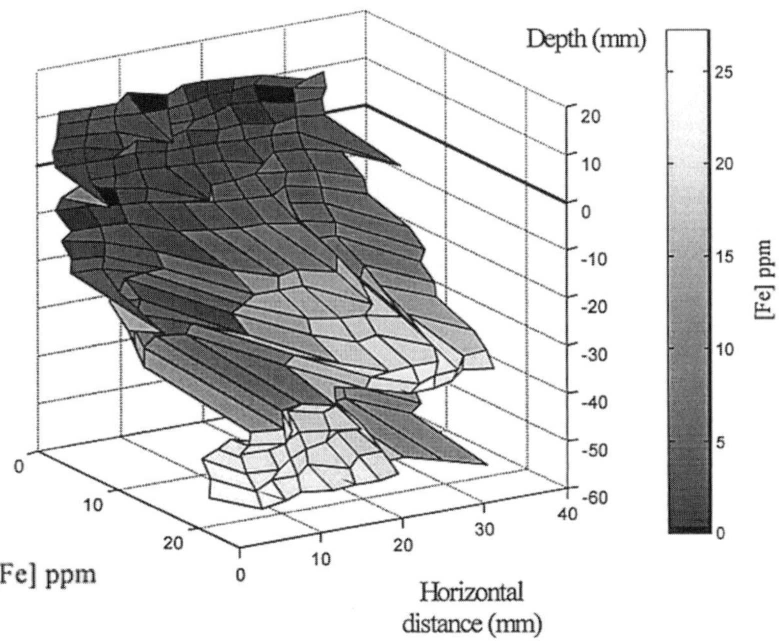

Figure 8.1   DET images of Mn and Fe distributions down a sediment profile. DET samplers were deployed at 15 m depth in Esthwaite Water in winter 1998. After retrieval, the solute distribution was fixed by precipitation in 1 mol L^{-1} NaOH. The gels were then sliced to 3 mm × 3 mm pieces, eluted and analysed. In addition to the expected solute distribution changes with depth, considerably horizontal concentration changes are visible. Adapted with permission from Shuttleworth et al. [23]. Copyright 1999 American Chemical Society.

Drawing from the same mechanised DGT processing method as used in the earlier sediment work [23], Luo et al. [64] illustrated the multiple advantages of using DGT to assess the detailed solute structure of porewaters in the near surface layer of soils historically treated by sewage sludge. Providing more than just a spatial view of highly localised trace-cation (Mn, Cd, Co, Cu, Ni and Zn) mobilisation, the study was able to show that the supply was substantial, highly localised and sustained. Whilst using element stoichiometry patterns, additional information relating to the processes of mobilisation were gleaned. In contrast, when a conventional soil sampling and extraction technique was employed on the same soils, not only was the availability of metals overestimated, but also the method failed to discriminate any localised concomitant release of Fe, Mn or Co or locate the zones of maximum chemical interest. The technique has also been validated in cores collected from a marine harbour [65], where metal release was found to be controlled not by the depositional history of the sediment, but rather by the decomposition of organic matter.

Measurements to assess the precision of the guillotine slicing method (3 mm × 3 mm, $n = 24$) in probes deployed for 24 h in synthetic seawater solution recorded a relative standard deviation for Mn, Fe, Co, Ni and Cu of between 4 and 7 per cent. While, the DGT measurement differed from an independent analytical spike QC by less than 10 per cent, validating the accuracy of the method [65]. Fine gel slicing can also be quickly and simply achieved, using a cutting template. By locking sample gels within perpendicular orientated horizontal and vertical cutting guides, cushioned by a spacer to avoid gel compression, it is possible to accurately cut 600, 4 mm × 4 mm squares with a thickness of 0.4 mm in less than thirty minutes [66].

8.2.2.3 Gel Slicing – Microscale Sectioning

Further improvements to the cutting methodology have now taken slicing to sub-millimetre scales, with the invention of a simple new sectioning tool, designed especially for DGT [56]. Comprising of 80 Teflon-coated razor blades of a thickness of 0.1 mm, with each pair of neighbouring blades separated by a 0.35-mm plastic spacer, this cutter has been shown to be able to precisely slice discrete 450 μm × 450 μm gel segments. Microscope examination of the distance between each pair of adjacent cutting edges, revealed an acceptable 2 per cent variability. Furthermore, random measurement of *ca.* 100 cut gel squares found the surface area variance to be within 5 per cent.

However, the technique is manually intensive. A 1 cm^2 binding layer section would require the transfer of >500 individual cut gel pieces; assuming an average transfer rate of 1 segment per minute, this would take over 8 h or the equivalent of a full working day. The other steps of the procedure though are relatively less demanding as elution and pipetting can be done on mass. However, to obtain a sizable two-dimensional measurement of 4–5 cm^2 would involve a whole weeks work, and as demonstrated later on in this chapter other methods are available that provide comparable results but on considerably larger spatial scales (>100 cm^2). Partnering high-resolution cutting with colorimetric analysis is extremely cost effective, but typically restricted to a single analyte parameter. With advances in small volume, high throughput sample introduction systems, such as with flow injection, the future of micro-cutting is likely to lie with increased automation coupled with multi-element micro-detection systems [67].

8.2.3 Colorimetric/Computer-Imaging Densitometry (CID)

8.2.3.1 CID – the Origins

At the same time that advances in the DGT technique were creating a new vantage point from which to explore the grouped behaviours of elements [23], a novel DGT application was undergoing its final validation. Using a conventional flat-bed scanner as the detector system and appropriate data processing software, colour densitometry measurement was emerging as a powerful and cost-effective tool for two-dimensional DGT imaging [51]. Harnessing the change in optical density induced by the quantitative conversion of pale yellow $AgI_{(s)}$ to black $Ag_2S_{(s)}$ in the presence of dissolved sulphide species (S^{2-}, HS^-, H_2S), a spatial scale of analysis from the sub-mm to potentially tens of cm became available.

In addition to being affordable, the AgI DGT technique was both rapid and accurate. Quality control trials of the DGTs in spiked solutions of sulphide yielded recoveries of 95 per cent and precisions close to 5 per cent. With an image quality of 4.2 pixels per millimetre, this offered a theoretical resolution of 238 µm, which, with the exception of PIXE (1 µm), was unparalleled by any other comparable technique at the time. Using multiple blank gels, the standard deviation (SD) of the sulphide density of the zero sulphide controls was determined as 1.41 nmol cm^{-2}, conferring a limit of detection (LOD = 3 × SD) of 4.23 nmol cm^{-2}. For a standard DGT deployment, with a device employing a typical 0.08 cm diffusive gel thickness, this corresponds to a LOD of about 0.26 µmol L^{-1}. These low levels of detection meant the sensitivity of the DGT technique was comparable to that of the most sensitive probes or microsensors [51].

The main limitation with the original AgI gel formulation was the relatively small dynamic range, with the maximum concentration measurable being ca. 60 µmol L^{-1} [51]. This was subsequently bettered, when $AgNO_3$ as opposed to AgI was cast into the polymerising gel and the iodine substituted for nitrate by soaking in a solution of KI; this not only ensured the binding phase was evenly distributed within the gel matrix but also doubled the binding capacity, simultaneously alleviating the measurement range problem [50, 68]. Using the new gel and improved scanning technologies, the spatial resolutions obtained were ca. 100 µm [68]. Higher concentrations can be measured by using shorter deployment times or thicker diffusion layers, but these approaches will also increase the LOD.

Arguably, the central disadvantage to using colour or density-based methods is that typically the analysis can only be used to quantify a single element. Yet, to decipher many localised geochemical processes a more holistic, multi-parameter approach is necessary. So it was a natural progression to start partnering AgI gels alongside other DGT binding layers within the same probes, to enable capture of the simultaneous dynamics of sulphide and metals. These early pairings were only able to cover relatively small areas of 1–10 cm^2 [69, 70], so for larger spatial zones (>100 cm^2) a new approach was required. This demand was met in 2007 when a large-scale (ca. 100 cm^2) DET colorimetric method for Fe(II) that used a ferrozine stain was developed. Using a polyacrylamide hydrogel equilibrated in a 0.01 mol L^{-1} ferrozine and 0.1 mol L^{-1} ammonium acetate solution, a 'reactive hydrogel' was produced that once placed on top of a complementary sample gel immediately retrieved from a sediment, develops a magenta coloration in proportion to the formation of ferrozine–Fe(II) complexes [24]. For determining concentration, the resulting colour images, recorded using a flat-bed scanner, were converted to green-monochrome, as this provided better sensitivities and a more linear calibration for Fe(II). Attempts to partner the ferrous iron measurements with sulphide were made, but the latter was only captured qualitatively [24].

Simultaneous Fe(II) and sulphide two-dimensional imaging was realised when AgI DGT was combined with the new colorimetric DET method for ferrous iron [25, 71]. Adopting a similar 'reactive hydrogel' or 'staining gel layer' approach to Jézéquel et al. [24], this work advanced the measurements in two major ways. Firstly scale: the measurement windows for the deployment encompassed an area of ca. 136 cm^2 representing a significant size gain. However, more importantly was the incorporation of a true measurement of total dissolved sulphide species within the same sampling device [50, 51, 68]. There were also a number of key improvements to the sulphide determination method. Ferric iron can precipitate within AgI gels, especially when probes are deployed across anaerobic-aerobic transition zones. The resultant darkening of the gel cannot be differentiated by the scanners from that caused by Ag$_2$S formation, but bathing the sample gels in a reductant solution of 0.01 mol L^{-1} hydroxylamine hydrochloride for 12 h [25] was shown to dissolve the Fe hydroxide and remove the interference.

The suite of analyte targets available for large-scale DET imaging has been extended further with methods for alkalinity (Figure 8.2) [19, 29] and reactive phosphate (Figure 8.3) [26], the latter being adapted from the classic blue stain that forms when orthophosphate interacts with molybdate-ascorbic acid [26]. Similar to the ferrozine method [24, 25], the indicator doped 'reactive hydrogel' is overlaid on the sample, enabling the reagent to diffuse into the sample gel, initiating a colour change. Optimal staining times of 18–24 min were based on a balance between maximising sensitivity and avoiding measurement biases/artefacts due to a propensity for colour changes to be concentrated at the peripheries of the gel [26]. With detection limits of 0.22 μmol L^{-1} and a large dynamic range up to ca. $1{,}000$ μmol L^{-1}, the method is versatile, but is not as sensitive as traditional solution based colorimetry and would be unable to measure very low phosphate concentrations in fresh and marine waters [72].

8.2.3.2 Hyperspectral Imaging (HSI) Methods

Desktop scanners typically employed for densitometry utilise three large spectral colour channels; recording red (620–740 nm), green (495–570 nm) and blue (450–495 nm).

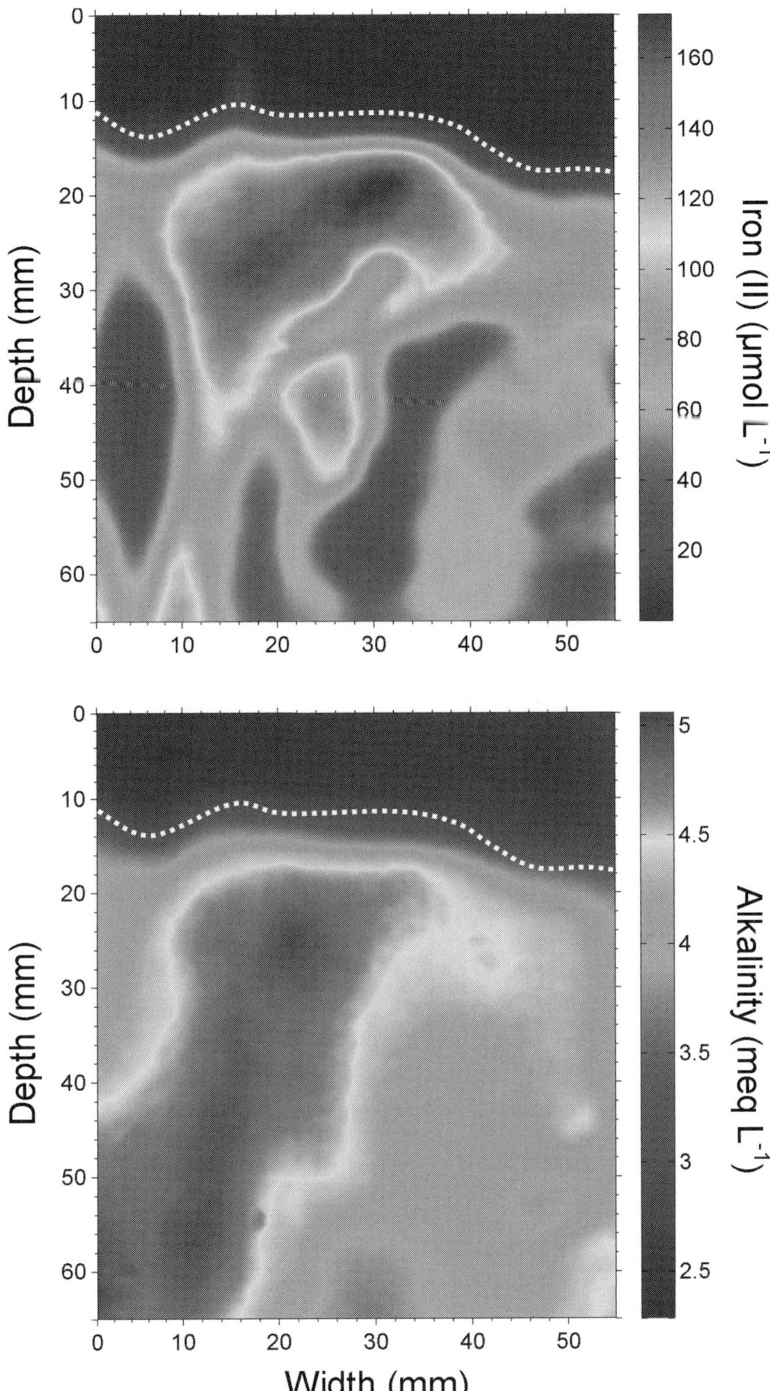

Figure 8.2 Simultaneous colorimetric DET images of the Fe(II) and alkalinity distributions in a coastal marine sediment. This sediment was colonised by the seagrass *Zostera capricorni*. The highly heterogeneous Fe(II) distribution might be linked to oxygen leakage of the seagrass roots, which leads to Fe(II) oxidation and precipitation as Fe(III) oxyhydroxides. The alkalinity distribution is less heterogeneous. In some parts, alkalinity and Fe(II) are clearly linked, possibly because Fe(II) and alkalinity are generated by microbial dissimilatory iron reduction during organic matter mineralisation. The dotted line represents the sediment-water interface. Adapted from Bennett et al. [29], with permission from Elsevier. A black and white version of this figure will appear in some formats. For the colour version, please refer to the plate section.

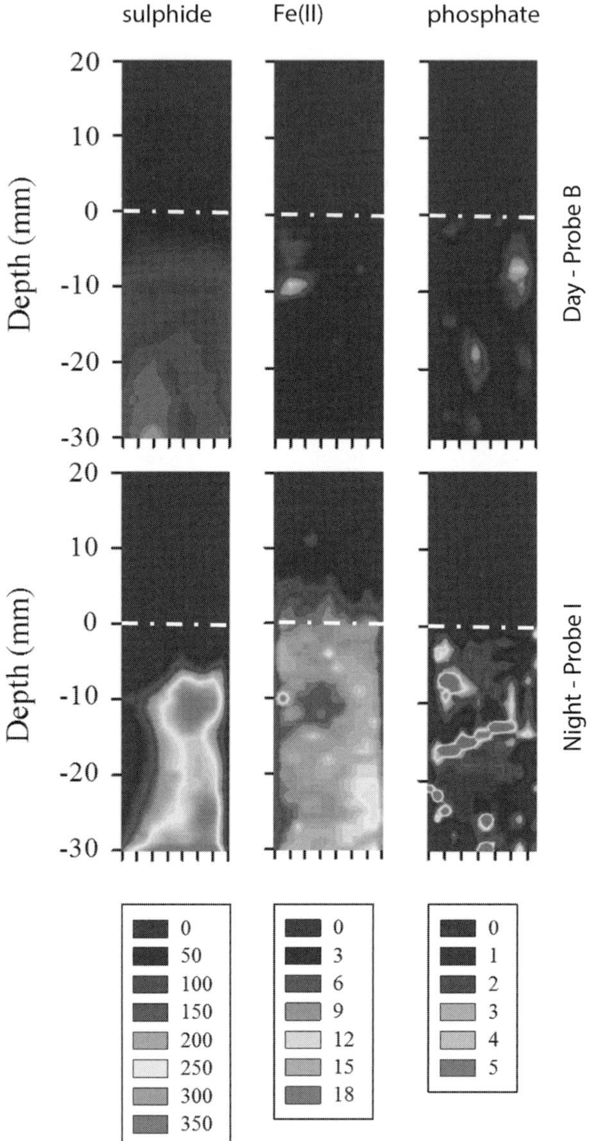

Figure 8.3 Distribution of sulphide, phosphate and Fe(II) in a microbial mat, measured by triple-layer DGT/DET probes. The width of all gels is 14 mm. Distinct differences in solute concentrations and distributions between the day (top) and the night (bottom) deployments are visible, while no clear element co-distribution features can be seen. The dotted line represents the sediment–water interface. Reprinted from Pagès et al. [30], with permission from Elsevier. A black and white version of this figure will appear in some formats. For the colour version, please refer to the plate section.

Therefore, the ability to deconvolve multiple element signals is very restricted. With HSI, the entire visible spectrum, from 380–780 nm, is captured with a resolution of ca. 4 nm using 100 wavelength channels [28, 73]. Therefore, as each pixel within an HSI image forms a detailed, high-resolution spectrum, it becomes possible to unmix analyte signals associated with different colorimetric stains from each other and from the background [28].

Quantification of the target analytes can be performed by comparing the spectra from the background (no colour), and those from the highest concentrations of each respective coloration response, which in the case of the Cesbron et al. [28] study was dissolved reactive phosphate (DRP) and Fe(II). Using linear spectra unmixing software such as ENVI (Environment for visualising images, Exelis Visual Information Solutions, Boulder, CO, USA), analyte concentrations can be quickly calculated. With each pixel representing an area of ca. 60 μm $\times$ 60 μm, the spatial constraints are governed/dictated by gel processing and relaxation effects. This results in the technique's actual spatial resolution being in the hundreds of micrometres rather than tens of micrometres [28].

8.2.3.3 CID DGT Methods

Quantification of DRP using DGT with a binding phase of ZrO [56, 57], which has a high binding capacity, works well and so it is desirable to combine it with the measurement of sulphide using AgI DGT with CID [51]. However, although appearing similar, the AgI and DRP methods are markedly different. With the former the colour change occurs on the surface of particles already incorporated into the gel, while the latter is much more akin to DET imaging where reagents must interact with target analytes to induce a colour response. For a colorimetric DGT method this poses a problem if either the target analyte or resulting colour pigments are mobilised by the reagent staining solution and are thus free to diffuse within the gel, especially if times of 40–45 mins are required for optimal colour development. This problem can be circumvented to some extent by increasing the binding strength of the phosphate-zirconium oxide bonds, by heating the prestained gel in a water bath at 85°C for five days. However, the benefits gained in maintaining the spatial integrity of the images, are somewhat offset by a decrease in the method's sensitivity, as the colorimetric quantification is performed on only a small fraction of the total analyte reservoir.

8.2.4 DGT Measurement by Laser Ablation-ICPMS

8.2.4.1 The Beginnings of DGT Measurement by LA-ICPMS

From its origins in the late 1980s [74, 75] laser ablation (LA) -ICPMS has developed into the dominant bio/geo-imaging technique [76–78]. Modern systems now boast relative ease of use, long-term stability, high sensitivities and a dynamic range encompassing ca. 12 orders of magnitude [78]. The method combination of DGT and ultra-high-resolution ($\leq$ 100 μm) multi-element analysis using LA-ICPMS, was first presented publically in 2000 [79] with the full paper being in print in 2003 [69]. The probes used were multi-layer, comprising of a membrane filter, 0.4 mm diffusive gel and two 0.4 mm binding gels. The first, a gel layer containing AgI, was underlain by a layer containing colloidal resin formed from a suspension of 0.2 μm iminodiacetate (suspended particulate reagent – iminodiacetate; SPR-IDA) beads. Using a raster (*line by line*) scan, a 100-μm diameter beam was drawn over dried gels at 50 μm s^{-1}. Element signals were recorded at 0.66 s intervals corresponding to a physical distance of 0.33 μm, which were then processed as 100 μm averages. The carbon content of the gels was used as the internal standard. Relative standard deviations obtained from probes deployed in spiked solutions of 10 μg L^{-1} for 4 h were 5 per cent for Fe and Mn and 10 per cent for the other metals.

The first two-dimensional analysis study followed shortly in May of 2004 [48], using a near identical set-up as the previous work [69, 79]. Some modification of the ablation cell was made to accommodate a 9 cm, SPR-IDA functionalised DGT binding-layer strip, which had been deployed in situ in the North-East Atlantic Ocean sediment using a benthic lander. The measurements were performed on a 15 × 5 mm section of the dried gel. However, due to instrument drift between standards and samples, the data was not calibrated, instead signal intensities were plotted as *m/z* counts per second [48]. A fully quantitative two-dimensional LA-ICPMS technique for the measurement of polyacrylamide gels, with a novel application of internal standardisation that overcame the technical difficulties associated with the long-term drift was published six months later [49]. Using the same laser base unit as the earlier one- and two-dimensional imaging, the detector sensitivity was improved by coupling to a Thermo X7 ICPMS equipped with a high-performance Pt-tipped sampler/skimmer cone set. Changing all argon gas and sample lines to Teflon-lined Tygon tubing enhanced the transport of ablated material. Another advance was realised by careful ICPMS optimisation using a sample desolvation system, for dry plasma tuning, maximum signal/noise and minimal interference species (CeO^+, Ba^{2+}). The benefits of this approach are evidenced with element sensitivities at ca. 400 Mcounts s^{-1} ppm^{-1}.

8.2.4.2 Advantages of DGT Analysis by LA-ICPMS

Aside from the general analyte pre-concentration and homogenous binding surface characteristics of the DGT gels, that are key attributes for high-resolution chemical imaging, the principal trait that makes the combined techniques of DGT and LA-ICPMS so mutually compatible is the ease and effectiveness that samples can be calibrated by reference to gels with known mass loadings. Matrix matched calibrations are a key requirement of many measurement methods, but they are especially pertinent with LA-ICPMS. Different matrices exhibit characteristic aerosol compositions of nanoparticles and larger aggregates. These unique mixes give rise to varying ionization patterns [80] and the non-stoichiometric transfer of target ions from the source to the detector, which generates a bias between measurement results [81]. The uniformity of the binding plane of the hydrogel matrix of the DGT gels enables a very close match in terms of the chemical and physical properties of the samples and standards, ensuring any measurement discrepancies are kept to a minimum. Another advantage of coupling these two techniques is the low laser energies required for

DGT gel ablation. This enables a faster analysis and improved signal stability in comparison to other denser samples [82]. Furthermore, the wide range of elements accumulated and the ability to easily and precisely control analyte concentrations within the DGT binding layer simplifies plasma tuning and signal optimisation procedures in comparison to other matrix types such as glass or biological tissue [49].

8.2.4.3 Corrections for Signal Variability

Measurement uncertainties associated with LA-ICPMS mainly stem from the signal variability of the ablation analysis [83], which can be due to the laser stability, sample transport and plasma conditions. Internal standardisation of the analyte signals with a reference output can be effective in compensating for any changes in the measurement condition. Signal normalisation based on Ca is the favoured approach for LA-ICPMS analysis of hard tissues such as bone [82], but with the high abundance of carbon in both biological samples and DGT gels the minor isotope of ^{13}C is the most common standardisation method for soft tissues and polyacrylamide gels. However, the atomic mass and/or first ionisation potential of ^{13}C is not an optimal match to that of many trace metals. Furthermore, each LA-ICPMS set-up has a unique ^{13}C signature/background due to impurities in the argon gas, atmosphere and walls of gas and sample lines. The latter can be resolved by monitoring and cleaning of the sample chamber and tubing after analytical runs, in addition to coating internal surfaces with a low friction composite (e.g. Teflon) [49]. However, if not checked, the signal intensity of ^{13}C may drift independently from the analytes in the sample [84].

Some abundance sensitivity effects may also be observed on ^{13}C on instruments with a low mass resolution (W-10 per cent, 0.80 amu), as adjacent ^{14}N mass peaks can overlap if the signal is sufficiently intense [84]. Despite polyacrylamide containing ca. 20 mass per cent of N, significant interferences formed from N liberated from the polyacrylamide gel matrix have been shown to be negligible under normal LA-ICPMS operating conditions [52]. However, N-mixed plasmas for LA-ICPMS analysis that are commonly used to suppress element oxide and hydride formations as well as to provide a general signal enhancement by a factor of 2–3 [85], would require more careful monitoring. Other internal standards that have been validated for DGT LA-ICPMS analysis include In, Ba, Ce and Tb [49], while for ferrihydrite gels, there is an option to use ^{57}Fe for signal normalisation [52–54, 86].

8.2.4.4 Sample Cell Optimisation

The performance of the LA-ICPMS is fundamentally dependent on the characteristics of the ablation cell. Signal stability is restricted by the distance of the sample from the gas inlet/outlet and uniformity of the gas flow within the cell, and this impacts on different elements to varying degrees, with the disruption found to be particularly problematic for ^{66}Zn and $^{56,57}Fe$ [84]. Recent advances in LA technologies, which have included the development of double volume cells [80, 87], have been of enormous benefit to DGT LA-ICPMS analysis. Rapid washout times, improvements in sensitivity and stability, along with the ability to analyse much larger gel samples have culminated in significant analytical gains and improved clarity of DGT images [59, 86, 88, 89].

8.2.4.5 Ablation Optimisation

Another key consideration for DGT analysis is whether the analysis should be carried out using a spot raster [49, 53, 54] or line scan procedure [52, 69]. A detailed comparison of the two techniques is provided in Gao and Lehto [90]. Continuous (line) scans are currently the most popular laser ablation approach. The main advantage being that with line scans the measurement times are reduced. Spot analysis requires a gas purge and laser warm-up cycle before each short ablation, which adds considerable time to the analysis. Other benefits include a simpler optimisation procedure, as careful defocusing of the laser is necessary for spot analysis to avoid burning through the gel layers. Line-scan ablation additionally allows for a higher maximum spatial resolution of ca. 10 μm [90].

8.2.5 Sampling Setup Design

A range of different sampling setups has been used in high-resolution DET and DGT studies, as described below. Figure 8.4 shows standard gel probe designs, as well as a schematic of the simultaneous application of DGT and planar optodes. For schematics of large sediment probes for imaging and a robust steel gel housing, the reader is referred to Robertson et al. [25] and Ullah et al. [20].

8.2.5.1 Sediment Probes

The standard sediment gel holder design consists of a backing plate onto which the gel layer(s) and the protective membrane are laid (Figure 8.4a). A second plastic plate, with a rectangular sampling window, is placed on top of this assembly. Such probes with a window size of 1.8 cm × 15 cm are commercially available (DGT Research Ltd, Lancaster, LA2 0QJ, UK).

This probe design has mainly been used for the determination of one-dimensional DGT and DET profiles. Sampling has either been done in retrieved sediment cores in the laboratory [39], by manual insertion in situ [51], or by benthic landers for deep sediments [31]. In addition, some colorimetric two-dimensional imaging applications have also been realised using standard sediment probes [26, 51, 57, 58, 68, 91]. Similar, custom-made probes with different sampling window sizes (10–40 cm length and 1–3 cm width) have been commonly used [15, 29, 37, 57, 92].

8.2.5.2 Large Format Sediment Probes for Imaging Applications

For simultaneous DET-DGT imaging applications, Robertson et al. [25] developed a sediment gel probe with a gel exposure window size of 170 mm × 80 mm, which was also used in several other studies [26, 71, 93]. With a total size of 275 mm × 110 mm × 14 mm (H × W × D), and especially as these probes have no wedge-shaped bottom to minimise

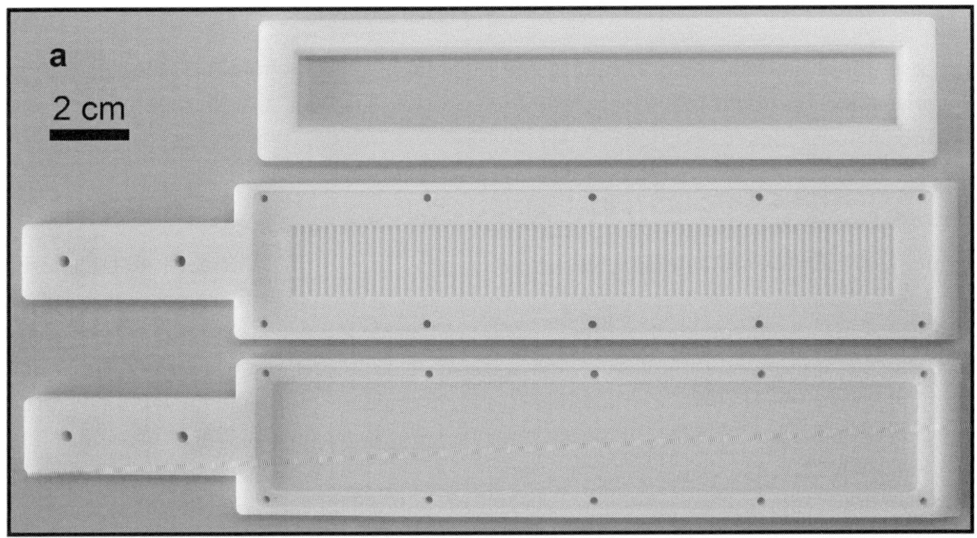

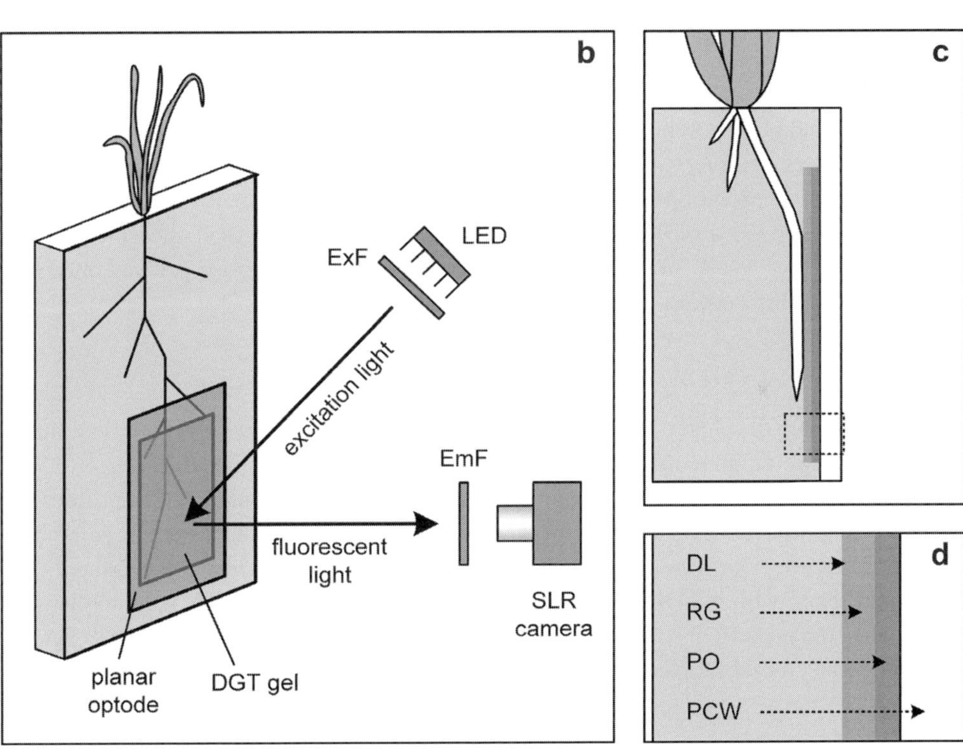

Figure 8.4 Sampling setup designs. (a) Standard size DGT and DET gel probes. Bottom: Backside plate of an open gel probe. The gel(s) and the covering membrane are cut to size, placed onto the backing plate and the assembly is closed with the front part containing the sampling window (top). Middle: Backside plate of a constrained probe. After filling the gel slits, the gel is hydrated, covered with protective membrane and closed with the front part. (b) Schematic of a simultaneous DGT – planar optode application on soil and roots. The DGT gel is overlaid by a planar optode, thereby serving as diffusion layer for the optode sensor. The fluorophore is excited by LED light, the emitted fluorescence light is imaged using a standard single lens reflex camera, ExF: excitation filter. EmF: emission filter. (c) Side view of the DGT-optode assembly inside the growth box. (d) Detail of (c) showing the membrane, gel and optode layers. DL: diffusion layer (membrane), RG: resin gel, PO: planar optode, PCW: plastic container wall. (b–d) reproduced from Santner et al. [11]. A black and white version of this figure will appear in some formats. For the colour version, please refer to the plate section.

the insertion resistance, the disturbance and material displacement in the sediment during sampler insertion might be considerable (See Section 8.3.1 for a discussion of sediment disturbance by sampler insertion).

8.2.5.3 Constrained Probes

While the analytes are immobilised on DGT gels, they diffusive freely in DET gels. After the retrieval of DET gels from sediments this continuous diffusion causes solute equilibration and blurring of the solute distribution inside the gel (see Section 8.3.8). If the analyte can be immobilised quickly after retrieval, e.g. by precipitation of Fe^{2+} or Mn^{2+} through immersion in alkaline solution [12], a high spatial resolution in the DET gel can be preserved, rendering gel slicing into ca. 1-mm thick pieces feasible. As such immobilisation is not possible for many solutes, 'constrained' DET probes have been developed (Figure 8.4a). In these samplers, an array of parallel slits is cut into the backing plate of a gel probe. The first constrained DET probe consisted of three columns of 0.2-mm wide and 2-mm long slits [94]. The commercially available constrained probes have seventy-five slits with a dimension of 1 mm × 1 mm × 18 mm arranged in one column with 1 mm distance from each other. These recesses are filled with agarose gel, which, unlike polyacrylamide gel, does not expand after polymerisation and are covered with a protective membrane. With constrained DET samplers it is therefore possible to exclude diffusional relaxation and achieve a vertical resolution of 2 mm or less. To avoid gel cutting before the elution step, Docekal and Gregusova [45] recently used constrained probes for DGT measurements of U, Fe and Mn profiles. A mixture of Spheron-Oxin® resin and agarose gel was filled into the recesses for this purpose.

8.2.5.4 Gel Introduction System for Coarse Sediments

Ullah et al. [20] developed a DET deployment system for coarse riverbed sediments, consisting of two neighbouring DET gels with a combined dimension of 2.5 cm × 30 cm that are contained in a stainless steel housing. First, a stainless steel plate, which is intended for protecting the DET gels during introduction into the sediment, is pushed or hammered into the riverbed. Afterwards, the steel gel housing is hammered into the sediment next to the steel plate with the DET gels facing towards the plate to avoid damage to the gels. Afterwards, the protective plate is removed and the collapsing gravel sediment falls against the DET probe, thereby ensuring close contact of the sediment and the probe.

8.2.5.5 DGT Application in the Rhizosphere

For DGT-based chemical imaging of solute distributions in the rhizophere, plants are usually grown in rhizotrons (Figure 8.4) [52, 59, 86, 88, 89]. In these flat growth containers that are kept at an inclination of 30°–45°, plant roots develop in the very surface layer of the contained soil. The plastic rhizotron wall adjacent to the rooted soil layer is removed just before DGT deployment. The resin gel, covered by a piece of membrane, is taped onto the rhizotron wall, which is then put back into place for the sampling period. This DGT deployment setup has been used for gels of sizes up to 3 cm × 6 cm.

8.2.5.6 Dual-Layer DGT and DET Imaging

For obtaining simultaneous chemical images of analytes that cannot be analysed with a single DET or DGT gel layer, dual-gel stacks have been used. The first dual resin gel applications were the simultaneous determination of S^{2-} and metals using AgI and either Chelex or SPR-IDA gels in lake and marine sediments [69, 70]. Subsequently, PVC foils [24] or AgI DGT gels [25] for imaging S^{2-} were combined with DET gels for imaging Fe(II). Stacks of two DET gels have also been used to measure Fe(II) and PO_4^{3-} [26] colorimetrically.

8.2.5.7 Combinations with Complementary Planar Imaging Techniques

Similar to dual-layer DGT/DET applications, DGT imaging can be combined with other two-dimensional imaging methods. Planar optodes, which offer a range of analytes that complement DGT measurements, show considerable promise [11]. This methodology is based on sensors containing fluorescence indicators immobilised in an analyte-permeable matrix. After contact of the optode and soil or sediment is established, the analyte diffuses into the optode and interacts with the fluorophore. Ultraviolet light or LEDs are then used to excite the indicator and the emitted fluorescence is captured by a camera. Since DGT, DET and planar optode imaging all rely on analyte diffusion into a thin (gel) layer it is possible to realise simultaneous applications of these methods. DGT gels have been placed onto the medium to be sampled and then overlain by the optode (Figure 8.4b). In this setup, the DGT resin gel acts as diffusive layer for the optode, requiring the resin gel thickness to be minimised to reduce diffusive relaxation on the optode images. Using numerical simulation, Stahl et al. [95] showed the relationship between DGT resin gel thicknesses and diffusive broadening of O_2 images. A resin gel thickness of ≤ 100 μm broadened O_2 images by 0.8–2.0 per cent, demonstrating that this resin gel thickness is adequate for planar optode–DGT sandwich sensors. Ultrathin DGT resin gels are so far only available for transition metals cations [96], oxyanions [55] and transition metal cations and oxyanions [59]. DGT–planar optode sandwich sensors have been used for the simultaneous imaging of trace metals and O_2 (Figure 8.5) [89, 95], trace metals and pH (Figure 8.6) [89] and phosphate, arsenate and O_2 [55].

A powerful method combination was recently introduced by Pagès et al. [93], who characterised the microbial communities in a modern stromatolite using DGT/DET images of the alkalinity, S^{2-}, PO_4^{3-} and Fe(II) in combination with lipid biomarker sampling. The observed heterogeneous solute distribution, linked with the information on microbial community composition in different depth layers of the mat provided a unique, detailed picture of the vertical, but also lateral heterogeneity of the biogeochemical cycling inside the mat.

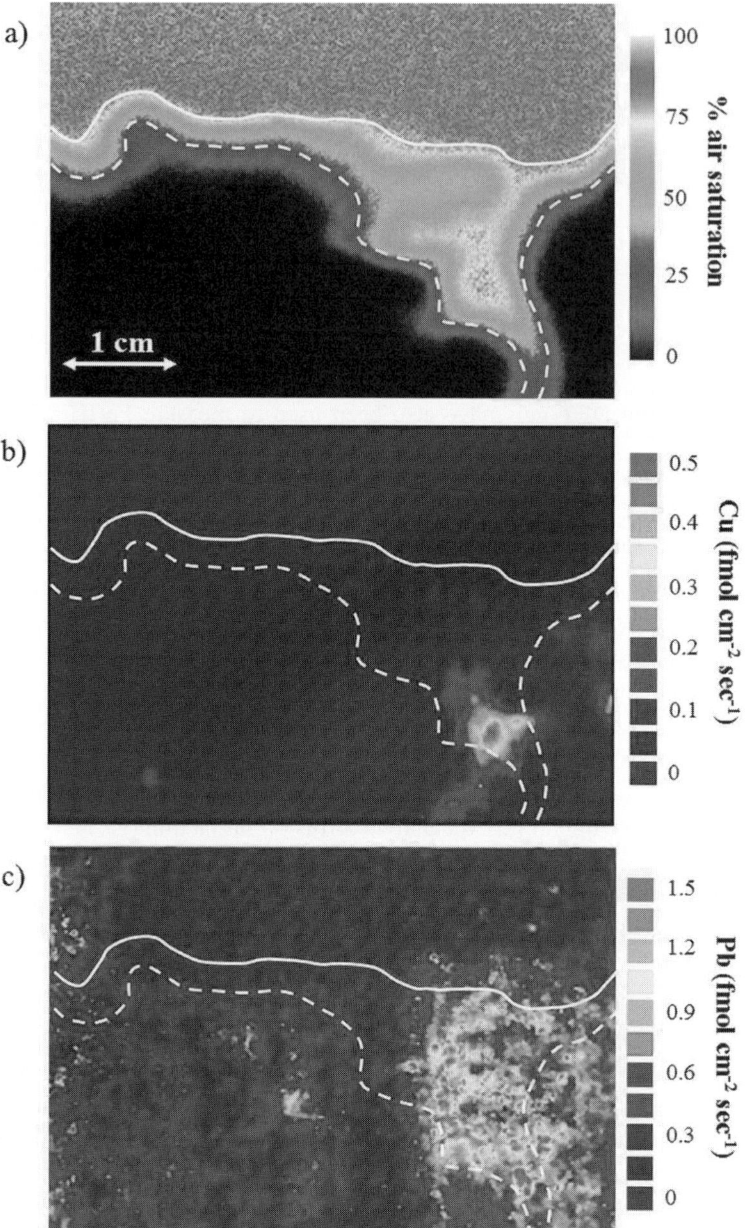

Figure 8.5 Simultaneous high resolution two-dimensional imaging of Cu, Pb and O_2, in polychaete (*Hediste diversicolor*) burrows. Large-scale flume mesocosms were filled with intact harbour sediment containing *H. diversicolor*. The distribution of (a) O_2, (b) Cu and (c) Pb inside and adjacent to an individual burrow was imaged with a simultaneous deployment of planar O_2 optodes and DGT gels. The majority of localised Cu and Pb hotspots were located within the irrigated burrow. However, some Pb release was also observed in the neighbouring sediment. Movement and burrowing activity of the polychaete, in and on top of the sediment, may have affected the metal distributions during the gel deployment period, which might have caused the differences between the near instantaneous concentration measurement of the optode compared with the integrated flux (12 h) of the DGT. Reproduced from Stahl et al. [95], with permission of John Wiley and Sons. A black and white version of this figure will appear in some formats. For the colour version, please refer to the plate section.

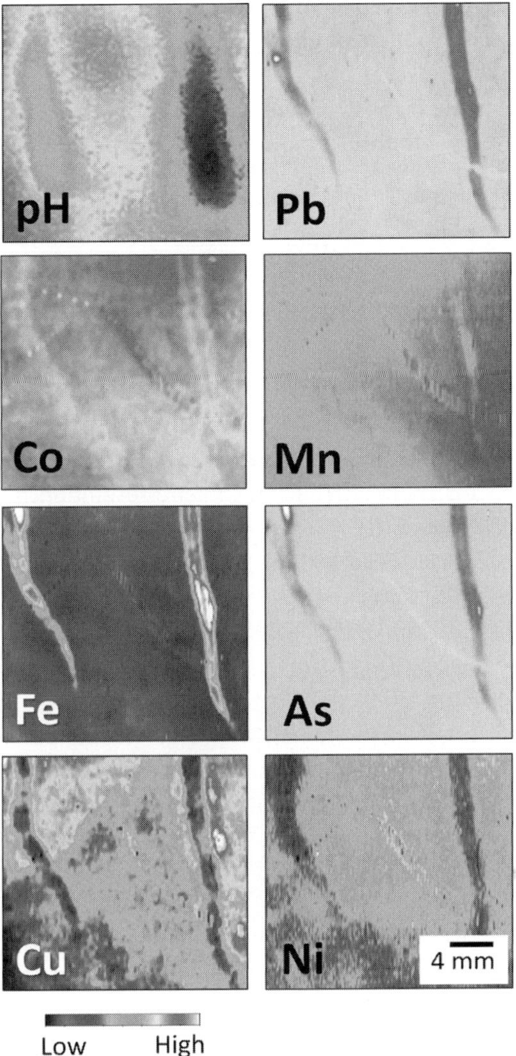

Figure 8.6   Zones of greatly enhanced As, Pb and Fe(II) mobilisation, located immediately adjacent to a set of rice root tips. Rice seedlings were grown for three weeks with deployment of an ultrathin 0.05-mm thick SPR-IDA DGT binding gel, partnered with a planar pH optode, 8 cm below the soil–water interface for 24 h. The blue to white colour scale represents a sequential increase in metal fluxes (f_{DGT}, pg cm^{-2} s^{-1}). Planar optode pH measurements demonstrate a localised acidification around the rice roots of $\sim$0.5 pH units. Adapted from Williams et al. [89]. A black and white version of this figure will appear in some formats. For the colour version, please refer to the plate section.

Another potential method combination is DGT/DET and soil zymography, which allows mapping of the distribution of enzyme activity in soil [97, 98]. In this technique, 4-methylumbelliferyl (MUF) tagged enzyme substrates, like MUF-phosphate or MUF-cellulose, are immobilised on a polyamide membrane. Upon enzymatic cleavage from the substrate, MUF is hydrolysed to 4-methylumbelliferone, which is fluorescent. After retrieving the membrane from the soil, this fluorescence is measured by excitation using UV light and photographing the fluorescent light. A range of MUF-labelled enzyme substrates is available; therefore it should be possible to image a range of enzyme activities in this way. As zymography also relies on diffusive contact of the sampled medium and the sensing layer, combinations of zymography and DGT/DET techniques, similar to planar optode – DGT combinations, should be possible.

8.2.6 Binding Gel Fabrication and Handling

8.2.6.1 Iminodiacetic Acid (Chelex, SPR-IDA) Binding Gels

Standard Chelex 100 binding gel. Chelex 100 (Bio-Rad), which binds transition metal cations via its iminodiacetic acid groups, is the most widely used resin in DGT applications [99]. It has been used for millimetre- to centimetre-resolution measurements of metal profiles in sediments [37, 100, 101].

SPR-IDA binding gel. The bead size of Chelex 100 (ca. 200 μm) is too coarse for sub-millimetre resolution measurements with LA-ICPMS analysis. For these applications suspended particle reagent – iminodiacetate (SPR-IDA; Teledyne CETAC, Omaha, NE, USA), which has a bead size of 0.2 μm, has been used. Warnken et al. [102] developed and characterised a 400-μm thick polyacrylamide gel containing SPR-IDA resin, and subsequently used it in LA-ICPMS studies [49]. This binding gel is made by mixing the resin into the gel solution prior to polymerisation and casting gels between glass plates.

Ultrathin SPR-IDA binding gel. For simultaneous DGT-planar optode imaging studies Lehto et al. [96] developed a polyacrylamide SPR-IDA resin gel that was only 50 μm thick. This gel has a significantly decreased metal capacity compared to the thicker gel and is very easily torn. However, image blurring on the optode sensor film by lateral diffusion is minimised with ultrathin resin gels [11, 95]. As pipetting the viscous gel solution between glass plates with a very small gap is not possible, this gel type is made by pipetting a quantity of gel solution-resin mixtures onto one glass plate that has 50-μm thick spacers near its edges, and placing a second plate on top. Interestingly, no gel expansion was observed in this gel, although the gel solution composition was very similar to pure diffusive gels.

8.2.6.2 Ferrihydrite Binding Gels

Ferrihydrite slurry binding gel. In some studies, standard ferrihydrite slurry gels, made by mixing freshly prepared ferrihydrite slurry into the gel solution prior to polymerisation [39], have been used in gel probes for measuring profiles of the P distribution in sediments [39, 103].

Precipitated ferrihydrite binding gel. Like Chelex 100, slurry ferrihydrite gels are too coarse for reliable sub-mm measurements. Based on the suggestion of Teasdale et al. [51] to directly precipitate mineral binding phases into pre-cast diffusion gels, Stockdale et al. [53] developed a combined AgI-ferrihydrite gel with a directly precipitated ferrihydrite phase, and successfully used it to measure P, V and As alongside sulphide. Similar precipitated ferrihydrite gels without AgI were developed and characterised for the analysis of P [52] and As, Se, V and Sb [104]. These gels are made by equilibrating diffusive gels in Fe(III) solutions and transferring the gels into a pH 6.7 buffer solution, which leads to the formation of a finely distributed ferrihydrite phase on and in the gel. Due to their easy preparation and higher P capacity than slurry gels, these gels have also been used in standard DGT samplers used for deployments in solutions and soils [105].

8.2.6.3 Ti-oxide Binding Gels

Metsorb, a Ti-oxide-based binding material, was used in DGT resin gels for sampling total inorganic arsenic, antimonate, molybdate, phosphate, selenate, vanadate, tungstate and dissolved aluminium species [106–109]. This gel type has been used for investigating As-Fe(II) co-distributions in freshwater, estuarine and marine sediments [91], as well as for studying the speciation of As mobilised in freshwater and estuarine sediments in relation to Fe(II) distributions [40].

8.2.6.4 Zr-oxide Binding Gels

Zr-oxide slurry binding gel. To increase the phosphate capacity, Ding et al. [38] developed a Zr-oxide bisacrylamide resin gel, which has a ∼50-fold increased P capacity compared to the original ferrihydrite slurry gel [39]. In a subsequent work, ultrasound treatment of the Zr-oxide particles was used to decrease the particle size and enable sub-mm resolution measurements [56], which were done by slicing the gel to 0.45 mm × 0.45 mm slices with a self-made cutting device. A further modification of this gel was the incorporation of AgI particles for the simultaneous analysis of sulphide and phosphate in sediments [58].

Precipitated Zr-oxide binding gel. Recently, a directly precipitated Zr-oxide resin gel was presented by Guan et al. [55], which was characterised for the measurement of P, V, As, Se, Mo and Sb, both after elution and directly by LA-ICPMS. Compared to gels containing Zr-oxide particles [38, 110], this gel has a somewhat smaller P and As capacity. However, the P and As capacity per unit of Zr-oxide is higher in the precipitated gel, indicating a larger surface area of the directly precipitated Zr-oxide phase compared to the Zr-oxide particles.

8.2.6.5 Combined ZrOH and SPR-IDA Binding Gel

In an effort to simultaneously map metal cations and oxyanions using DGT, Kreuzeder et al. [59] developed an ultrathin (ca. 100 µm) resin gel incorporating SPR-IDA and Zr-hydroxide. The common acrylamide hydrogel matrix polymerised rapidly after contacting Zr-hydroxide, leading to a very inhomogeneous distribution of binding material in the gel, making them unsuitable for sub-millimetre imaging. Therefore Hydromed D4 (Advan-Source biomaterials, Wilmington, MA, USA), a urethane-based hydrogel, was adopted as the hydrogel matrix for this binding material combination. Hydromed D4 does not require a polymerisation reaction, but is formed upon solvent (ethanol) evaporation. An additional advantage of this material is, that the resulting ultrathin gels are highly stable and tear-proof, therefore handling between supporting membrane layers is not required. This gel was used to image P, Mn, Zn, Co, As, Cd and Cu in the vicinity of a maize (*Zea mays* L.) root [59].

8.2.6.6 AgI Binding Gels

AgI powder binding gel. For the sub-mm resolution measurement of sulphide in sediments Teasdale et al. [51] mixed AgI powder into a polyacrylamide gel solution prior to polymerisation. The pale-white AgI phase is converted to black Ag_2S in the presence of sulphide, which can be quantified by CID using simple office flat-bed scanners.

Precipitated AgI binding gel. Based on this first AgI gel, Devries and Wang [50] presented a method to directly precipitate AgI into a plain polyacrylamide gel. Unlike other direct precipitation methods, $AgNO_3$ was mixed into the bisacrylamide gel solution at a concentration of ca. 0.1 mol L^{-1} prior to polymerisation. After setting, this gel was immersed in a 0.2 mol L^{-1} KI solution to precipitate AgI. As this method was not always reliable, a similar AgI precipitation method, which utilizes a gel solution based on the agarose-derived DGT cross-linker, was developed and is now the method used most often [53]. AgI gels have to be kept in the dark to avoid photoreduction of AgI to black Ag.

8.2.6.7 3-Mercaptopropyl-Silica Binding Gels

Binding gels incorporating 3-mercaptopropyl-functionalised silica have been developed and tested for the selective sampling of As(III) and methylmercury (MeHg) in the presence of other As and Hg species [111, 112]. With the aid of this binding gel, the speciation of As(III) and MeHg was investigated in freshwater, estuarine and marine sediment profiles [40, 43, 44].

8.2.6.8 Spheron-Thiol and Spheron-Oxin Binding Gels

Diviš et al. [42] developed a binding gel for the measurement of Hg containing Spheron-Thiol, which immobilizes Hg via its thiol groups. In exemplary measurements of Hg in sediment profiles, a much larger Hg uptake was observed for the gel containing Spheron-Thiol rather than Chelex-100, and attributed to the higher affinity of Spheron-Thiol for Hg. However, although an increased total Hg uptake of Spheron-Thiol DGT units is possible as this resin binds both Hg^{2+} and methylmercury (MeHg), while Chelex-100 only binds Hg^{2+}, a fivefold larger Hg uptake by Spheron-Oxin DGT samplers seems very large, given that the percentage of MeHg in total sediment Hg scarcely exceeds a few per cent [43, 113]. A detailed investigation of the performance of Spheron-Thiol binding gels, including its selectivity for different Hg species, should be carried out before further applying this binding gel.

Gregusova and Docekal [45, 114] presented a gel incorporating Spheron-Oxin, which binds U by its 8-hydroxyquinoline groups, as an alternative to binding U using Chelex-100 gels. Spheron-Oxin was slightly more efficient in binding U at high CO_3^{2-} concentrations in the sampled solution. The increase in carbonate concentration led, however, to a decrease in the U uptake for both, Spheron-Oxin and Chelex-100 containing gels. There was only a small range of carbonate concentrations where Spheron-Oxin was more efficient in binding U, while the c_{DGT}/c^{soln} ratio was still >0.9. At still higher carbonate concentrations both Spheron-Oxin and Chelex-100 ceased to take up U quantitatively; therefore it is doubtful whether the use of Spheron-Oxin rather Chelex-100 provides a worthwhile gain in performance.

8.3 Method Characteristics and Limitations

8.3.1 Material Displacement During Sampler Insertion

The displacement of sediment material by pushing gel probes into the sediment, which might lead to artefacts in the solute distribution, has been considered negligible in the past [41, 48]. Mn and Co DGT profiles were obtained after insertion of probes into a constructed sediment (homogenised sand overlain by ca. 2 cm layer of contaminated sediment) [48]. The penetration of elevated Mn and Co into the sand layer below the sediment was considered to be consistent with diffusion into the sand medium. In another study, Zhang et al. [41] also reported the absence of indications for material displacement. As a precautionary measure, the lower 5 mm of the DGT resin gels were not analysed in these studies, as it was assumed that displaced material would by retained by the lower lip of the sampling window and thus impact the adjacent gel region. Using fluorescent sand particles and a sediment corer, Santner et al. [11] however reported considerable displacement of particles down a sediment profile. To show this effect, the sediment was covered with a ca. 1 mm thick layer of fluorescent particles. Subsequently, a bevelled corer was pushed into the sediment and removed with the contained sediment. The distribution of the fluorescent sand down the sediment profile was then imaged under UV light excitation. The originally thin particle layer was widened to ca. 15 mm vertical thickness. In this region the density of displaced material from the sediment surface was high. In addition to the relatively homogeneous distribution of displaced particles in this layer, more than 350 individual fluorescent particles were carried further down the whole insertion depth of ca. 120 mm. Although the geometries of the corer and gel probes differ considerably, this experiment demonstrated that particle displacement during sampler insertion may be a source of artefacts when measuring porewater solute distributions. This phenomenon should thus be more thoroughly tested for DGT probes, with the aim of avoiding artefacts wherever possible, and acknowledging potential problems in data interpretation where displacements of sediments may have occurred.

8.3.2 Ensuring Complete Contact of Gel Probe and Sediment

Gel probe insertion into compact sediment may often require considerable force, which could lead to cavitation and downward flow of overlying water, while soft sediments will flow into small gaps, should any be created [41]. Modified gel probes have been tested to reduce the risk of cavitation and ensure close contact between probes and the sediment [92, 115]. Bevelled probe backs, which create forward pressure that pushes the sampling window and the sediment closely together, have been found to lead to slightly higher Fe and more scattered Fe and Mn concentration profiles compare to probes that were not wedge-shaped, indicating that oxic overlying water might be introduced into the sediment and lead to precipitation of redox-sensitive elements if no wedge is used [115]. However, these effects were small for Fe and Mn, and were absent for U, Re and Mo.

8.3.3 Locating the Sediment–Water Interface

The precise location of the sediment-water interface (SWI) is often hard to determine after retrieving gel probes from the sediment. Although some sediments leave a visible stain on the protective membrane, which can be used to infer the SWI position [37], many sediments do not leave any visible trace. To facilitate the precise localisation of the SWI, Ding et al. [92] attached a sheet of sponge to the upper part of the back of their probe design. After removing the sampler from the sediment, the sponge was clearly stained by sediment particles, making it easy to locate the SWI.

8.3.4 Preventing Damage During Sampler Insertion

In coarse, stony sediments, insertion of gel probes by pushing them into the sediment may lead to tears in the membrane and gel layer, or, due to the physical resistance, may not by possible at all. Ullah et al. [20] presented a steel gel probe design that effectively protects the membrane and gel layer. See Section 8.2.5.4 for details.

8.3.5 Capacity Artefacts in Long-term Deployments

The ideal DGT resin gel has a very high binding capacity combined with a high affinity for the sampled solutes. If the binding sites of the resin gel approach saturation, however, solutes with low affinities may be displaced from the resin by solutes with higher affinity, or by solutes with similar affinities that are present at very high concentrations. Tankéré-Muller et al. [116] tested Chelex-100 resin gels for capacity and displacement effects in seawater and marine sediments. In solutions with Ca and Mg concentrations resembling those of seawater, Mn uptake by Chelex-100 DGT samplers was only ca. 25 per cent of the Mn uptake from solutions without Ca and Mg. This effect was probably due to relatively slow Mn binding associated with the resin gel becoming saturated with Ca and Mg within ca. 2 h, and its affinity for Mn being only slightly higher than that for Ca and Mg. Co uptake, for which Chelex-100 has a much higher affinity, was only slightly reduced in the presence of high Ca and Mg. Furthermore, displacement of Mn from resin gel discs preloaded with various metals, which were deployed in solutions with high Fe(II) concentrations, was observed. Mn and Fe sediment profiles were measured in a mesocosm with deployment times of 3, 6 and 21 h. For the shorter deployment periods, increasing Mn and Fe porewater concentrations were measured down to a depth of 60 mm below the SWI. In the 21-h deployment, however, the Mn concentration profile showed an increase down to ca. 10 mm depth and decreased below this point almost to the concentration of the overlying water, while the Fe profile was similar to that observed for the shorter periods. This experiment showed that Mn can be displaced from Chelex-100 gels at the very high porewater Fe(II) concentrations found in highly reducing sediments. Such artefacts in Mn sediment profiles can be avoided by using short deployment times. However, lower Mn concentrations associated with high Ca and Mg in marine systems cannot be avoided.

8.3.6 Potential Artefacts in Imaging Studies in Unsaturated Soil

In imaging studies in soils that are not naturally water-saturated, saturation during DGT sampling has to be avoided to prevent artificial anaerobia and reducing conditions. Working with multiple layers of gels and membranes, which ideally contact each other and the adjacent soil via a continuous water film, is however challenging and may lead to artefacts if not performed carefully. In the first DGT imaging study on plant roots [86], the plant growth container was fully water saturated just before the application of the DGT gels and was maintained saturated throughout the sampling period. In this setup continuous soil–gel contact throughout the entire gel surface area was accomplished, and air bubbles between individual layers could be prevented by careful positioning. However, for the very redox-sensitive Mn, reductive dissolution following water saturation in an imaging setup, was observed (data not shown). Even if direct alterations of solute distributions are not expected, retaining natural growth conditions is important to avoid indirect effects like root stress. Therefore, more recent DGT imaging studies in soils worked in unsaturated conditions [88, 117]. In this situation, air bubbles are easily entrapped between membrane, resin gel and soil, sometimes leading to only a small fraction of soil and gel being in close contact (Figure 8.7). Reducing the number of stacked thin-films (membrane, resin gels, planar optode sensors) can help to reduce the number of air bubbles and thereby the number of gel applications that have to be discarded. It should also be noted that even small temperature changes can lead to the outgassing of air from the water films between the thin-films, so keeping the temperature constant also helps to minimise the number of air bubbles.

8.3.7 Image Generation

Reconstruction of solute porewater images, obtained from DGT LA-ICPMS analysis becomes more complex as the number of individual measurements, the size of the gels, and the number of both target and monitored analytes (i.e. interference ions or signal stability checks) increases. A typical gel analysis of 4 cm^2 can generate around six million data points [89], which all require internal standard correction. Complete on-line, real-time data processing is possible with some LA-ICPMS software packages, such as Glitter (GEMOC,

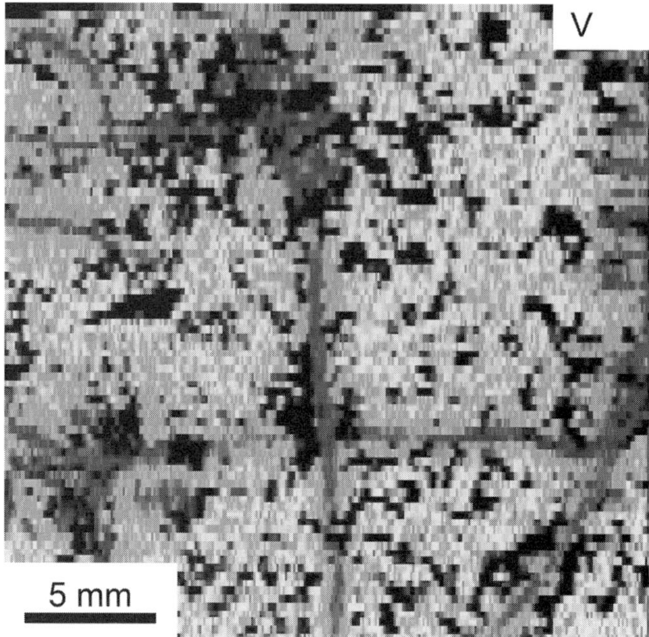

Figure 8.7   Gel–soil contact artefacts in DGT imaging applications in unsaturated soil. Vanadium distribution around lupin roots (blue, elongated features). The black areas, where no V was taken up by the DGT gel, are caused by numerous small entrapped air bubbles. Vanadium was found to be a very good indicator of soil-gel contact discontinuities in this study, as the soil background V concentration yielded a good signal in LA-ICPMS analysis, while the blank V concentration on the resin gel was negligible. Unpublished data of Andreas Kreuzeder, Jakob Santner, Vanessa Scharsching and Walter W. Wenzel. A black and white version of this figure will appear in some formats. For the colour version, please refer to the plate section.

Macquarie University, Sydney). However, to date, all the published two-dimensional LA-DGT studies adopt a post-analysis visualisation approach. This means the success of the sample analysis can only be immediately gauged by on-line one-dimensional profiles of individual line scans of the gel and not in two dimensions, which would be the ideal view.

Other considerations centre on the processing of obtained data, which requires the careful selection/mining of regions of interest (ROI) as complete measurement files also contain signal information from sample washout and laser warm-up cycles, and encompass the time taken for gels to be moved into position for analysis; a sequence which is repeated with each spot or line ablation. Although there is at present no dedicated software to facilitate this, most regular DGT-LA-ICPMS users have developed their own spreadsheets automated with macros, to complete what is otherwise a time-intensive task. As a general rule, datasets are formatted in two basic schemes.

The first are three-dimensional graphs, where individual ion intensities (z) are represented on a two-dimensional format, and allied with their corresponding (x,y) axis coordinates in a three-column layout. There are a wide number of scientific graphing and statistical analysis platforms capable of processing this data format. SigmaPlot (Systat Software) has found favour with many users [96, 118], while other examples include Matlab routines (Mathworks Natick, MA), Origin (Originlab Corporations, Northampton), IMAGENA (C++ software), Pmod (Zürich, Switzerland), ISIDA (Python) and the SMAK-package. One of the advantages of columnating data for three-dimensional graphs is that it is possible to display the complex spatial data in partnered three-dimensional x,y,z graphs and two-dimensional plots, as elegantly demonstrated in the sediment imaging study of Gao et al. [118] (Figure 8.8).

The second scheme involves simply creating data-driven pictures from a grid-matrix of ion intensities, with each cell corresponding to an image pixel. Using commonly available digital photography software such as the free application ImageJ (National Institute of Health, Bethesda, MD, USA; http://imagej.nih.gov/ij/), imported comma-separated value files (.csv) are transformed into 32-bit images for subsequent colour rendering, scaling and information visualisation [11]. Especially, when dealing with smaller area images and when specific ROI are interrogated, care must be taken to avoid pixel interpolation during image resizing, as this may unintentionally smooth the data. This is often unavoidable in contour plots, but in programmes like Photoshop or ImageJ there is the option not to interpolate. As a guide, images with a pixel size of ca. 100 μm × 150 μm will, and should look pixelated when not interpolated.

8.3.8 Spatial Resolution Constraints

Detailed interpretation of DET and DGT measurements has to consider the resolution constraints that are imposed by the techniques, which are both affected by lateral solute diffusion.

In DET, two effects of lateral diffusion can be distinguished: (1) Diffusive relaxation of the chemical imprint of the sediment solute distribution in the DET gel during sediment–gel contact. This effect was demonstrated in laboratory experiments by Davison et al. [22], who showed that the solute distribution at the gel side that is not in direct contact with the sampled sediment is widened compared to the opposite gel side. As this effect depends on the thickness of the DET gel, thinner gels should be chosen to maximise accuracy of the solute distribution within the sediment. (2) When DET gels are retrieved from the sediment and the diffusional contact is lost, the maxima in the solutes distribution inside the gel begin to relax, blurring the image. Quick processing of DET gels is therefore essential for retaining high spatial resolution. Precipitation of Fe and Mn should be done within 1–2 s for a resolution of ca. 250 μm, while 20 s will suffice for a 1 mm resolution [22]. Given this need for very short processing periods, the spatial resolution limit of colorimetric DET methods, which is considered to be in the range of 1 mm [25], could actually be larger. Thorough investigation of the diffusivity of the coloured complex and the staining reagent in the DET gel would help to elucidate the resolution limits of colorimetric DET applications.

Lateral diffusion has also an impact on DGT application, however only while the solutes diffuse through the diffusion layer. As soon as immobilised by the binding material, the

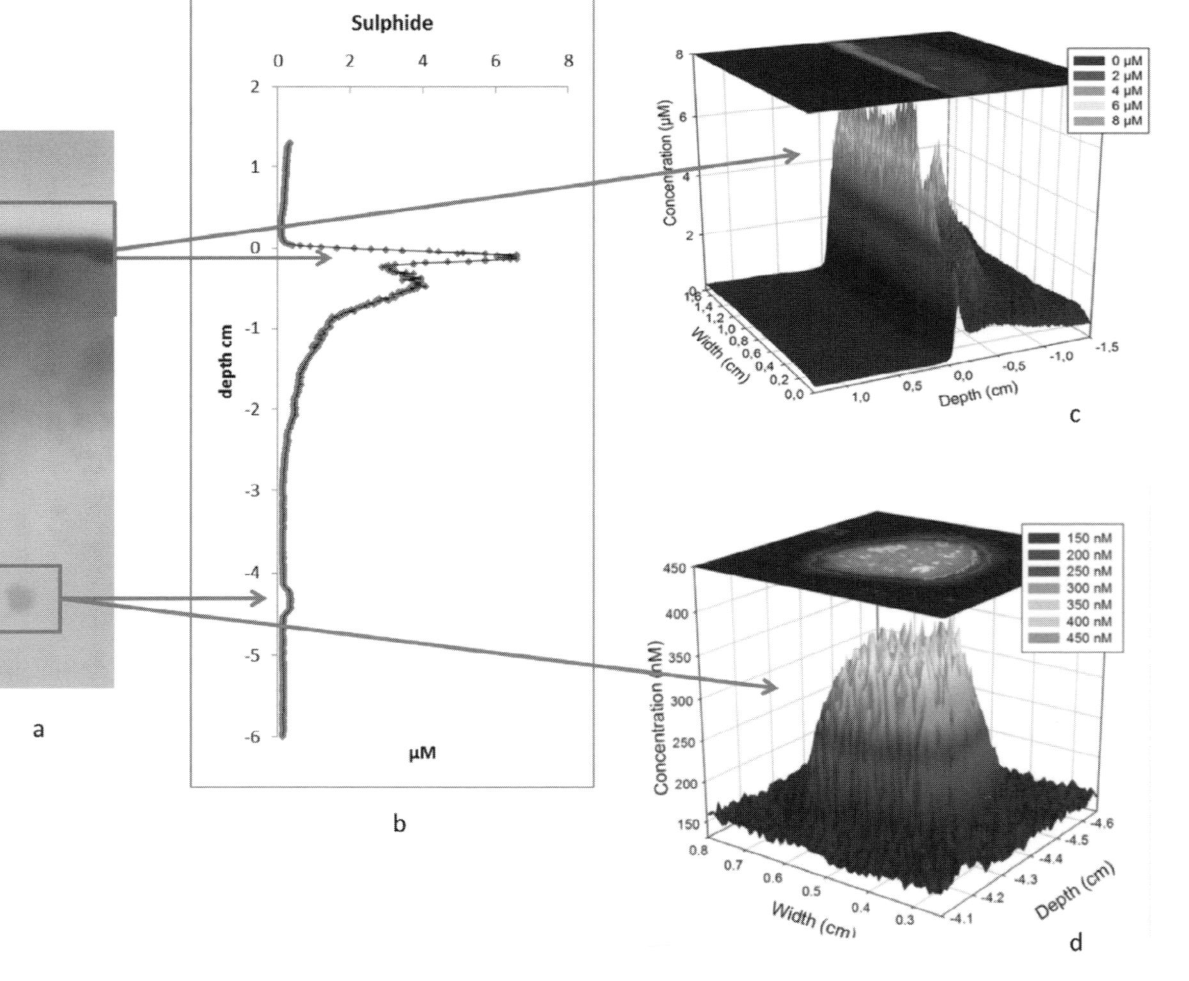

Figure 8.8   Two-dimensional density plot superimposed on a three-dimensional graphic. Combination of visual formats to display sulphide release in estuarine sediment: (a) greyscaled scanned DGT-CID image of silver iodide (AgI) gel. (b) 1-DGT representation of sulphide release. (c, d) Combined two- and three-dimensional plots of the sediment–water interface and microniche zones. Reprinted from Gao et al. [118], with permission from Elsevier. A black and white version of this figure will appear in some formats. For the colour version, please refer to the plate section.

solute distribution is retained. Modelling studies on the reproduction geometry of porewater solute concentration peaks on DGT binding gels yielded contrasting results. Harper et al. [62] reported a minimum porewater peak width of 1.2 mm for good reproduction (DGT peak width divided by porewater peak width >0.8) on DGT images, whereas simulated DGT peak widths were similar to porewater peak widths in the study of Sochaczewski et al. [63]. Because the latter model was based on three-dimensional diffusion and considered different mechanisms that may generate concentration maxima, the results of Sochaczewski et al. [63] should reflect the real situation in porewaters and during sampling more closely. However, experimentalists now tend to use diffusion layer thicknesses of 10–100 μm [59, 86, 88, 89], which should reduce lateral diffusion to a degree where it becomes negligible.

8.3.9 The Interaction of DGT with the Sampled Medium

The capture of micro-scale geochemical processes obtained by high-resolution DET/DGT imaging has been the catalyst to breakthroughs in our understanding of many solid and solution phase phenomena in sediments and soils [12, 48, 54, 65, 68, 69, 71, 88, 89, 118]. Visualising the dynamics of solute exchange has been a key in recognising that flux maxima and minima can occur at discrete sites, all within small spatial scales [119]. Detailed interpretation of the DGT measurement must consider the local dynamic equilibrium of solute between solid phase and solution, and the diffusional transport of solute through the porewaters and DGT device [63], as discussed in detail in Chapter 7. A major first step in modelling the processes captured by DGT was made by Harper et al. [120, 121], who created the DIFS (DGT Induced Fluxes in Sediment and Soils) model, which simulates solute uptake by a DGT device using a 1D diffusion geometry. Subsequently, a two-dimensional implementation of DIFS was published, which additionally accounts for lateral diffusion during the solute uptake by the DGT device [119]. A three-dimensional variant of DIFS [122] was used to assess the interaction of microniches with the DGT setup and to understand the reproduction of microniches on DGT chemical images [63].

These simple, idealised systems serve as proxies for processes that may occur in sediment [119], such as solute transport [120], mineral formation [123] and bioavailability [124]. Modelling showed that the pronounced small-scale maxima of metals captured by DGT can represent a local elevation in the concentration in the porewater. The width (one-dimensional) or diameter (two-dimensional) of these concentration maxima are likely to be only slightly less than those of the maxima that were actually present in the porewater at the same location, without the DGT present. Furthermore, the concentrations measured by DGT at the maxima (corrected for the concentration background) will only be slightly lower than the true concentrations at the same locations (within 62–87 per cent) [63]. However, the background concentration measured by DGT is likely to be more markedly reduced compared with the unperturbed concentration, giving the feature measured by DGT a somewhat sharper appearance than the true feature in the sediment. Perhaps most importantly for the interpretation of porewater solutes by high-resolution DGT imaging, was that Sochaczewski et al. [63] were able to demonstrate that although the soil/sediment surface concentration will be slightly modified by the DGT sink, the device cannot induce a localised maximum or minimum. Rather, localised maxima in the flux reflect localised maxima in the concentration, indicating a localised process of mobilisation [63].

8.3.10 General DGT and DET-related Limitations

Besides the discussed limitations that are specific to high-resolution DGT and DET setups, the formation of biofilms on the protective membrane of DGT samplers has been found to alter solute uptake [125]. Moreover, analyte binding to the hydrogel matrix has been found to be a potential artefact in both, DGT and DET measurements [126–128]. As these phenomena are not specific to high-resolution setups, the reader is referred to Chapters 3 and 5 of this book.

8.4 Applications

The emphasis of this chapter has been on the methodological development of high-resolution imaging rather than the scientific breakthroughs *per se*. In this final section, a selection of case studies are presented to highlight how these technical improvements have resulted in advances in our understanding of key geochemical processes.

8.4.1 Sediment

The highly localised distribution and intertwined mobilisation of metals and sulphide in sediment, within microniches, was discovered using DGT and has now become a widely accepted part of geochemical theory. Microniches are characterised by small-scale localisation where geochemical behaviour differs significantly from the average for that depth. Local geochemistry is very dependent on location and also the size and reactivity of any aggregate of organic material. At greater depths where nitrate is absent, sulphide would be expected to be produced in small microniches, while the existence of sulphidic microniches in otherwise non-sulphidic zones of the sediment may have implications for the formation of metal sulphides [46]. This improved understanding garnered by chemical imaging has changed approaches to modelling element behaviour, directing attention away from over simplistic one-dimensional reaction transport models to the more comprehensive three-dimensional reaction transport model schemes, which accommodate microniches [28, 30, 48, 53, 65, 68, 69, 122]. However, detailed diagenetic predictions still require the characterisation of a wide range of solutes and solid phases.

Variations, in two dimensions, in the S isotopic ratio of porewater sulphide is a powerful new parameter that can be measured successfully at a spatial resolution of ca. 100 μm by combining DGT sampling with CID and laser ablation with high-resolution multicollector inductively coupled plasma mass spectrometry (LA-HR-MC-ICPMS) [46]. Measurements of sulphur isotopes were performed on DGT AgI binding layers after deployment in sediment from a eutrophic lake contained in laboratory mesocosms. Bacterial sulphate reduction and sulphide formation in this sediment are predominantly localised to discrete, mm-sized microniches, where oxidation of labile organic matter drives the reduction of sulphate. The results emphasise the importance of microniches as localised, highly dynamic reaction sites in sediment, where significant shifts in $\delta\,^{34}S$ of up to $+20‰$ relative to the local background were measured across microniches. This first observation of small-scale Rayleigh fractionation in microniches has potential implications for the interpretation of sedimentary isotope data more generally, such as for C, N and Fe, if the phenomenon proves to be common (Figure 8.9) [46].

8.4.2 Rhizosphere

In wetland adapted plants, such as rice, it is typically root apices, sites of rapid entry for water/nutrients where radial oxygen losses (ROLs) are highest. Nutrient/toxic metal uptake largely occurs through oxidised zones and pH microgradients. However, the processes controlling the acquisition of trace elements have been difficult to explore experimentally because of a lack of techniques for simultaneously measuring labile trace elements and oxygen/pH. When a sandwich sensor based on DGT and planar optodes was first deployed in situ on rice roots a new geochemical niche was discovered. Located immediately adjacent to the root tips, a zone of greatly enhanced As, Pb and Fe(II) mobilisation (Figure 8.6) was observed, with this phenomenon also occurring co-incidentally with an oxygen enrichment and low pH. Fe(II) mobilisation was congruent to that of the peripheral edge of the aerobic root zone, demonstrating that the Fe(II) mobilisation maximum only developed in a narrow oxygen range as the oxidation front penetrated the reducing soil. The Fe flux to the DGT resin at the root apexes was threefold higher than the anaerobic bulk soil and twenty-seven times greater than the aerobic rooting zone. These results provided new evidence for the importance of coupled diffusion and oxidation of Fe in modulating trace metal solubilisation, dispersion and plant uptake [89].

A metal-accumulating willow was grown under greenhouse conditions on a Zn/Cd polluted soil to investigate the effects of sulphur (S^0) application on metal solubility and plant uptake. Porewater was sampled throughout the sixty-one days of experimental observations, while DGT-measured metal flux and oxygen were chemically mapped at selected times. Sulphur oxidation resulted in soil acidification and related mobilisation of Mn, Zn and Cd (Figure 8.10). This phenomenon was more pronounced in the rhizosphere compared with the bulk soil and highlights sustained microbial S^0 oxidation and associated metal mobilisation close to root surfaces. The localised depletion of oxygen along single roots upon S^0 addition indicated the contribution of reductive Mn (oxy)hydroxide dissolution with Mn eventually becoming a terminal electron acceptor after depletion of oxygen and nitrate. The S^0 treatments increased the foliar metal concentrations markedly, but had no significant impact on biomass production. The key point is that lower metal solubilisation in the bulk soil should translate into reduced leaching, offering opportunities for using S^0 as an environmentally friendly amendment for enhancing the phytoextraction of metal polluted soils [88].

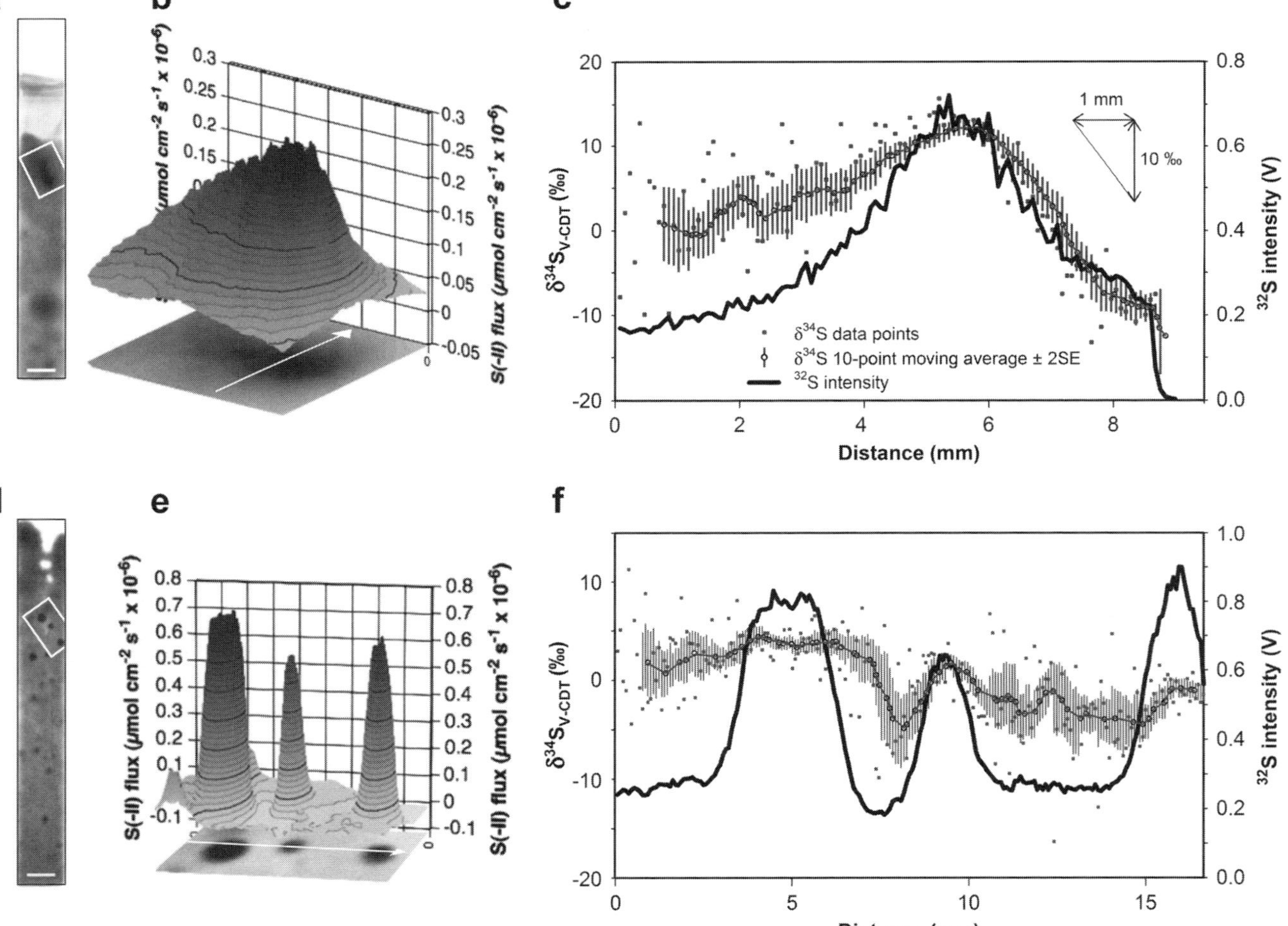

Figure 8.9   Dissolved sulphide flux and sulphur isotopic composition across a subsurface redox-transition zone. (a, d) Scanned DGT-CID greyscale image of AgI gel with sub-surface zone of maximum bacterial sulphate reduction (BSR) and sulphidic microniches (dark spots). White scale bar is 10 mm. (b, e) three- and two-dimensional plots of the subsurface zone of maximum BSR obtained by DGT-CID, with a maximum horizontal S(-II) flux of 0.6×10^{-6} μmol cm^{-2} s^{-1}. (c, f) Vertical LA-HR-MC-ICPMS profiles across the zones indicated in (b, e) by white arrows showing ^{32}S intensity and δ^{34}S in Ag$_2$S precipitated in AgI gel. Gels deployed in mesocosm with sediment from

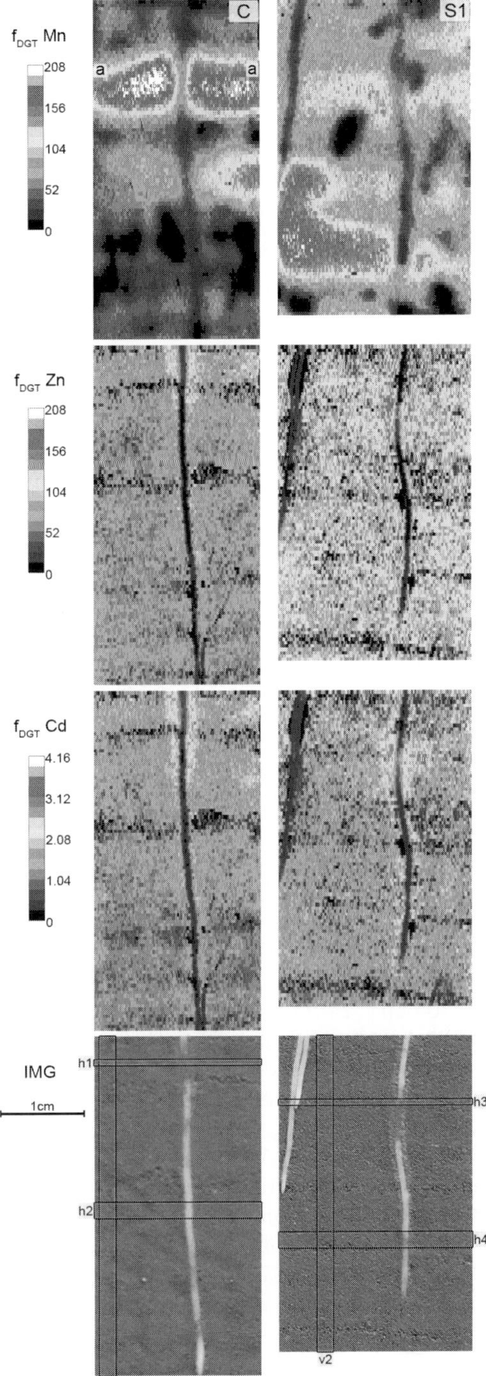

Figure 8.10 Distribution of Mn, Cd and Zn around a *Salix x smithiana* root. C and S1 indicate soil treatments without (C) and with (S1) elemental sulphur addition. For this image, the DGT resin gel was applied for 20 h on soil saturated to 80 per cent maximum water holding capacity. At the start of the experiment, the soil was filled into the growth container and carefully compacted layerwise. DGT-labile Mn is clearly elevated in two of these soil layers, indicating that the bulk soil density is higher than in the layers above and below. As a result, the water saturation of the soil pores is increased, leading to anaerobia and a localised reductive dissolution of Mn. Zinc, which is not redox-sensitive, does not show elevated concentrations in these more strongly compacted layers. Reproduced from Hoefer et al. [88]. A black and white version of this figure will appear in some formats. For the colour version, please refer to the plate section.

Acknowledgments

We thank Dr. William W. Bennett, Dr. Anaïs Pagès, Andreas Kreuzeder and Christoph Hoefer for providing high-resolution figures for this chapter. The contribution of Jakob Santner was funded by the Austrian Science Fund (FWF): P23798-B16, Paul N. Williams received financial support from the Newton Fund/Royal Society (R1504GFS).

References

1. R. C. Aller, 8.11 – Sedimentary diagenesis, depositional environments, and benthic fluxes, In *Treatise on geochemistry*, ed. H. D. Holland and K. K. Turekian (Oxford: Elsevier, 2014), pp. 293–334.
2. R. A. Berner, *Early diagenesis: A theoretical approach* (Princeton: Princeton University Press, 1980).
3. H.-P. Blume, G. W. Brümmer, U. Schwertmann et al., *Lehrbuch der Bodenkunde.* 15th edn. (Heidelberg, Berlin: Spektrum Akademischer Verlag, 2002).
4. M. Kühl and N. P. Revsbech, Biogeochemical microsensors for boundary layer studies, In *The benthic boundary layer*, ed. B. P. Boudreau and B. B. Jørgensen (Oxford: Oxford University Press, 2001), pp. 180–210.
5. M. H. Weisenseel, A. Dorn and L. F. Jaffe, Natural H+ currents traverse growing roots and root hairs of barley (Hordeum vulgare L.), *Plant Physiol.* 64 (1979), 512–518.
6. M. Taillefert, G. W. Luther III and D. B. Nuzzio, The application of electrochemical tools for in situ measurements in aquatic systems, *Electroanalysis.* 12 (2000), 401–412.
7. W. W. Wenzel, G. Wieshammer, W. J. Fitz and M. Puschenreiter, Novel rhizobox design to assess rhizosphere characteristics at high spatial resolution, *Plant Soil.* 237 (2001), 37–45.
8. P. R. Teasdale, G. E. Batley, S. C. Apte and I. T. Webster, Pore water sampling with sediment peepers, *TrAC-Trend Anal Chem.* 14 (1995), 250–256.
9. B. Jaillard, L. Ruiz and J. C. Arvieu, pH mapping in transparent gel using color indicator videodensitometry, *Plant Soil.* 183 (1996), 85–95.
10. K. K. S. Bhat and P. H. Nye, Diffusion of phosphate to plant roots in soil – I. Quantitative autoradiography of the depletion zone, *Plant Soil.* 38 (1973), 161–175.
11. J. Santner, M. Larsen, A. Kreuzeder and R. N. Glud, Two decades of chemical imaging of solutes in sediments and soils – A review, *Anal Chim Acta.* 878: (2015), 9–42.
12. W. Davison, G. W. Grime, J. A. W. Morgan and K. Clarke, Distribution of dissolved iron in sediment pore waters at submillimetre resolution, *Nature* 352 (1991), 323–325.
13. H. Dočekalová, O. Clarisse, S. Salomon and M. Wartel, Use of constrained det probe for a high-resolution determination of metals and anions distribution in the sediment pore water, *Talanta.* 57 (2002), 145–155.
14. M. Koschorreck, I. Brookland and A. Matthias, Biogeochemistry of the sediment-water interface in the littoral of an acidic mining lake studied with microsensors and gel-probes, *J Exp Mar Biol Ecol.* 285–286 (2003), 71–84.
15. M. O. Krom, P. Davison, H. Zhang and W. Davison, High-resolution pore-water sampling with a gel sampler, *Limnol Oceanogr.* 39 (1994), 1967–1972.
16. R. J. G. Mortimer, M. D. Krom, P. O. J. Hall, S. Hulth and H. Stahl, Use of gel probes for the determination of high resolution solute distributions in marine and estuarine pore waters, *Mar Chem.* 63 (1998), 119–129.

17. K. T. Yu, M. H. W. Lam, Y. F. Yen and A. P. K. Leung, Behavior of trace metals in the sediment pore waters of intertidal mudflats of a tropical wetland, *Environ Toxicol Chem.* 19 (2000), 535–542.

18. E. J. Palmer-Felgate, R. J. G. Mortimer, M. D. Krom and H. P. Jarvie, Impact of point-source pollution on phosphorus and nitrogen cycling in stream-bed sediments, *Environ Sci Technol.* 44 (2010), 908–914.

19. E. Metzger, E. Viollier, C. Simonucci et al., Millimeter-scale alkalinity measurement in marine sediment using DET probes and colorimetric determination, *Water Res.* 47 (2013), 5575–5583.

20. S. Ullah, H. Zhang, A. L. Heathwaite et al., In situ measurement of redox sensitive solutes at high spatial resolution in a riverbed using diffusive equilibrium in thin films (DET), *Ecol Eng.* 49 (2012), 18–26.

21. M. Gregusova and B. Docekal, High resolution characterization of uranium in sediments by DGT and DET techniques ACA-S-12–2197, *Anal Chim Acta.* 763 (2013), 50–56.

22. W. Davison, H. Zhang and G. W. Grime, Performance characteristics of gel probes used for measuring the chemistry of pore waters, *Environ Sci Technol.* 28 (1994), 1623–1632.

23. S. M. Shuttleworth, W. Davison and J. Hamilton-Taylor, Two-dimensional and fine structure in the concentrations of iron and manganese in sediment pore-waters, *Environ Sci Technol.* 33 (1999), 4169–4175.

24. D. Jézéquel, R. Brayner, E. Metzger et al., Two-dimensional determination of dissolved iron and sulfur species in marine sediment pore-waters by thin-film based imaging. Thau lagoon (France), *Estuar Coast Shelf Sci.* 72 (2007), 420–431.

25. D. Robertson, P. R. Teasdale and D. T. Welsh, A novel gel-based technique for the high resolution, two-dimensional determination of iron (II) and sulfide in sediment, *Limnol Oceanogr Methods.* 6 (2008), 502–512.

26. A. Pagès, P. R. Teasdale, D. Robertson et al., Representative measurement of two-dimensional reactive phosphate distributions and co-distributed iron(II) and sulfide in seagrass sediment porewaters, *Chemosphere.* 85 (2011), 1256–1261.

27. A. J. Kessler, R. N. Glud, M. B. Cardenas and P. L. M. Cook, Transport zonation limits coupled nitrification-denitrification in permeable sediments, *Environ Sci Technol.* 47 (2013), 13404–13411.

28. F. Cesbron, E. Metzger, P. Launeau et al., Simultaneous 2D imaging of dissolved iron and reactive phosphorus in sediment porewaters by thin-film and hyperspectral methods, *Environ Sci Technol.* 48 (2014), 2816–2826.

29. W. W. Bennett, D. T. Welsh, A. Serriere, J. G. Panther and P. R. Teasdale, A colorimetric DET technique for the high-resolution measurement of two-dimensional alkalinity distributions in sediment porewaters, *Chemosphere.* 119 (2015), 547–552.

30. A. Pagès, D. T. Welsh, P. R. Teasdale et al., Diel fluctuations in solute distributions and biogeochemical cycling in a hypersaline microbial mat from Shark Bay, WA, *Mar Chem.* 167 (2014), 102–112.

31. G. R. Fones, W. Davison, O. Holby, B. B. Jorgensen and B. Thamdrup, High-resolution metal gradients measured by in situ DGT/DET deployment in black sea sediments using an autonomous benthic lander, *Limnol Oceanogr.* 46 (2001), 982–988.

32. Y. Gao, W. Baeyens, S. De Galan, A. Poffijin and M. Leermakers, Mobility of radium and trace metals in sediments of the Winterbeek: Application of sequential extraction and DGT techniques, *Environ Pollut.* 158 (2010), 2439–2445.

33. Y. Gao, M. Leermakers, M. Elskens et al., High resolution profiles of thallium, manganese and iron assessed by DET and DGT techniques in riverine sediment pore waters, *Sci Total Environ.* 373 (2007), 526–533.

34. D. C. Gillan, W. Baeyens, R. Bechara et al., Links between bacterial communities in marine sediments and trace metal geochemistry as measured by in situ DET/DGT approaches, *Mar Pollut Bull.* 64 (2012), 353–362.

35. M. Leermakers, Y. Gao, C. Gabelle et al., Determination of high resolution pore water profiles of trace metals in sediments of the Rupel River (Belgium) using DET (diffusive equilibrium in thin films) and DGT (diffusive gradients in thin films) techniques, *Water Air Soil Poll.* 166 (2005), 265–286.

36. Z. Wu, M. He, S. Wang and Z. Ni, The assessment of localized remobilization and geochemical process of 14 metals at sediment/water interface (SWI) of Yingkou coast (China) by diffusive gradients in thin films (DGT), *Environmental Earth Sciences.* 73 (2015), 6081–6090.

37. H. Zhang, W. Davison, S. Miller and W. Tych, In situ high resolution measurements of fluxes of Ni, Cu, Fe, and Mn and concentrations of Zn and Cd in porewaters by DGT, *Geochim. Cosmochim. Acta.* 59 (1995), 4181–4192.

38. S. Ding, D. Xu, Q. Sun, H. Yin and C. Zhang, Measurement of dissolved reactive phosphorus using the diffusive gradients in thin films technique with a high-capacity binding phase, *Environ. Sci. Technol.* 44 (2010), 8169–8174.

39. H. Zhang, W. Davison, R. Gadi and T. Kobayashi, In situ measurement of dissolved phosphorus in natural waters using DGT, *Anal. Chim. Acta.* 370 (1998), 29–38.

40. W. W. Bennett, P. R. Teasdale, J. G. Panther et al., Investigating arsenic speciation and mobilization in sediments with DGT and DET: A mesocosm evaluation of oxic-anoxic transitions, *Environ. Sci. Technol.* 46 (2012), 3981–3989.

41. H. Zhang, W. Davison, R. J. G. Mortimer et al., Localised remobilization of metals in a marine sediment, *Sci. Total Environ.* 296 (2002), 175–187.

42. P. Diviš, M. Leermakers, H. Dočekalová and Y. Gao, Mercury depth profiles in river and marine sediments measured by the diffusive gradients in thin films technique with two different specific resins, *Anal. Bioanal. Chem.* 382 (2005), 1715–1719.

43. O. Clarisse, B. Dimock, H. Hintelmann and E. P. H. Best, Predicting net mercury methylation in sediments using diffusive gradient in thin films measurements, *Environ. Sci. Technol.* 45 (2011), 1506–1512.

44. J. Liu, X. Feng, G. Qiu et al., Intercomparison and applicability of some dynamic and equilibrium approaches to determine methylated mercury species in pore water, *Environ. Toxicol. Chem.* 30 (2011), 1739–1744.

45. B. Docekal and M. Gregusova, Segmented sediment probe for diffusive gradient in thin films technique, *Analyst.* 137 (2012), 502–507.

46. A. Widerlund, G. M. Nowell, W. Davison and D. G. Pearson, High-resolution measurements of sulphur isotope variations in sediment pore-waters by laser ablation multicollector inductively coupled plasma mass spectrometry, *Chem Geol.* 291 (2012), 278–285.

47. W. Davison, G. R. Fones and G. W. Grime, Dissolved metals in surface sediment and microbial mat at 100-μm resolution, *Nature* 387 (1997), 885–888.

48. G. R. Fones, W. Davison and J. Hamilton-Taylor, The fine-scale remobilization of metals in the surface sediment of the North-East Atlantic, *Cont. Shelf Res.* 24 (2004), 1485–1504.

49. K. W. Warnken, H. Zhang and W. Davison, Analysis of polyacrylamide gels for trace metals using diffusive gradients in thin films and laser ablation inductively coupled plasma mass spectrometry, *Anal. Chem.* 76 (2004), 6077–6084.

50. C. R. Devries and F. Wang, In situ two-dimensional high-resolution profiling of sulfide in sediment interstitial waters, *Environ. Sci. Technol.* 37 (2003), 792–797.

51. P. R. Teasdale, S. Hayward and W. Davison, In situ, high-resolution measurement of dissolved sulfide using diffusive gradients in thin films with computer-imaging densitometry, *Anal. Chem.* 71 (1999), 2186–2191.

52. J. Santner, T. Prohaska, J. Luo and H. Zhang, Ferrihydrite containing gel for chemical imaging of labile phosphate species in sediments and soils using diffusive gradients in thin films, *Anal. Chem.* 82 (2010), 7668–7674.

53. A. Stockdale, W. Davison and H. Zhang, High-resolution two-dimensional quantitative analysis of phosphorus, vanadium and arsenic, and qualitative analysis of sulfide, in a freshwater sediment, *Env. Chem.* 5 (2008), 143–149.

54. A. Stockdale, W. Davison and H. Zhang, 2D simultaneous measurement of the oxyanions of P, V, As, Mo, Sb, W and U, *J. Environ. Monit.* 12 (2010), 981–984.

55. D.-X. Guan, P. N. Williams, J. Luo et al., Novel precipitated zirconia-based DGT technique for high-resolution imaging of oxyanions in waters and sediments, *Environ. Sci. Technol.* 49 (2015), 3653–3661.

56. S. Ding, F. Jia, D. Xu et al., High-resolution, two-dimensional measurement of dissolved reactive phosphorus in sediments using the diffusive gradients in thin films technique in combination with a routine procedure, *Environ. Sci. Technol.* 45 (2011), 9680–9686.

57. S. Ding, Y. Wang, D. Xu, C. Zhu and C. Zhang, Gel-based coloration technique for the submillimeter-scale imaging of labile phosphorus in sediments and soils with diffusive gradients in thin films, *Environ. Sci. Technol.* 47 (2013), 7821–7829.

58. S. Ding, Q. Sun, D. Xu et al., High-resolution simultaneous measurements of dissolved reactive phosphorus and dissolved sulfide: The first observation of their simultaneous release in sediments, *Environ. Sci. Technol.* 46 (2012), 8297–8304.

59. A. Kreuzeder, J. Santner, T. Prohaska and W. Wenzel, Gel for simultaneous chemical imaging of anionic and cationic solutes using diffusive gradients in thin films, *Anal. Chem.* 85 (2013), 12028–12036.

60. W. Davison and H. Zhang, In situ speciation measurements of trace components in natural waters using thin-film gels, *Nature* 367 (1994), 546–548.

61. M. P. Harper, W. Davison and W. Tych, Temporal, spatial, and resolution constraints for in situ sampling devices using diffusional equilibration: Dialysis and DET, *Environ. Sci. Technol.* 31 (1997), 3110–3119.

62. M. P. Harper, W. Davison and W. Tych, Estimation of pore water concentrations from DGT profiles: A modelling approach, *Aquat. Geochem.* 5 (1999), 337–355.

63. Ł. Sochaczewski, W. Davison, H. Zhang and W. Tych, Understanding small-scale features in DGT measurements in sediments, *Env. Chem.* 6 (2009), 477–485.

64. J. Luo, H. Zhang, W. Davison et al., Localised mobilisation of metals, as measured by diffusive gradients in thin-films, in soil historically treated with sewage sludge, *Chemosphere* 90 (2013), 464–470.

65. S. Tankéré-Muller, H. Zhang, W. Davison et al., Fine scale remobilisation of Fe, Mn, Co, Ni, Cu and Cd in contaminated marine sediment, *Mar. Chem.* 106 (2007), 192–207.

66. P. N. Williams, High resolution 2D imaging of trace elements in soil and sediments by DGT and LA-ICP-MS. Thermo Fisher Scientific ICP-MS user group meeting, Hemel Hempstead, UK, 2014.

67. A. L. Fabricius, L. Duester, D. Ecker and T. A. Teres, New microprofiling and micro sampling system for water saturated environmental boundary layers, *Environ. Sci. Technol.* 48 (2014), 8053–8061.

68. A. Widerlund and W. Davison, Size and density distribution of sulfide-producing microniches in lake sediments, *Environ. Sci. Technol.* 41 (2007), 8044–8049.

69. M. Motelica-Heino, C. Naylor, H. Zhang and W. Davison, Simultaneous release of metals and sulfide in lacustrine sediment, *Environ. Sci. Technol.* 37 (2003), 4374–4381.

70. C. Naylor, W. Davison, M. Motelica-Heino, G. A. Van Den Berg and L. M. Van Der Heijdt, Simultaneous release of sulfide with Fe, Mn, Ni and Zn in marine harbour sediment measured using a combined metal/sulfide DGT probe, *Sci. Total Environ.* 328 (2004), 275–286.

71. D. Robertson, D. T. Welsh and P. R. Teasdale, Investigating biogenic heterogeneity in coastal sediments with two-dimensional measurements of iron(II) and sulfide, *Env. Chem.* 6 (2009), 60–69.

72. F. C. Stephens, E. M. Louchard, R. P. Reid and R. A. Maffione, Effects of microalgal communities on reflectance spectra of carbonate sediments in subtidal optically shallow marine environments, *Limnol Oceanogr.* 48 (2003), 535–546.

73. M. Manley, Near-infrared spectroscopy and hyperspectral imaging: Non-destructive analysis of biological materials, *Chem. Soc. Rev.* 43 (2014), 8200–8214.

74. P. Arrowsmith and S. K. Hughes, Entrainment and transport of laser ablated plumes for subsequent elemental analysis, *Appl. Spectrosc.* 42 (1988), 1231–1239.

75. A. L. Gray, Solid sample introduction by laser ablation for inductively coupled plasma source mass spectrometry, *Analyst.* 110 (1985), 551–556.

76. J. Becker, A. Matusch and B. Wu, Bioimaging mass spectrometry of trace elements – Recent advance and applications of LA-ICP-MS: A review, *Anal. Chim. Acta.* 835 (2014), 1–18.

77. A. Kindness, C. N. Sekaran and J. Feldmann, Two-dimensional mapping of copper and zinc in liver sections by laser ablation – Inductively coupled plasma mass spectrometry, *Clin. Chem.* 49 (2003), 1916–1923.

78. J. Koch and D. Günther, Review of the state-of-the-art of laser ablation inductively coupled plasma mass spectrometry, *Appl. Spectrosc.* 65 (2011), 155A–162A.

79. M. Motelica-Heino and W. Davison, In situ trace metals distribution in lake sediment pore waters: High spatial resolution depth profiling an 2D-mapping, *Goldschmidt Abstracts.* 5 (2000), 724.

80. J. Pisonero, D. Fliegel and D. Günther, High efficiency aerosol dispersion cell for laser ablation-ICP-MS, *J. Anal. Atom. Spectrom.* 21 (2006), 922–931.

81. N. Kivel, H. D. Potthast, I. Günther-Leopold, F. Vanhaecke and D. Günther, Modeling of the plasma extraction efficiency of an inductively coupled plasma-mass spectrometer interface using the direct simulation Monte Carlo method, *Spectrochim Acta B.* 93 (2014), 34–40.

82. G. Caumette, S. Ouypornkochagorn, C. M. Scrimgeour, A. Raab and J. Feldmann, Monitoring the arsenic and iodine exposure of seaweed-eating North Ronaldsay sheep from the gestational and suckling periods to adulthood by using horns as a dietary archive, *Environ. Sci. Technol.* 41 (2007), 2673–2679.

83. A. Kreuzeder, J. Santner, H. Zhang, T. Prohaska and W. W. Wenzel, Uncertainty evaluation of the diffusive gradients in thin films technique, *Environ. Sci. Technol.* 49 (2015), 1594–1602.

84. C. Austin, F. Fryer, J. Lear et al., Factors affecting internal standard selection for quantitative elemental bio-imaging of soft tissues by LA-ICP-MS, *J. Anal. Atom. Spectrom.* 26 (2011), 1494–1501.

85. Z. Hu, S. Gao, Y. Liu et al., Signal enhancement in laser ablation ICP-MS by addition of nitrogen in the central channel gas, *J. Anal. Atom. Spectrom.* 23 (2008), 1093–1101.

86. J. Santner, H. Zhang, D. Leitner et al., High-resolution chemical imaging of labile phosphorus in the rhizosphere of Brassica napus L. cultivars, *Environ. Exp. Bot.* 77 (2012), 219–226.

87. Y. Liu, Z. Hu, H. Yuan, S. Hu and H. Cheng, Volume-optional and low-memory (VOLM) chamber for laser ablation-ICP-MS: Application to fiber analyses, *J. Anal. Atom. Spectrom.* 22 (2007), 582–585.

88. C. Hoefer, J. Santner, M. Puschenreiter and W. W. Wenzel, Localized metal solubilization in the rhizosphere of Salix smithiana upon sulfur application, *Environ. Sci. Technol.* 49 (2015), 4522–4529.

89. P. N. Williams, J. Santner, M. Larsen et al., Localized flux maxima of arsenic, lead, and iron around root apices in flooded lowland rice, *Environ. Sci. Technol.* 48 (2014), 8498–8506.

90. Y. Gao and N. Lehto, A simple laser ablation ICPMS method for the determination of trace metals in a resin gel, *Talanta.* 92 (2012), 78–83.

91. W. W. Bennett, P. R. Teasdale, D. T. Welsh et al., Inorganic arsenic and iron(II) distributions in sediment porewaters investigated by a combined DGT colourimetric DET technique, *Env. Chem.* 9 (2012), 31–40.

92. S. Ding, C. Han, Y. Wang et al., In situ, high-resolution imaging of labile phosphorus in sediments of a large eutrophic lake, *Water Res.* 74 (2015), 100–109.

93. A. Pagès, K. Grice, M. Vacher et al., Characterizing microbial communities and processes in a modern stromatolite (Shark Bay) using lipid biomarkers and two-dimensional distributions of porewater solutes, *Environ. Microbiol.* 16 (2014), 2458–2474.

94. G. R. Fones, W. Davison and G. W. Grime, Development of constrained DET for measurements of dissolved iron in surface sediments at sub-mm resolution, *Sci. Total Environ.* 221 (1998), 127–137.

95. H. Stahl, K. W. Warnken, L. Sochaczewski et al., A combined sensor for simultaneous high resolution 2-D imaging of oxygen and trace metals fluxes, *Limnol Oceanogr. Meth.* 10 (2012), 389–401.

96. N. J. Lehto, W. Davison and H. Zhang, The use of ultra-thin diffusive gradients in thin-films (DGT) devices for the analysis of trace metal dynamics in soils and sediments: a measurement and modelling approach, *Env. Chem.* 9 (2012), 415–423.

97. M. Spohn and Y. Kuzyakov, Distribution of microbial- and root-derived phosphatase activities in the rhizosphere depending on P availability and C allocation – Coupling soil zymography with 14C imaging, *Soil Biol. Biochem.* 67 (2013), 106–113.

98. M. Spohn and Y. Kuzyakov, Spatial and temporal dynamics of hotspots of enzyme activity in soil as affected by living and dead roots-a soil zymography analysis, *Plant Soil.* 379 (2014), 67–77.

99. H. Zhang and W. Davison, Performance characteristics of diffusion gradients in thin films for the in situ measurement of trace metals in aqueous solution, *Anal. Chem.* 67 (1995), 3391–3400.

100. R. J. G. Mortimer, M. D. Krom, S. J. Harris et al., Evidence for suboxic nitrification in recent marine sediments, *Mar. Ecol. Prog. Ser.* 236 (2002), 31–35.

101. J. W. M. Wegener, G. A. van den Berg, G. J. Stroomberg and M. J. M. van Velzen, The role of sediment-feeding oligochaete Tubifex on the availability of trace metals in sediment pore waters as determined by diffusive gradients in thin films (DGT), *J. Soils Sed.* 2 (2002), 71–76.

102. K. W. Warnken, H. Zhang and W. Davison, Performance characteristics of suspended particulate reagent- iminodiacetate as a binding agent for diffusive gradients in thin films, *Anal. Chim. Acta.* 508 (2004), 41–51.

103. P. Monbet, I. D. McKelvie and P. J. Worsfold, Combined gel probes for the in situ determination of dissolved reactive phosphorus in porewaters and characterization of sediment reactivity, *Environ. Sci. Technol.* 42 (2008), 5112–5117.

104. J. Luo, H. Zhang, J. Santner and W. Davison, Performance characteristics of diffusive gradients in thin films equipped with a binding gel layer containing precipitated ferrihydrite for measuring arsenic(V), selenium(VI), vanadium(V), and antimony(V), *Anal. Chem.* 82 (2010), 8903–8909.

105. J. Santner, M. Mannel, L. D. Burrell et al., Phosphorus uptake by Zea mays L. is quantitatively predicted by infinite sink extraction of soil P, *Plant Soil.* 386 (2015), 371–383.

106. W. W. Bennett, P. R. Teasdale, J. G. Panther, D. T. Welsh and D. F. Jolley, New diffusive gradients in a thin film technique for measuring inorganic arsenic and selenium(IV) using a titanium dioxide based adsorbent, *Anal. Chem.* 82 (2010), 7401–7407.

107. J. G. Panther, W. W. Bennett, P. R. Teasdale, D. T. Welsh and H. Zhao, DGT measurement of dissolved aluminum species in waters: Comparing Chelex-100 and titanium dioxide-based adsorbents, *Environ. Sci. Technol.* 46 (2012), 2267–2275.

108. J. G. Panther, R. R. Stewart, P. R. Teasdale et al., Titanium dioxide-based DGT for measuring dissolved As(V), V(V), Sb(V), Mo(VI) and W(VI) in water, *Talanta* 105 (2013), 80–86.

109. J. G. Panther, P. R. Teasdale, W. W. Bennett, D. T. Welsh and H. Zhao, Titanium dioxide-based DGT technique for in situ measurement of dissolved reactive phosphorus in fresh and marine waters, *Environ. Sci. Technol.* 44 (2010), 9419–9424.

110. Q. Sun, J. Chen, H. Zhang et al., Improved diffusive gradients in thin films (DGT) measurement of total dissolved inorganic arsenic in waters and soils using a hydrous zirconium oxide binding layer, *Anal. Chem.* 86 (2014), 3060–3067.

111. W. W. Bennett, P. R. Teasdale, J. G. Panther, D. T. Welsh and D. F. Jolley, Speciation of dissolved inorganic arsenic by diffusive gradients in thin films: Selective binding of As III by 3-mercaptopropyl-functionalized silica gel, *Anal. Chem.* 83 (2011), 8293–8299.

112. O. Clarisse and H. Hintelmann, Measurements of dissolved methylmercury in natural waters using diffusive gradients in thin film (DGT), *J. Environ. Monit.* 8 (2006), 1242–1247.

113. K. Kannan, R. G. Smith Jr, R. F. Lee et al., Distribution of total mercury and methyl mercury in water, sediment, and fish from South Florida estuaries, *Arch. Environ. Contam. Toxicol.* 34 (1998), 109–118.

114. M. Gregusova and B. Docekal, New resin gel for uranium determination by diffusive gradient in thin films technique, *Anal. Chim. Acta.* 684 (2011), 142–146.

115. J. Morford, L. Kalnejais, W. Martin, R. Francois and I. M. Karle, Sampling marine pore waters for Mn, Fe, U, Re and Mo: Modifications on diffusional equilibration thin film gel probes, *J. Exp. Mar. Biol. Ecol.* 285–286 (2003), 85–103.

116. S. Tankéré-Muller, W. Davison and H. Zhang, Effect of competitive cation binding on the measurement of Mn in marine waters and sediments by diffusive gradients in thin films, *Anal. Chim. Acta.* 716 (2012), 138–144.

117. A. Kreuzeder, J. Santner, V. Scharsching et al. In situ investigation of localized plant induced rhizosphere processes related to phosphorus deficiency. In preparation.

118. Y. Gao, S. van de Velde, P. N. Williams, W. Baeyens and H. Zhang, Two-dimensional images of dissolved sulfide and metals in anoxic sediments by a novel diffusive gradients in thin film probe and optical scanning techniques, *TrAC-Trend Anal. Chem.* 66 (2015), 63–71.

119. Ł. Sochaczewski, W. Tych, B. Davison and H. Zhang, 2D DGT induced fluxes in sediments and soils (2D DIFS), *Environ. Modell. Softw.* 22 (2007), 14–23.

120. M. P. Harper, W. Davison and W. Tych, DIFS – A modelling and simulation tool for DGT induced trace metal remobilisation in sediments and soils, *Environ. Modell. Softw.* 15 (2000), 55–66.

121. M. P. Harper, W. Davison, H. Zhang and W. Tych, Kinetics of metal exchange between solids and solutions in sediments and soils interpreted from DGT measured fluxes, *Geochim. Cosmochim. Acta.* 62 (1998), 2757–2770.

122. Ł. Sochaczewski, A. Stockdale, W. Davison, W. Tych and H. Zhang, A three-dimensional reactive transport model for sediments, incorporating microniches, *Env. Chem.* 5 (2008), 218–225.

123. W. Stumm and J. J. Morgan, *Aquatic chemistry. Chemical equilibria and rates in natural waters* (New York: Wiley, 1996).

124. M. J. McLaughlin, E. Smolders and R. Merckx, Soil-root interface: Physicochemical processes, In *Soil chemistry and ecosystem health*, ed. P. M. Huang (Madison: Soil Science Society of America, 1998), pp. 233–277.

125. W. Li, H. Zhao, P. R. Teasdale, R. John and F. Wang, Metal speciation measurement by diffusive gradients in thin films technique with different binding phases, *Anal. Chim. Acta.* 533 (2005), 193–202.

126. K. W. Warnken, H. Zhang and W. Davison, Trace metal measurements in low ionic strength synthetic solutions by diffusive gradients in thin films, *Anal. Chem.* 77 (2005), 5440–5446.

127. Ø. A. Garmo, W. Davison and H. Zhang, Interactions of trace metals with hydrogels and filter membranes used in DET and DGT techniques, *Environ. Sci. Technol.* 42 (2008), 5682–5687.

128. Ø. A. Garmo, W. Davison and H. Zhang, Effects of binding of metals to the hydrogel and filter membrane on the accuracy of the diffusive gradients in thin films technique, *Anal. Chem.* 80 (2008), 9220–9225.

9

DGT and Bioavailability

FIEN DEGRYSE AND ERIK SMOLDERS

9.1 Introduction

This chapter discusses the relationship between DGT measurements and bioavailability. Several definitions for the term bioavailability have been put forward [1], but an exact definition is not essential in the context of this chapter. The main idea behind bioavailability is that the uptake or the effect of an element in and on an organism depends not only on its total concentration in the environment, but also on its speciation and mobility. For almost twenty years, the DGT technique has been used to measure in situ speciation and fluxes of different elements in the environment. Since 1999, an increasing number of studies have related DGT-measured fluxes to uptake, growth responses (in the case of nutrients) and toxicity. Table 9.1 presents a non-exhaustive list of such studies in soils, waters and sediments. The DGT technique has been most developed for toxic cationic divalent trace metals but has also been expanded to test bioavailability of plant nutrients (phosphate, potassium), uranium, methylmercury, arsenate, molybdate and has recently also been used to understand effects of nanoparticles in the environment. Figure 9.1 shows two examples where DGT better predicted bioavailability than other physico-chemical measurements and an example where DGT appeared less successful.

Rather than giving a description of the numerous empirical tests on the predictive power of DGT, the emphasis in this chapter is on the understanding of why and when DGT measurements may or may not relate to bioavailability. No model or assay of bioavailability is perfectly applicable to all elements, biota and environments and DGT is no exception to this. The main merit of DGT is that it is a relative simple research tool that allows testing of bioavailability concepts and helps understanding of how biota interact with their environment. For these reasons, this chapter is mainly devoted to the mechanisms of bioavailability and to what extent the DGT technique mimics these.

The chapter focuses on divalent trace metals (Cd, Co, Cu, Ni, Pb, Zn) and phosphorus, since most studies relating DGT measurements to bioavailability have investigated one or several of these elements. As speciation is essential to the concept of bioavailability, we first discuss the speciation of these elements in natural environments. This is followed by an overview of models/frameworks that have been used to explain/predict bioavailability, before we move to the discussion of studies that used DGT in bioavailability assessments.

216

Table 9.1 *Selected overview of studies discussed in this chapter in which the DGT technique has been used to test bioavailability relative to other indices of bioavailability*

Compartment	Endpoint for bioavailability	Organism	Elements	Major findings	Ref
Soil	Uptake	Plants	Cu, Zn, Co, Cd, Ni, Pb, As, Cr, CH_3-Hg, P, U	Superior predictive power of DGT relative to other tests when diffusion limits uptake; similar predictions as by soil solution concentrations in most other cases	[2–12]
	Growth increase (nutrient response)	Plants	P, K	DGT adequately predicts nutrient responses in most cases	[13–18]
	Growth decrease (toxicity)	Plants	Zn, Cu, Mo, As	DGT predicts toxicity better than total concentration but similar to soil solution concentrations	[19–24]
Sediment	Uptake	Plants/invertebrates	Cd, Cu, Pb, Cr, Ni, Zn	DGT sometimes predicts better than AVS/SEM but no generalization possible	[25–28]
	Toxicity	Invertebrates	Ni, Cu	No generalizations possible	[29, 30]
Water	Uptake	Algae/periphyton, invertebrates, fish	Cu, Zn, Cd, Ni, Pb, CH_3-Hg,	Diffusion probably only limits uptake in particular situations (e.g., periphyton), in which case DGT is a good predictor; otherwise predictive power is variable	[31–37]
	Toxicity	Micro-algae	Cu, Al, Ag-nanoparticles	Toxic effects of Al are well predicted by DGT (but needs more data); predictions in other cases often not successful since DGT does not account for ion competition effects	[38–40]

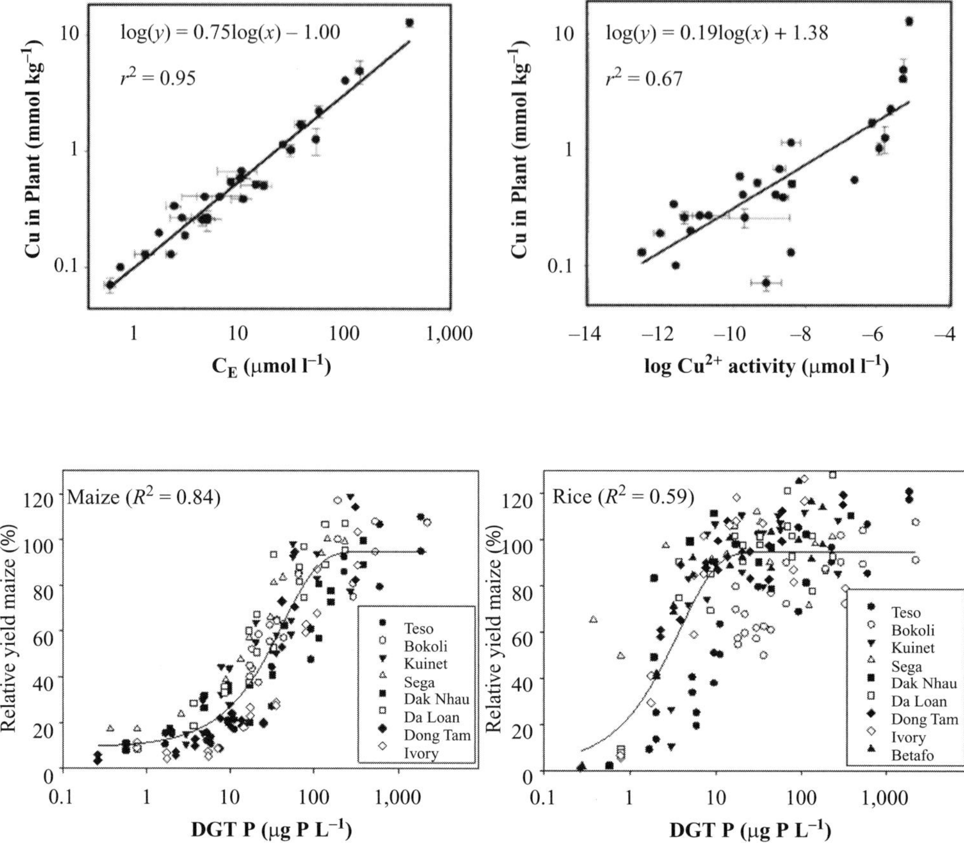

Figure 9.1   The DGT technique predicting bioavailability of elements in the environment. (a) Shoot Cu concentrations of plants grown in soils with contrasting properties and contamination levels as a function of DGT-measured concentration (expressed as effective concentration c_E) or as a function of free Cu^{2+} ion activity, illustrating that DGT gives better predictions than free ion activities. Reprinted with permission from ref. [3]. Copyright 2001 American Chemical Society. (b) Growth response of maize (left) or rice (right) seedlings to fertilizer P in several P-deficient soils. The prediction by DGT was better than the conventional Olsen-P test on the same soils for maize but the reverse was true for rice plants (not shown). Reprinted from ref. [16] with kind permission of Springer Science and Business Media.

9.2 Elemental Speciation in Water, Sediments and Soil

The speciation of an element is the distribution of an element among defined chemical species within a system. Chemical compounds that differ in their conformation, oxidation or electronic state, covalently bound substituents, or in their isotopic composition, can be regarded as distinct chemical species [41].

Table 9.2 *Median and range of total dissolved (c_{total}) and free ion concentrations (c_{free}; expressed as $-\log$ of the concentration in M) and the ratio of free ion to total dissolved concentration observed in seven river and lake waters with pH between 7.6 and 8.7. Data from ref. [47]*

		Cd	Ni	Pb	Cu	Zn
$-\log c_{total}$ (M)	Median	9.7	8.0	9.3	7.6	6.8
	range	10.2–9.1	8.2–7.2	9.9–9.0	7.9–7.1	7.6–6.6
$-\log c_{free}$ (M)	Median	11.2	9.6	12.1	10.4	8.1
	range	11.9–10.7	9.8–8.5	12.4–10.7	10.9–10.1	8.3–7.3
c_{free} / c_{total}	Median	0.03	0.03	0.002	0.001	0.09
	range	0.01–0.09	0.02–0.05	0.0007–0.02	0.0002–0.01	0.04–0.17

9.2.1 Trace Metals

9.2.1.1 Natural Waters and Soil Solution

Trace metals in water are present as free ions (or technically as aquo-complexes), as complexes with inorganic or organic ligands, or associated with colloids. The term colloids usually denotes macromolecules or microparticles in suspension with a size between 1 and 1000 nm [42]. These colloids may be either inorganic (e.g., hydroxides), organic (e.g., humic acid) or a combination of both. Their chemical composition may vary widely and is often not well defined [43]. A distinction between 'dissolved' and 'particulate' metal is often made by filtration through a 0.45-μm filter. However, this is a purely operational distinction, as the dissolved fraction may contain many colloidal metal species.

Determining speciation of metals in waters is not straightforward. Use of speciation programs (e.g., WHAM [44] or NICA-Donnan [45]) requires a complete chemical characterization of the sample including concentration and chemical composition of the complexing ligands. Usually assumptions are made (e.g., a certain fraction of measured dissolved organic C behaving like fulvic or humic acid), which results in inherent uncertainties. Direct measurement of free ion concentrations in natural waters is also challenging. Ion selective electrodes and the Donnan membrane technique measure the free ion concentration, but detection is often an issue as the free ion concentrations are often very low [46]. Table 9.2 gives an example of total dissolved (<0.45 μm) and free ion concentrations determined for seven natural waters (rivers or lakes) [47]. The free ion fraction was much smaller for Cu and Pb than for Cd, Ni and Zn, which can be explained by the much higher affinity for dissolved organic matter (and possibly other colloids) of these metals. Total solution concentrations are at nM (Cd) to μM (Zn) levels. In oceanic waters, metal concentrations are even lower, with total Zn concentrations only at nM levels and free Zn concentrations only around 0.01 nM [48].

Table 9.3 *Range of total dissolved metal concentrations (c_{total}; expressed as $-$log of the concentration in M) and ratio of free ion concentration (c_{free}) or DGT-measured concentration (c_{DGT}) to total dissolved concentration for three freshwaters. Data from ref. [54]*

	Cd	Ni	Pb	Cu
$-$log c_{total} (M)	10.4–9.7	8.4–7.7	9.7–8.8	7.8–7.6
c_{free}/c_{total}	0.01–0.13	0.03–0.06	1.9×10^{-3}–1.3×10^{-2}	1.5×10^{-3}–4.1×10^{-3}
c_{DGT}/c_{total}	0.36–0.75	0.34–0.70	0.03–0.82	0.13–0.75

Humic substances are the main organic ligand in natural waters. Also low-molecular weight organic compounds are present in natural waters, but these have usually little affinity for trace metals, and therefore have little influence on the trace metal speciation. However, in anthropologically affected waters, also EDTA-type ligands may be present [49] and these may have a strong effect on the speciation of trace metals [50, 51].

Several techniques used to measure speciation in waters do not measure a well-defined species (e.g., free ion), but rather 'labile' metal [52], as is for instance the case with the DGT technique. Complexes of metals usually contribute to the DGT measurements (Chapter 5), though they are often not detected to the same extent as the free ion due to the lower diffusion coefficient of the complex [53]. Table 9.3 illustrates this contribution of complexes to the DGT-measured concentration for three freshwaters [54]. The DGT-measured concentrations were higher than the free ion concentrations as determined with the Donnan technique. In another study with seven seawaters [31], it was found that the ratio of DGT-measured to total dissolved concentration was around 70% for Cd, around 50% for Ni and around 25% for Pb and Cu, with little variation among waters. The free ion fractions were much smaller, also demonstrating that DGT measurements include metal complexes with dissolved organic matter (DOM) or inorganic ligands. In general, the variation in the ratio of DGT-measured to total dissolved concentration is small compared to the range in total dissolved concentrations across natural waters, explaining why total dissolved concentrations and DGT-measured concentrations often are strongly correlated within datasets [55].

Soil solutions or interstitial water of sediments contain the same type of metal species as natural waters. The complexation of metals with DOM is highly pH dependent. Metals with low affinity for DOM (Co, Ni, Cd, Zn) are usually mainly as free ion in solution at pH < 6.5, while metals with high affinity for DOM (Pb, Cu) are mainly present complexed with DOM, even in acidic soil solutions (pH < 5) [56]. DGT-measured concentrations of elements in soils mostly relate to their total dissolved concentration in the soil solution, but also reflect the buffering of the metals in solution by the soil solid phase. If the metal in solution is mostly present as free ion (as is expected for Cd, Co, Ni and Zn at pH < 6), the ratio between c_{DGT} and porewater concentration (R_c) can theoretically be calculated

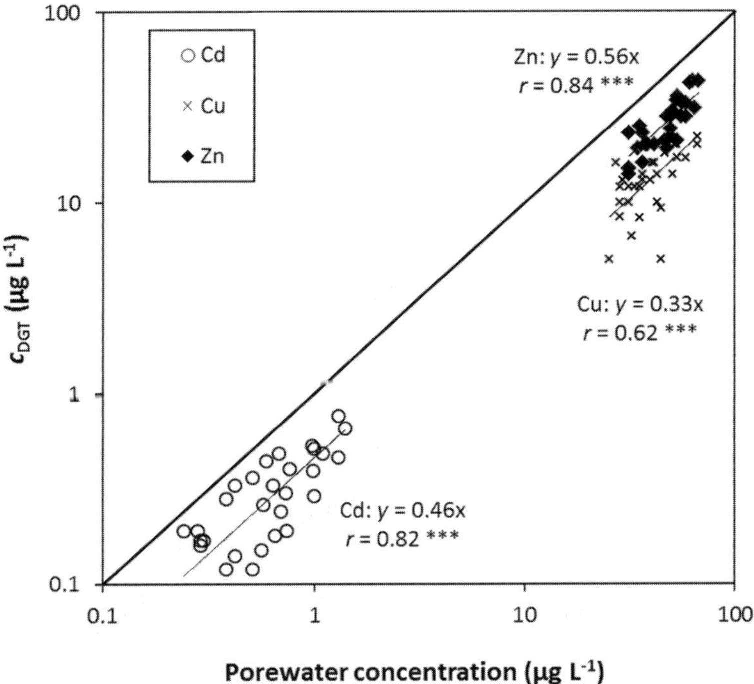

Figure 9.2   DGT-measured concentration of Cd, Zn and Cu versus porewater concentration for a dataset of thirty mine-affected soils with pH between 6.7 and 7.8. *** indicates a significant correlation at $P < 0.001$. Data from ref. [60].

[57, 58] and roughly ranges from ~0.2 in case of weak buffering (e.g., low pH sandy soil) to >0.5 for strong buffering for a 24-h deployment time [59]. However, several studies have found DGT-measured concentrations considerably lower than expected given the soil solution concentration. This is likely due to the presence of quasi-inert colloids in the soil solution. For instance, in a study with fifteen soils covering a wide range of properties [59], for seven out of the fifteen soils DGT-labile Zn concentrations were much lower than expected if all Zn would have been present as DGT-labile species in solution.

Despite the effect of solid-phase buffering and presence of inert colloids on the ratio of c_{DGT} to porewater concentration, strong correlations between c_{DGT} and porewater concentration are usually observed across a dataset of various soils, since the variation in porewater concentrations is usually larger than the variation in R_c. This is illustrated in Figure 9.2, which shows the relationship between porewater concentration and c_{DGT} in thirty garden soils from a mining-affected area in Romania [60]. Porewater and DGT-measured concentrations were highly correlated ($P < 0.001$). The correlation was the weakest for Cu, likely because Cu is mostly present as a complex with organic matter in soil solutions and the variable diffusion coefficient and degree of lability of the complexes poses an additional source of variability. The value of R_c for Cu was, on average, around 0.3, which is similar to the ratio of diffusion coefficients of a fulvic acid complex to that of the free metal ion [61], again suggesting that Cu-DOM complexes are in general DGT-labile.

9.2.1.2 Soils and Sediment

This section gives a very short overview of metal speciation in soils and sediments. More detailed discussions can be found in the literature [56, 62]. Metal ions adsorb to organic matter, oxyhydroxides and clay minerals. Metals can also be precipitated as pure or mixed solids. In uncontaminated soils, adsorption reactions rather than precipitation reactions are generally assumed to control the ion activity of trace metals in the solution. Organic matter and oxyhydroxides are the main adsorbents, with the relative importance of oxyhydroxides predicted to increase with increasing pH [63, 64]. Soil pH is the most important soil property determining the adsorption of free metal ions on the solid phase. For metal ions that show relatively little solution complexation, this results in an increase in the solid:liquid distribution with increasing soil pH. However, for metals that have high affinity for dissolved organic matter such as Cu, the solid:liquid distribution shows less dependence on pH because the increasing adsorption of the free ion is offset by the increasing complexation with DOM at higher pH [65].

Speciation is not constant over time; transformations may occur at various time scales, and these changes affect the solubility and availability. For instance, changes in soil properties over time (e.g., acidification) affect the speciation of trace metals [66]. Long-term 'ageing' reactions may render metals less soluble [56], while conversely, mineral weathering can result in metal release [67].

The speciation of trace metals in sediments is complex and may show great temporal and spatial variation. The redox condition of the sediment strongly influences the metal speciation. Under oxic conditions, the reactions determining speciation are similar to those in soils, with Fe/Mn hydroxides, organic matter, carbonates and clay minerals acting as main sorbents. Several studies have indicated that Fe/Mn hydroxides are the main sorbent for Cd, Zn and Ni in most sediments, whereas Cu is mostly bound to organic matter [68]. Under anoxic conditions, the reduction of sulphate yield sulphides that precipitate metals. As a result, porewater concentrations of metals are often higher in the oxic zone of a sediment than in the underlying anoxic zone [68]. It should be kept in mind that describing sediments as one-dimensional entities with homogenous layers is an oversimplification. There is considerable horizontal and vertical heterogeneity in sediments. For instance, sulphide-producing microniches in the oxic zone have been shown and are likely common [69].

9.2.2 Phosphorus

Under pH conditions normally encountered in natural waters, dissolved P occurs mainly as the orthophosphate ions $H_2PO_4^-$ or HPO_4^{2-}, but it may also contain condensed phosphates (pyro- or polyphosphate). Colloidal P may account for a large part of P in surface waters or soil solutions [70, 71]. This colloidal P is usually not well characterized and may consist of P bound to (hydr)oxides, organic P, or organo-mineral colloids. Most often, P in solutions is characterized in operationally defined fractions, such as dissolved (molybdate) reactive P (DRP), dissolved non-reactive P (difference between total dissolved P and DRP) and particulate P. The separation between dissolved and particulate P is usually made by filtering through a 0.45-μm filter. Often, DRP is regarded as a measure of orthophosphate, though it has been well recognized that other species may also be included, since the colorimetric method includes an acid step that may hydrolyze labile organic P or dissolve inorganic colloids [72].

As with metals, most of the P in soils and sediments is not in solution, but is associated with the solid phase. Phosphate sorbs strongly on Fe and Al (hydr)oxides. The sorption depends on pH, but the effect of pH within the normal pH range (4–8) is moderate. The sorption affinity and maximum binding capacity therefore mainly depends on the oxide content of the soil and the nature of these oxides. For instance, Sakadevan et al. [73] found that the P sorption capacity of soils and other materials was very strongly correlated with the oxalate-extractable (amorphous) Fe and Al oxide content of the samples. Börling et al. [74] showed that P sorption in cultivated Swedish soils was well correlated with the oxalate-extractable Fe and Al content (~amorphous Fe and Al oxides). Organic matter is negatively charged and does not sorb P directly. However, it has been suggested that phosphate sorption may occurs on humic-Fe/Al surfaces, i.e., the trivalent cation (Fe or Al) acts as a bridge between humic acid and phosphate, and that this may be an important P pool in certain soils [75].

Phosphorus may also be in precipitated forms in soils, especially at high pH where Ca-phosphates may form. The formation of Ca-phosphates such as dicalcium phosphate dihydrate or octacalcium phosphate has been inferred from solubility considerations [76] and more recently been confirmed using in-situ speciation measurement such as X-ray absorption spectroscopy (XANES) [77]. It has sometimes been suggested that Fe/Al phosphates may form in low pH soils (e.g., strengite), but there is little direct evidence for this.

In cultivated mineral soils, most P is usually in mineral form, but there is often a large proportion of organic P in pasture soils. Typical organic P compounds include phosphomonoesters (compounds with a single ester linkage to orthophosphate), phosphodiesters (compounds with two ester linkages to orthophosphate), deoxyribonucleic acid (DNA), lipoteichoic acid, phospholipid fatty acids (e.g., lecithin), and organic polyphosphates [78]. Most studies using ^{31}P nuclear magnetic resonance (NMR) spectroscopy have indicated that monoesters generally account for the largest proportion of organic P.

9.3 Uptake and Toxicity – Concepts and Models

9.3.1 Relationship among Uptake, Internal Concentrations and Effect

In many cases, uptake of a solute by biota can be described by a Michaelis–Menten-type equation which relates the uptake flux (J_{upt}) to the concentration of the solute (or of the species taken up) in the surrounding solution:

$$J_{upt} = J_{max} \frac{c}{c + K_m} \tag{9.1}$$

The 'Michaelis constant' K_m equals the concentration at which the uptake flux is half of the maximal uptake flux (J_{max}). At low concentrations ($c \ll K_m$), this equation can be reduced to a linear relationship:

$$J_{upt} = \frac{J_{max}}{K_m} c \qquad (9.2)$$

The slope of this linear relationship between uptake flux and solution concentration has usually been termed the root absorption power (α) in plant uptake research and, more generally, the (membrane) permeability (P_m). The Michaelis-Menten parameters are often considered to be physiological characteristics [79]. However, this relies on the assumption that the bulk solution concentration equals the concentration at the site of uptake, which is not the case if diffusion limitation prevails, as discussed below (see Section 9.3.3).

Uptake fluxes and internal concentrations in the organism are related through the growth rate. Growth can often be described by an exponential equation:

$$W_t = W_0 . \exp\left(\mu_g t\right) \qquad (9.3)$$

where W_t and W_0 are the plant weights at time t and time 0, respectively, and μ_g is the specific (or relative) growth rate. In case of exponential growth, the uptake rate under steady-state conditions equals the product of the specific growth rate, μ_g, and the internal concentration in the biota, c^{bio}:

$$J_{upt} = c^{bio} . \mu_g \qquad (9.4)$$

Under non-deficient, non-toxic conditions, the growth rate will be at its maximal value (e.g., around $0.25\ d^{-1}$ for many plant species). In this case, a decrease in the solution concentration in the external medium, resulting in a decrease in the uptake flux (equation 9.1), will result in a decrease in the internal concentration in the biota. However, for nutrients, a minimum internal concentration (c^{bio}_{min}) is required for growth. As a result, under deficient conditions, the growth rate, rather than the internal concentration, will decrease with decreasing supply [80]:

$$\mu_g = \frac{J_{upt}}{c^{bio}_{min}} \qquad (9.5)$$

Conversely, if the internal concentration exceeds a toxic threshold, c^{bio}_{tox}, growth will also be affected. Figure 9.3 illustrates the relationship among nutrient supply, uptake, internal concentrations and growth rate.

Growth or other effects (deficiency or toxicity symptoms) can usually be related to the internal concentrations, with the effects only appearing below or above a certain threshold concentration. However, there are exceptions, for instance when complexes enter the cell and do not fully dissociate and hence are not assimilated in the same way as the free ion [81].

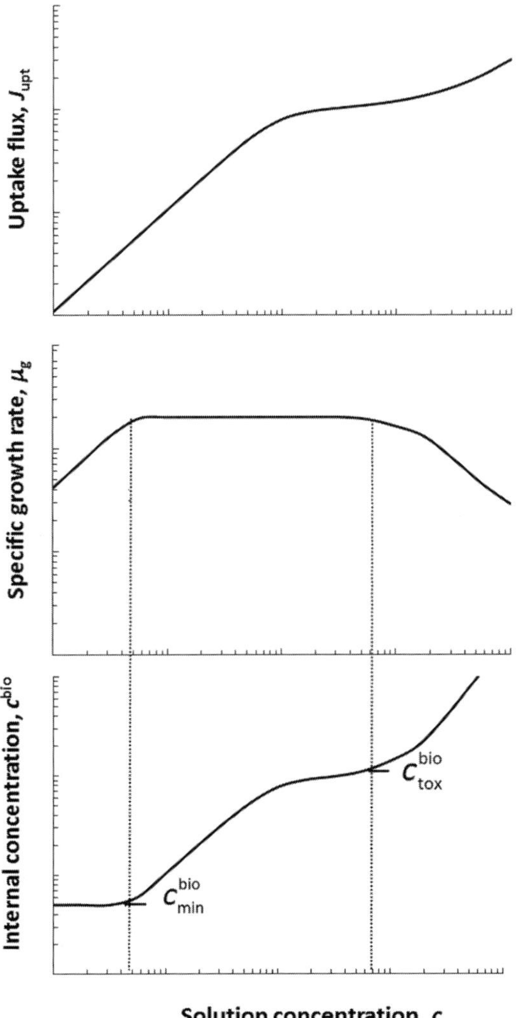

Figure 9.3 Schematic diagram showing uptake flux, specific growth rate and internal concentrations as functions of solution concentration in the external medium. The critical internal concentration for deficiency (c^{bio}_{min}) and toxicity (c^{bio}_{tox}) below or above which the growth rate is reduced is indicated. Simplified from ref. [80].

9.3.2 *Free Ion Activity Model (FIAM) and Biotic Ligand Model (BLM)*

The free ion activity model (FIAM) relates metal availability and toxicity to the free ion activity of the metal in the surrounding solution [82]. In the FIAM, the free metal ion is assumed to be in rapid equilibrium with the binding site at the cell surface [83]. The FIAM was developed based on experimental observations that bioavailability to aquatic organisms was mainly related to free ion activities and not to total dissolved metal concentrations [82].

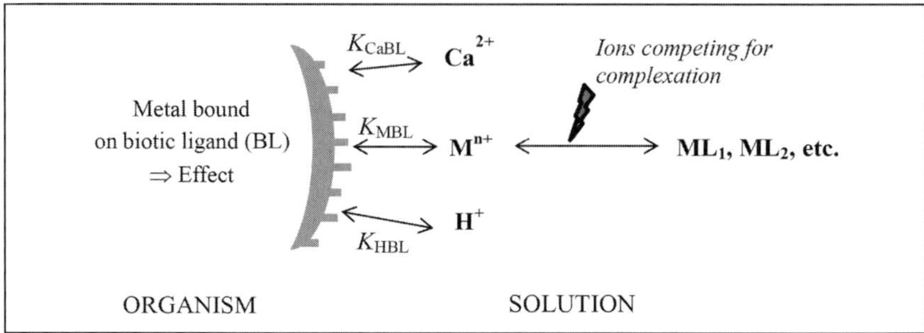

Figure 9.4 Schematic overview of the concepts of the biotic ligand model (BLM). The amount of metal bound to the biotic ligand determines the biological effect. The free ion (M^{n+}) of the element of interests binds to the biotic ligand. Other competing ions may also bind and thus affect the binding of the ion to the biotic ligand. In solution, the metal may form complexes with organic (e.g., DOM) or inorganic ligands. The extent of complexation is affected by the presence of other ions which also complex with these ligands.

This is related to the fact that most complexes of trace metals are too large to cross the cell membrane, ion channels or be taken up by membrane transporters.

 The biotic ligand model (BLM) is an extension of the FIAM [84]. Like the FIAM, the BLM is an equilibrium model that relates the effect of the metal to the activity of the metal ion in the surrounding solution, but it also takes into account the effect of competing ions in the uptake process (Figure 9.4). For instance, at the same free ion activity, the binding of the metal to the uptake site will be less in hard than in soft water if Ca^{2+} and Mg^{2+} compete with the metal ion for binding at the so-called biotic ligand. Similarly, at given metal activity, toxic effects are less at low pH than at high pH if protons compete for binding. This is mathematically expressed using constants that express the binding of the metal ion of interest and of competing ions to the biotic ligand. These binding constants are most commonly determined by measuring toxic effects at different concentrations of the potentially competing ions. To apply the BLM, the solution speciation (specifically the free ion activity of toxic ion and of competing ions) needs to be known. The speciation is calculated with speciation programs that take into account the complexation with inorganic and organic aquatic ligands, such as WHAM.

 The concepts implemented in the BLM have been around long before the BLM term was coined, for instance in the Michaelis-Menten equation with competitive inhibition [85], which can be written as follows [86]:

$$J_{\text{upt}} = J_{\text{max}} \frac{K_M \cdot c_M}{1 + K_M \cdot c_M + \sum_i K_i \cdot c_{X_i}} \tag{9.6}$$

where c_M and c_{X_i} are the free ion concentrations of the metal ion of interest and the competing ions X_i, and K_M and K_i the binding constants at the uptake site for the free metal ion and for competing ions. In the absence of competing ions, this equation can be simplified to the normal Michaelis-Menten equation (equation 9.1), with K_m, the Michaelis constant, the reciprocal of K_M. This type of model was for instance used to describe Ni biouptake by a green alga in the presence of competing ions [87].

The BLM was originally developed to predict toxicity of metals in natural waters to aquatic biota. More recently, the BLM framework has been further developed to describe metal toxicity to terrestrial plants. As plants are predominantly exposed to metals through the soil solution, it is assumed that the same concepts can be used as in aquatic BLMs. These BLMs to predict toxicity to plants can be developed using hydroponic solutions. For instance, in the development of a BLM to predict Ni toxicity for barley root elongation, Mg was found to be the main competitive ion [88]. Effects of Co toxicity on root growth of barley (*Hordeum vulgare*) in hydroponic solutions could be described using a BLM taking into account Mg and K competition. However, validation with two soils showed that the EC50 was only predicted within a factor of 4 [89]. These authors indicated that rhizosphere soil is different from bulk soil, which may (partly) explain the poor predictions.

In some studies, terrestrial BLMs (t-BLM) were developed based on measurements in soils, with free metal activities in soil solution either calculated from the soil solution composition [89] or from the total metal concentration in soil [90]. The number of studies that calibrated and validated t-BLMs is limited, but in general, it was found that the BLM framework predicted metal toxicity in soils reasonably well. For instance, Co toxicity to potworm (*Enchytraeus albidus*) was reasonably described taking into account competition of Ca^{2+}, Mg^{2+} and H^+ [91].

9.3.3 Deviations from the FIAM/BLM and the Role of Complexes on Bioavailability

The FIAM and BLM predict that complexes do not contribute to the uptake or toxic effect of a metal. Thus, if the FIAM concept applies, addition of a metal-complexing ligand to a solution in which all metal is present as free ion should reduce the uptake to the same extent as the decrease in free ion concentration. In a medium in which the free ion activity is buffered (e.g., soil), addition of a complexing ligand will increase the total dissolved concentration, but will have no effect on the free ion concentration unless the quantity of metal on the solid phase is considerably depleted and should, therefore, have no effect on the uptake if the FIAM applies. As discussed below, departures from the FIAM and BLM can in many cases be related to transport processes outside the organisms, which are the same processes that also affect DGT fluxes, and, therefore, these biophysical mechanisms are elaborated here.

Several studies with aquatic organisms have demonstrated exceptions to the FIAM. For instance, the presence of citrate complexes increased uptake of Cd and Zn by a freshwater alga, likely due to accidental or piggy-back uptake of the complex [92]. Similarly, enhanced uptake of Ag in presence of thiosulphate has been observed [93], most likely via a (thio)sulphate transporter. Also exceptions with lipophilic complexes have been reported, as these can diffuse across the cell membrane [94]. Furthermore, failures of the FIAM/BLM concept for Zn uptake by a green alga have been reported [95]. No first-order uptake flux was observed, and the uptake flux for algae preconditioned under sub-optimal Zn conditions was constant over a wide range of Zn^{2+} concentration in solution and much larger than for algae preconditioned under optimal conditions, which was attributed to up-regulation of high-affinity transporters. Due to active transport, both uptake fluxes and receptor-bound Zn may be independent of solution chemistry, and it has therefore been suggested that the FIAM and BLM are most useful for predicting the uptake of non-essential metals under reasonably controlled conditions, but have serious drawbacks to predict the uptake of essential metals which is also controlled by homeostatic processes [96].

Strong deviations from the FIAM/BLM concept have also observed for uptake of rare earth elements by freshwater algae [97]. The uptake of Tm^{3+} was up to almost two orders of magnitude higher in the presence of complexes than at the same Tm^{3+} activity in absence of complexes. This could not be attributed to diffusion limitations (see below), as the observed uptake flux was more than two orders of magnitude lower than the calculated maximum diffusive flux. The authors hypothesized that in this case, the contribution of the complexes was due to the formation of ternary complexes at the metal uptake site. As pointed out by the authors, organic ligands are ubiquitous in natural waters, and rare earth elements are usually mostly present as complexes with these ligands. Estimating their bioavailability based on free ion-based equilibrium models calibrated using simple solutions would likely result in strong underestimates of their availability in natural systems.

For metal uptake by higher plants, many exceptions to the FIAM have been reported, as shown by contribution of complexes to the uptake [98–101]. Such contribution may be due to direct uptake of the complex. Alternatively, the increased uptake in presence of complexes may be due to diffusion limitations in the uptake [102]. The FIAM and BLM assume that the metal is in equilibrium with the biotic ligand and with any ligands in the surrounding medium. In order for this equilibrium assumption to be fulfilled, the potential rate of uptake has to be slow compared to the rate at which the free ion can be supplied to the uptake site. If this is not the case, the uptake will be diffusion limited, and the free ion concentration at the uptake site will be lower than that in the bulk medium, invalidating the assumption of equilibrium between (bulk) solution and biotic surface. Under such conditions, complexes that dissociate quickly enough contribute to the uptake, as they dissociate within the diffusive layer and hence enhance the flux of the free ion (Figure 9.5). Degryse et al. [103] measured Cd uptake by spinach at constant free ion concentration (1 µM) in solutions with varying concentrations of complexes of various dissociation rates. Despite the fact that all solutions had the same Cd^{2+} activity, pH and Ca concentration, the uptake varied over 1,000-fold, demonstrating the FIAM/BLM concept did not hold. The uptake increased with increasing concentration and increasing dissociation rate of the complex and was highly correlated with the diffusive flux as measured by the DGT technique, demonstrating that uptake was diffusion limited. Similar results were found for other plant species and for Zn [104]. Similarly, Wang et al. [105] also indicated that Zn uptake by wheat grown in solution culture was diffusion limited and enhanced in presence of complexes. Gramlich et al. [106] determined uptake of Zn in presence of citrate or

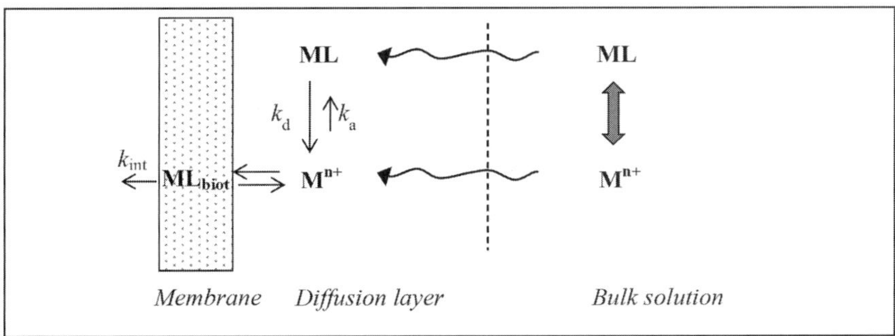

Figure 9.5   Diagram illustrating diffusion-limited uptake of a metal in the presence of complexes (k_{int}, k_d and k_a are first-order rate constants for the internalization and for the dissociation and association of the complex). In the bulk solution, the complex (ML) and free ion (M^{n+}) are in equilibrium. In the diffusion layer, the fast internalization compared to the diffusive supply of the free ion results in depletion of the free ion, which disturbs the equilibrium between complex and free ion and promotes the dissociation of the complex. Adapted from ref. [102].

histidine complexes at a complex:free ion ratio of 400, with radiolabelling of both Zn and the ligands. The citrate complexes increased Zn uptake 3.5-fold and the histidine complexes 9-fold. The enhanced uptake in presence of citrate appeared to be mainly due to increased diffusive supply, while for the histidine complexes, direct uptake also played a role.

This concept of diffusion-limited uptake has been theoretically well developed [102, 107]. To take into account the effect of diffusion limitation on biouptake, the Michaelis–Menten equation (equation 9.1) can be combined with the diffusion flux:

$$J_{diff} = \frac{D}{\delta} \cdot (c_M - c_{M0}) \tag{9.7}$$

where J_{diff} is the area based diffusion flux, and c_M and c_{M0} the (free) metal ion concentration in the bulk solution and at the biosurface, respectively, and δ is the diffusion layer thickness. This results in the following equation for the uptake flux [108]:

$$J_{upt} = \frac{D}{\delta} 0.5 \left(K_m + \frac{J_{max}.\delta}{D} + c_M - \sqrt{\left(K_m + \frac{J_{max}.\delta}{D} - c_M\right)^2 + 4.c_M.K_m} \right) \tag{9.8}$$

At low concentrations, in the linear part of the curve, a simpler equation can be derived [108]:

$$J_{upt} = \frac{c_M}{\dfrac{\delta}{D} + \dfrac{K_m}{J_{max}}} = \frac{c_M}{\dfrac{1}{P_{diff}} + \dfrac{1}{P_m}} \tag{9.9}$$

where P_{diff} is the diffusional permeability (cm s^{-1}), corresponding to D/δ, and P_m the membrane permeability (cm s^{-1}), corresponding to J_{max}/K_m. If $P_m \gg P_{diff}$, the uptake is diffusion limited, which also results in a bias of the experimentally derived K_m (Figure 9.6). If $P_m \ll P_{diff}$, the uptake is internalization limited. Based on observations of the effects of

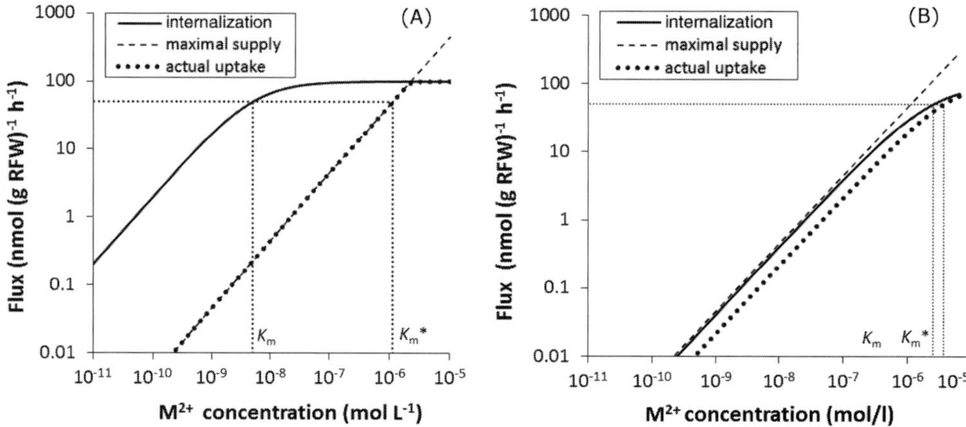

Figure 9.6 Conceptual diagram of the internalization flux (Michaelis–Menten curve; full line), maximal diffusive supply from solution to plant roots (dashed line) and actual uptake flux (dotted line) as a function of free ion concentration for two theoretical cases (left representative for Cd/Zn uptake by plants; right representative for Ni uptake). In (A), the potential internalization flux is much larger than the maximal diffusive supply at low concentrations, i.e., the uptake is strongly limited by the transport of the free ion to the root. The plant acts as a near-zero sink, and the actual plant uptake equals the maximal diffusive flux. The apparent K_m ($K_m{}^*$) is much larger than the 'true' K_m. In (B), the maximal diffusive flux is larger than the potential internalization flux. The uptake is not limited by diffusive transport, and the apparent K_m ($K_m{}^*$) and 'true' K_m are almost equal. (From ref. [109]. Copyright American Society of Plant Biologists, www.plantphysiol.org)

complexes and competing ions on metal uptake in hydroponic solutions, it was hypothesized that uptake of Cd and Zn by plants is strongly limited by diffusion and estimated that the physiological Michaelis constant is around 1 nM, i.e., 1,000-fold lower than determined in unbuffered solutions [109]. For Ni, in contrast, the uptake was estimated to be on the border between transport and internalization control. Soil-based experiments also suggested that diffusion limitations applied for uptake of Cd by both hyper- and non-hyper-accumulating plants and for Ni by hyperaccumulators, but that Ni uptake by non-hyperaccumulators was not limited by diffusion [110]. The effect of complexes on the diffusive supply depends on the concentration of the complex and on its dissociation rate. This can be expressed using reaction layer theory [111]. The reaction layer is the zone near the surface where the equilibrium between complex and free ion cannot be maintained and there is a strong decrease in free ion concentration compared to the bulk solution. Provided the complexes are not fully inert or fully labile [112], the thickness of the reaction layer, μ, can be calculated as

$$\mu = \sqrt{\frac{D_M}{k_d} \cdot \frac{c_M}{c_{ML}}} \tag{9.10}$$

where k_d is the first order association dissociation rate constant (s⁻¹), c_{ML} and c_M are the concentrations of complex and free ion, and D_M the diffusion coefficient of the free ion.

Table 9.4 *Estimated association rate constant with the biotic ligand (k_L) [116], membrane permeability (P_m; equation 9.12) at high (10^{-11} mol cm^2) or low (10^{-14} mol cm^2) surface ligand density (L_T) and the overall estimated permeability (P) for a diffusion layer thickness (δ) of 2 μm, representative for unicellular algae, or 300 μm, representative for plant roots. The permeabilities given in bold show the diffusion limited cases ($P_{diff} < P_m$, i.e., $P \rightarrow P_{diff}$).*

| | | | | P (cm s^{-1})[a] | | | |
| | | P_m (cm s^{-1}) | | $\delta = 2$ μm | | $\delta = 300$μm | |
Metal	k_L (M^{-1} s^{-1})	High L_T	Low L_T	High L_T	Low L_T	High L_T	Low L_T
Ni	1×10^5	1.0×10^{-3}	1.0×10^{-6}	9.6×10^{-4}	1.0×10^{-6}	**1.4×10^{-4}**	9.9×10^{-7}
Co	4×10^6	4.0×10^{-2}	4.0×10^{-5}	**1.5×10^{-2}**	4.0×10^{-3}	**1.7×10^{-4}**	3.2×10^{-5}
Mn	3×10^7	3.0×10^{-1}	3.0×10^{-4}	**2.3×10^{-2}**	3.0×10^{-4}	**1.7×10^{-4}**	**1.1×10^{-4}**
Zn	4×10^7	4.0×10^{-1}	4.0×10^{-4}	**2.4×10^{-2}**	3.9×10^{-4}	**1.7×10^{-4}**	**1.2×10^{-4}**
Cd	4×10^8	4.0×10^0	4.0×10^{-3}	**2.5×10^{-2}**	3.4×10^{-3}	**1.7×10^{-4}**	**1.6×10^{-4}**
Cu	1×10^9	1.0×10^1	1.0×10^{-2}	**2.5×10^{-2}**	7.1×10^{-3}	**1.7×10^{-4}**	**1.6×10^{-4}**

[a] The permeability P was calculated as $P = 1/(1/P_{diff} + 1/P_m)$, assuming a diffusion coefficient (D) of 5×10^{-6} cm^2 s^{-1} for the calculation of P_{diff} ($= D/\delta$)

The diffusion flux increases as the reaction layer thickness decreases, i.e., with increasing degree of buffering (c_{ML}/c_M) and with increasing dissociation rate. The uptake flux in presence of complexes can be calculated using the reaction layer thickness instead of the diffusion layer thickness [109]:

$$J_{upt} = \frac{D}{\mu}0.5 \left(K_m + \frac{J_{max} \cdot \mu}{D} + c_M - \sqrt{\left(K_m + \frac{J_{max} \cdot \mu}{D} - c_M \right)^2 + 4.c_M.K_m} \right) \quad (9.11)$$

While this analytical solution is only an approximation, it provides results in very good agreement with those obtained using numerical simulations [103, 113].

Several studies have indicated that diffusive transport may limit the uptake of metals [9, 105, 106, 109] and phosphorus [114] by plants, even in well-agitated hydroponic solutions. This situation is even more likely in soils because the high tortuosity in the soil porous network reduces the diffusive flux compared to a solution. Also for metal uptake by fish and mussels, diffusion limitations have been suggested [115]. For unicellular organisms, there are far fewer studies that have suggested that uptake is diffusion limited, which can be related to the smaller thickness of the diffusion layer. The diffusive layer thickness for radial diffusion towards a spherical organism approximates the radius of the organism in steady state. Thus, for unicellular algae, the value of δ is usually around a few μm. In contrast, the diffusive layer thickness near plant roots has been estimated to be several 100 μm in thickness [109]. Thus, P_{diff} in absence of complexes is about two orders of magnitude higher for unicellular algae than for higher plants, and uptake therefore is less likely to be diffusion limited. This is illustrated in Table 9.4, in which the membrane permeability P_m

is estimated according to the following equation (which assumes that the internalization rate is fast compared to the dissociation rate of the metal-surface ligand complex) [80]:

$$P_m = L_T \cdot k_L \tag{9.12}$$

where L_T is the density of surface ligands and k_L is the association rate with the surface ligand. Since the loss of a water molecule from the hydration sphere of the metal ion is the rate limiting step for association with the surface ligand, k_L is proportional to the water exchange constant (k_{-w}). These theoretical considerations rely on many assumptions, but they show that for small unicellular organisms, uptake is generally expected to be internalization limited ($P_m < P_{diff}$), except at high surface ligand density for ions with fast water exchange. Indeed, it has been reported that at low Zn concentrations, Zn internalization can be up-regulated to the point where Zn diffusion becomes limiting [95]. For plant roots, for which the thickness of the diffusion layers is much larger, diffusion limitation is much more likely, with the exception of Ni, and to lesser extent Co, due to their slow water exchange kinetics.

9.3.4 Barber–Cushman Model and Related Models

The Barber–Cushman model is a mechanistic model developed to describe nutrient uptake by plants from soil. It is a dynamic model that combines Michaelis–Menten uptake kinetics with nutrient supply to the roots through diffusive supply as well as transpiration-driven mass flow. The model input includes plant-related parameters (root length and diameter, growth rate, water influx and Michaelis–Menten parameters) and soil-related parameters (water content, initial soil solution concentration and buffer power). The Barber–Cushman model and related models have mostly been used to model uptake of macronutrients, such as P [117, 118] and K [118, 119], but has in recent decades also been used for metals, such as Cd [120] and Zn [121]. However, as the model was initially developed for macronutrients, which are mostly present as free ion in the soil solution, it does not take into account solution speciation. Except in high pH soils, Cd and Zn are usually mainly present as free ion, justifying the use of the model. However, for other trace elements (e.g., Cu), often only a minor part of the metal in solution is present as free ion, and use of a model that considers that the element in solution is in free ionic form would not be appropriate. Expansion of the model to take into account the effects of solution speciation has already been carried out for simple ligands [122], but not for complexes with dissolved organic matter which cover a range of complexation strength and dissociation kinetics. Furthermore, the Barber–Cushman model assumes that there is an equilibrium between the solid phase and solution phase. However, desorption from the solid phase may be kinetically limited, and this type of plant uptake model could hence be further expanded to take into account the desorption kinetics of solid-phase bound metals (as is for instance done in the DIFS model [58]). However, obtaining reliable input parameters for such a model will remain a challenge, and this type of model is therefore unlikely to be used for general practice, but it is useful as a research tool to assess the importance of different variables and identify ways to modify nutrient uptake.

9.3.5 *Empirical Relationship Between Toxicity Thresholds and Soil Properties*

The BLM framework gives a mechanistic description of toxicity of metals and can be used to predict or explain metal toxicity in soil, but as discussed further on, the assumptions inherent to the FIAM/BLM may not always hold. Furthermore, not only does the development of a (terrestrial) BLM for a given species and metal require careful calibration and validations, its application is also not straightforward, as it requires measurements that are not routinely done (soil solution composition) as well as calculation of free ion activities. This requires speciation modelling which involves inherent uncertainties. In a more pragmatic approach, empirical models have been developed that relate toxicity threshold to soil properties. For instance, Cu toxicity thresholds for soil microbial processes across seventeen soils with widely varying properties were found to increase with increasing cation exchange capacity (CEC) and Ni toxicity thresholds increased with increasing CEC, clay content and Ni background values [123]. In a summarizing study, it was reported that for Cu, Ni, Zn and Co, toxicity thresholds based on total soil metal concentrations generally rise proportionally to the effective cation exchange capacity of soil. The fact that pH was not a consistent factor, despite its strong effect on metal adsorption in soil, may be due to protons having a similar effect on the adsorption of the metal ion on the soil surface and the complexation of the metal ion with the biotic ligand [123].

9.3.6 *Sediment-Specific Models*

The exposure of benthic (sediment-dwelling) organisms may occur via the water phase, but also through ingestion of sediment particles. The relative contribution of each exposure route depends both on sediment conditions (e.g., concentration in interstitial and overlying water and in the sediment) and on the organism as burrowing and feeding behaviour greatly varies [124].

Speciation of elements in sediments is similar to that in soils, with elements in interstitial water present as free ion, complex or colloid and elements associated with the solid phase either in adsorbed or precipitated form. However, redox processes play an important role in comparison with (aerobic) topsoils, which contributes to large temporal and spatial variability [125]. Sediment quality criteria have been empirically derived, most often from frequency distributions relating total concentrations to field or laboratory biological effects data [126]. However, as with soils, total metal concentrations bear little relationship to the availability of the metals. Models have been developed that attempt to account for differences in bioavailability. The SEM/AVS concept has been found useful to explain availability of metals in sediment [127]. The acid-volatile sulphide (AVS) and simultaneously extracted metals (SEM) in the same cold hydrochloric acid extract are used as measures of reactive sulphide and metal in the sediment. As the precipitates of sulphides with potentially toxic metals (typically Cd, Cu, Ni, Pb, Zn) are more stable than iron sulphide, it is assumed that these metals are present as insoluble metal sulphides if the molar AVS concentration exceeds the molar sum of SEM concentrations (i.e., SEM/AVS < 1). In further development of this concept, the excess SEM (SEM-AVS) is normalized against the organic C content of the sediment. Organic matter is often the main sorbent for trace elements in sediments, and hence controls the activity of metals that are not precipitated. Thus, depending on the organic C content, metal activities in solution may still be below a given toxic threshold even if SEM/AVS > 1 [128]. As an extension of the SEM/AVS theory, a sediment BLM was developed, which assumes equilibrium partitioning between organic C in the sediment and the excess SEM (SEM-AVS) [129]. Using BLM parameters derived for algae in aquatic systems, it was estimated that the toxic threshold of organic C-normalized SEM depends mainly on pH, as the effect of pH on the solid-solution partitioning of the metal was predicted to be greater than the effect of binding of the free metal ion to the biotic ligand.

The equilibrium partitioning approach of the sediment BLM and other equilibrium models only consider the exposure through water. It has been suggested that this is usually the main exposure pathway, chiefly based on observations made in metal-spiked sediments. However, as also observed for soils [130], porewater concentrations are usually much higher for laboratory-spiked than for field-contaminated sediments [124]. Thus, use of laboratory-spiked sediments likely results in an overestimate of the toxicity and an underestimate of the importance of the ingestion route. Other models that have considered the exposure via sediment ingestion have suggested that this pathway is often a major one. However, these type of models, such as the exposure-effect model (EEM) described by Simpson [131] or biodynamic models, pose the challenge that uptake and efflux rates for each exposure route need to be known, and this type of information is not easily obtained. At the moment, no extensive datasets are available for better development of these kinetic-based exposure and effect models [124].

9.4 DGT in Relation to Bioavailability

9.4.1 General Considerations

The DGT technique is a dynamic speciation technique, and conceptually similar to voltammetric speciation techniques, as it measures the diffusion flux toward a device. These kind of techniques do not measure a defined species (e.g., free ion), but they measure the species that are 'labile' within the effective time scale [111]. This time scale is not related to the duration of the measurement (deployment in case of DGT), but rather to the time available for complexes to dissociate. For voltammetry this is determined by the thickness of the diffusion layer (δ) while for DGT the thickness of both the diffusion and binding layers play a part (see Chapter 5). The contribution of complexes to the flux will be larger as these thicknesses increase, due to the longer residence time of the complex within the depletion layer. To quantify the contribution of complexes to the diffusion flux, the (degree of) lability, ξ, has been defined (see equation 5.5). A value of 1 for ξ indicates that the complexes are fully labile (and diffuse as quickly as the free ion), while a value of 0 implies that the complexes are inert. We refer to Chapter 5 for an in-depth analysis of DGT as speciation tool.

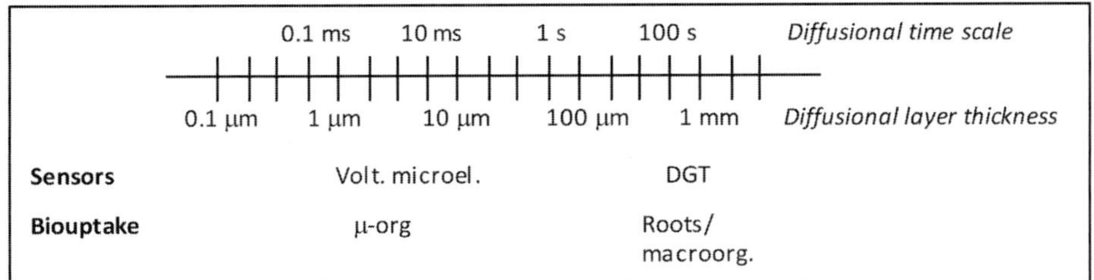

Figure 9.7   Thickness of the diffusion layer for analytical sensors and diffusion-limited biouptake processes, and the corresponding diffusional time scale. Adapted from ref. [102].

As discussed above, if uptake of an ion by biota is fast compared to the rate of supply by diffusion, the uptake will be diffusion limited. In such a case, the uptake will be determined by the species that contribute to the diffusion flux, and there is a clear parallel with the measurements by dynamic sensors, such as DGT. Ideally, if these sensors are used as predictors of diffusion-limited biouptake, the diffusional time scale should be similar for the measurement technique as for the biouptake process (Figure 9.7) [111]. The thickness of the diffusion layer during measurements with dynamic sensors is determined by the method used. In diffusion-limited biouptake, this thickness is less clearly defined and depends both on the organism and the hydrodynamic conditions. In the case of planar or cylindrical diffusion (e.g., to macroorganisms, roots, biofilms), the value of δ is highly dependent on hydrodynamic conditions and can vary from 10–100 µm under strongly agitated conditions to >1 mm under stagnant conditions [132]. For microorganisms, the radius of the organism is smaller than the hydrodynamic diffusive boundary layer, resulting in spherical diffusion, in which case the diffusion layer thickness effectively becomes the radius of the organism [86]. Thus, if uptake by unicellular organisms is considered, a technique with a diffusion layer thickness in the µm-range is theoretically most appropriate. A technique with larger dimension, such as DGT, would also measure the contribution of relatively slowly dissociating complexes that would not contribute to the biouptake. For instance, Tusseau-Vuillemin et al. [133] found that Cu-NTA complexes were almost fully DGT-labile, but did not contribute to uptake by *Daphnia magna*. Similarly, Meylan et al. [134] found that Cu uptake by periphyton correlated well with 'weakly complexed Cu,' including inorganic and weak Cu-organic complexes (not Cu-NTA), but did not show a good correlation with total dissolved Cu, which included Cu-NTA. No DGT measurements were carried out, but these would have most likely included Cu-NTA to a large extent, and hence would likely not have correlated well with the biouptake, even though the uptake was most likely diffusion limited. Thus, biosensors are expected to be a good predictor of diffusion-limited biouptake if the diffusional time scale is similar. However, if the diffusional time scale is larger/smaller for the sensor than for the biota, the contribution of complexes will be over-/underestimated.

If biouptake is not diffusion but internalization limited, the processes governing uptake are different from those governing the DGT flux. For instance, the presence of competing

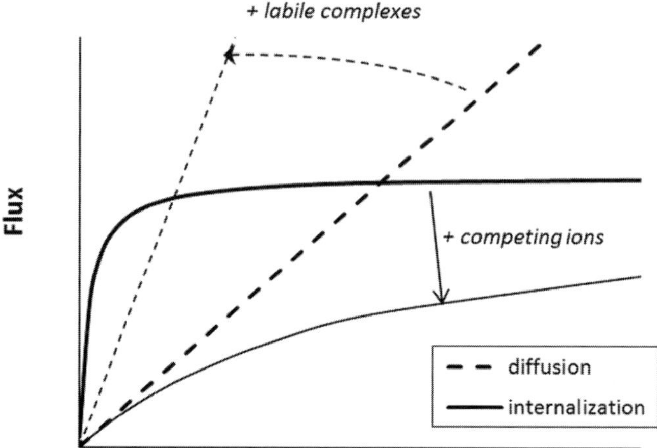

Figure 9.8   Schematic diagram showing the internalization flux and diffusion flux as a function of free ion concentration and how they are affected by the presence of labile complexes (diffusion flux) or the presence of ions that compete for uptake (internalization flux). The uptake is diffusion limited if the diffusion flux is smaller than the internalization flux and internalization limited if the internalization flux is smaller than the diffusion flux.

ions and the transporter (surface ligand) density affect internalization-controlled uptake. Whether uptake is diffusion or internalization controlled depends on the organism, the element and the external conditions. The biouptake may transition from diffusion to internalization limited if the concentration of competing ions increases (Figure 9.8) or if the transporter density decreases (Table 9.4). Furthermore, uptake is much more likely to be limited by internalization at high concentrations as the uptake starts to saturate (Figure 9.8). Thus, uptake is unlikely to be diffusion limited under toxic conditions.

Despite the obvious differences between DGT measurements and internalization-limited biouptake, DGT measurements may still correlate well with biouptake even if the uptake is internalization limited. For instance, across a wide range of waters, internalization-limited uptake may be mostly related to free ion concentrations (as competitive effects may be minor or similar across waters). If the free ion is the major species or if free ion fractions are fairly similar across waters (and free ion and total concentrations hence strongly correlated), DGT measured concentrations would also give a strong correlation with biouptake. A strong correlation between DGT fluxes and biouptake fluxes is therefore no proof of diffusion-limited uptake.

Natural or manufactured inorganic nanoparticles, with typical sizes 10–100 nm, present a special case in the considerations about the relationship between DGT and bioavailability. These particles can enter DGT systems that have the conventional 0.45 μm (450 nm) filter between the diffusive gel and the outer solution. However, the particles have a much lower diffusion coefficient than the true solutes and, therefore, may hardly diffuse toward the binding layer. With the exception of humic substances, there is little definitive information about the contribution of nanoparticles to DGT measurements as discussed in ref. [135]. Overall, available results suggest that both for DGT measurements and biouptake, nanoparticles generally have little contribution, though some contribution of P associated with natural or synthetic nanoparticles to DGT and to uptake fluxes in plant roots has been observed [12, 114].

9.4.2 DGT as Predictor of Bioavailability of Elements in Aquatic Systems

Compared to soils, there are only few studies relating metal bioavailability in water to measurements by DGT or other passive sampler measurements [136]. As discussed above, uptake by aquatic microorganisms is probably more often than not under internalization control, in which case DGT does not offer an advantage over other speciation techniques from a mechanistic point of view. Moreover, in cases where uptake is diffusion controlled, the diffusion layer thickness in the DGT technique is much larger than that surrounding the microorganism and voltammetric microelectrodes may be a more appropriate choice as biosensor microorganisms (cf. Figure 9.7). Overall, few studies have given strong indications whether uptake is limited by internalization or by diffusive transport [111]. Most studies on biouptake by aquatic organisms have assessed correlations between uptake/availability and metal concentrations as measured by different techniques, but this does not allow conclusions to be drawn about the processes involved. Furthermore, studies that have related uptake of elements from waters or sediments to DGT fluxes have used different methodologies (Table 9.5) and reached different conclusions, making any generalizations difficult.

Surprisingly few studies have related DGT measurements to metal uptake by aquatic microorganisms under nontoxic conditions. Slaveykova et al. [31] measured metal uptake by green microalga (*Chlorella salina*) in Black Sea and artificial seawaters and related this to free ion concentrations (measured by hollow fiber permeation liquid membrane) and DGT-measured concentrations. The uptake fluxes were very strongly correlated with free ion concentrations and were smaller than the calculated diffusive fluxes, suggesting that uptake was internalization limited. In contrast, Bradac et al. [32] found that uptake of Cd by periphyton in freshwaters correlated strongly to DGT-measured concentration but not to free ion concentration for chemically modified freshwaters. Addition of NTA decreased the free ion concentration, but increased both DGT-labile Cd and Cd uptake, likely because NTA displaced Cd from slowly exchanging colloidal complexes. This result suggests that Cd uptake was diffusion limited, which might be related to the larger diffusion layer thickness for periphyton compared to that for single microorganisms. The diffusive boundary layer thickness for periphytic biofilms has been estimated to be several 100 μm under conditions of moderate flow and in the millimeter-range under stagnant conditions [137–139]. Indeed, it was estimated that the diffusive flux (assuming $\delta = 1$ mm) was of similar order of magnitude to the measured internalization flux, indicating that diffusion may have been the rate-limiting process [32].

Table 9.5 *Selected studies looking at the relationship between metal bioavailability and DGT measurements in waters or sediments*

Metals (conc.[a])	Species	Description	Results	Ref.
Cd ($\sim$0.1 nM), Cu ($\sim$8 nM), Ni ($\sim$8 nM), Pb ($\sim$0.2 nM)	Green algae	Black Sea waters. Metal uptake by green algae determined	Uptake correlated most strongly with and increased proportional to free ion concentration	[31]
Cd (20–40 nM)	Periphyton	Cd-spiked natural freshwaters with or without NTA	Cd-NTA contributed to uptake and DGT flux. Hence, uptake was better correlated with c_{DGT} than with free ion concentrations	[32]
Cu (0–5 μM)	Trout	Trout exposed to soft waters amended with different DOM	Uptake correlated better with free Cu than with DGT-labile or total dissolved Cu	[33]
Cu (not reported, c_{DGT} up to 8 nM)	Oyster	Oyster cages deployed at fourteen sites and Cu uptake determined	Positive correlation ($r = 0.79$) between oyster available and c_{DGT}	[34]
Cu (0.08–0.24 μM)	Aquatic moss	Flow-through field microcosm with natural Cd-spiked waters	Moderate correlation between uptake and DGT flux, improved by taking into account cation competition effects	[35]
CH_3Hg^+ (0.1 or 100 ng/L)	Baltic clam	Reconstituted seawaters spiked with labeled CH_3Hg^+. Accumulation in DGT and clam over time measured	Changes in salinity equally affected uptake by clam and DGT	[36]
Cd (0.1–0.2 nM), Cu (4–14 nM), Zn (2–24 nM)	Bivalves	Metal concentrations in sediments and bivalves monitored in five coastal sites	Correlations between uptake by bivalve and metal concentrations species dependent.	[37]
Cu (up to 12 μM)	Water flea	Toxicity tested in Cu-spiked water sampled at three different times	Less variation in toxic threshold when expressed as c_{DGT} than as total dissolved concentration	[38]

Al ($\sim$7 μM)	Brown trout	Gill accumulation of Al in acid chemically modified surface water	Strong correlation between uptake/physiological effects and c_{DGT}.	[39]
Cu (up to 200 mg/kg)	Aquatic plant	Artificial Cu-spiked sediments. Cu uptake and growth inhibition determined	Relatively weak and non-linear correlation ($r^2 = 0.6$) between uptake and DGT flux	[25]
Cd (0.7–40 mg/kg)	Asian clam	Cd-spiked sediment. Uptake and antioxidant enzyme activities of clam measured.	Body Cd correlated to c_{DGT} (and to total Cd)	[26]
Cu, Cd, Pb (up to 965, 12 and 1,172 mg/kg resp.)	Chironomid	Laboratory microcosms with six freshwater sediments. Metal uptake by chironomids determined	c_{DGT} significantly correlated with uptake for Cu and Pb, but total metal concentrations best predictor	[27]
Cr, Ni, Zn, Cd, Pb (polluted sediments)	Freshwater snail	Laboratory microcosms with thirty freshwater sediments. Metal uptake by snail determined	Tissue concentrations of Cr, Cu, Ni and Pb positively correlated with c_{DGT}. No positive correlation with SOM-normalized SEM-AVS	[28]
Cu (up to 1,000 mg/kg)	Bivalve	Three Cu-paint spiked sediments. Bivalve survival determined	Overlying water Cu concentration and DGT (probe) flux gave similar dose-response curves for different sediments. Total concentrations did not	[29]
Ni (up to $\sim$10,000 mg/kg)	Macro-invertebrates	Ni-spiked sediments in mesocosm or in situ. Abundance or diversity of invertebrates determined	SOM-normalized SEM-AVS explained differences in toxicity. c_{DGT} (probe, 0–1 cm) did not	[30]

^{a} Concentrations in solution or total sediment concentration (in mg/kg)

Diffusion layers at the surface of aquatic macroorganisms are larger than for single microorganisms, and DGT might therefore be a better predictor under diffusion-limited conditions than for microalgae. However, most studies have assessed uptake or response to metals in the higher/toxic concentration range, where uptake is likely internalization limited. Luider et al. [33] determined accumulation of Cu by trout gills in Cu-spiked soft waters amended with natural organic matter (NOM) of different sources. Allochthonous, highly colored NOM complexed Cu more strongly and also reduced Cu accumulation by gills more than low-colored autochthonous NOM. Since NOM–Cu complexes contribute to the DGT flux, the DGT-measured concentration showed less difference than gill accumulation between the waters with the different sources of NOM. Hence, uptake was better correlated with free ion concentrations than with c_{DGT}. In a flow-through field microcosm experiment with four contrasting Cu-spiked waters, the correlation between DGT measured Cu and bioaccumulation by aquatic mosses (*Fontinalis antipyretica*) was poor, as the bioavailability of Cu was much higher for the two soft waters, but this was not reflected in the DGT measurements [35]. The authors used the DGT-measured concentration as input for a BLM-type model to take into account this protective effect of the hardness cations. However, as the BLM is an equilibrium concept while DGT measures a dynamic flux, we do not encourage this approach. The BLM concept should technically only use free ion concentrations as input, and while DGT-measured concentrations would be closely related to free ion concentrations if there is little complexation, this is not the case for Cu in natural waters, which is mostly present as complexes with organic matter.

Despite the fact that uptake is unlikely to be diffusion limited under toxic conditions, some studies have found good relationships with DGT-measured concentrations. Jordin et al. [34] found a positive correlation between oyster-available Cu and DGT-labile Cu across fourteen sites. Martin et al. [38] found that c_{DGT} better explained variation in Cu toxicity of Cu-spiked water sampled from the same place at different times compared to total dissolved Cu. In a field experiment with acid surface waters [39], accumulation of Al by trout gill and Al-induced physiological stress responses correlated well with c_{DGT}. The concentration of labile monomeric Al was lower than c_{DGT} and appeared to underestimate Al toxicity. These results suggest that organically complexed Al contributed both to the DGT flux and the bioaccumulation by fish gills.

In anoxic waters and sediments, inorganic Hg can be converted into methylmercury (CH_3Hg^+) by reducing bacteria. This organic form is highly toxic and readily biomagnified in aquatic food webs, and is therefore of environmental concern. A DGT technique has been developed to measure methylmercury, using a thiol resin as binding layer [140]. Accumulation of CH_3Hg^+ by the Baltic clam was determined in reconstituted seawaters with different salinity, and it was found that changes in salinity or mercury speciation (with or without DOM) affected the uptake by clam and by DGT in a similar way [36]. It was concluded that DGT is a promising tool to describe the first step of CH_3Hg^+ biomagnification, i.e., the uptake from water into biota.

The toxicity of silver (Ag) nanoparticles (NP) on microalgae has been measured and compared to DGT measurements [40]. The DGT-measured concentrations corresponded to measurements with ultra-filtration or ion selective electrode, indicating that the AgNP were not detected with DGT. A small fraction (1%) was present as Ag^+ ions, which is considered the most toxic fraction. However, the ionic Ag^+ could not fully explain the toxicity of AgNP, since toxicity was higher in a solution with AgNP than in a $AgNO_3$ solution at same Ag^+ concentration. The reasons are unclear but the authors speculated that the release rate of Ag^+ (due to oxidation of AgNP) was higher in presence of the algae than in the solutions used for speciation measurement.

9.4.3 DGT as Predictor of Bioavailability of Elements in Sediments

In studies with surface-dwelling biota, DGT measurements are usually made on the sediment, either using a DGT device horizontally deployed on the sediment surface or using a DGT probe that is vertically inserted into the sediment. For instance, Caillat et al. [25] determined Cu accumulation and toxicity in an aquatic plant after exposure to Cu-spiked sediments and measured DGT fluxes at the sediment surface. Even though uptake correlated better with c_{DGT} than with other metal concentrations (free, total dissolved), the correlation was weak and the relationship non-linear, suggesting that the uptake was not diffusion limited due to saturation in the Cu uptake. Ren et al. [26] showed a positive correlation between body Cd of Asian clam exposed to Cd-spiked sediments and c_{DGT} determined at the sediment surface. However, only a single sediment was spiked and c_{DGT} strongly co-varied with other metal measurements (e.g., total sediment concentration and total dissolved concentration) which also correlated well with body Cd. Roulier et al. [27] conducted microcosm experiments with six freshwater sediments and compared DGT concentrations of Cd, Cu and Pb measured at the sediment surface to bioaccumulation in a chironomid. A significant correlation was found between bioaccumulation and c_{DGT} for Cu and Pb, but the total concentrations in the sediment correlated even more strongly. Yin et al. [28] measured bioaccumulation of metals by freshwater snail in microcosm experiments with thirty freshwater sediments. The DGT-measured concentrations at the surface showed a significant positive correlation with bioaccumulation for Cr, Cu, Ni and Pb, but not for Cd and Zn. No correlation was found with organic matter normalized excess SEM (SEM-AVS) for any of the metals, which led the authors to conclude that DGT devices make a promising tool for assessing metal bioavailability in sediments. Also Simpson et al. [29] suggested that DGT measurements may provide a more reliable indication of metal bioavailability than AVS-SEM measurements. They determined Cu toxicity to bivalves in three sediments spiked with Cu antifouling paint and found that concentrations in the overlying water and DGT-measured concentrations (using a probe) gave similar dose-response curves for the three sediments, while this was not the case for total Cu concentrations (or excess Cu, as AVS was negligible). In contrast, Costello et al. [30] found that Ni toxicity to macroinvertebrates in Ni-spiked sediments could be well explained based on organic matter normalized SEM-AVS, but not by DGT-labile concentrations determined with a probe inserted in the sediment. The authors hypothesized that DGT mobilized a solid-phase fraction not available to the invertebrates. However, this hypothesis seems unlikely. The DGT-measured concentration corresponds to the time-averaged concentration of DGT-labile species at the sediment:DGT interface. Especially in spiked sediments, where most Ni would likely be as free ion in the interstitial water, DGT-measured concentrations would strongly correlate with dissolved concentrations, which in turn, according to the SEM/AVS theory, should correlate with organic matter normalized SEM-AVS. A more likely explanation for discrepancy between the SEM-AVS and DGT results may be that insertion of the DGT probe increased Ni solubility in some sediments due to perturbation.

Metal uptake is not just determined by the speciation of metals in the surrounding medium, but also by the organism itself, through factors such as internal regulations, the importance of different exposure routes (e.g., ingestion) and the effect that the organism may have on its surroundings, e.g., through bioturbation. For instance, Sakellari et al. [37] determined bioaccumulation of Cd, Cu and Zn by four marine bivalves at five coastal sites. The accumulation was both site- and species dependent. For Cd, two of the four organisms tested showed a positive correlation between bioaccumulation and DGT-measured concentration in the dissolved phase, while bioaccumulation of Cu and Zn for these same two organisms showed positive correlations with total or HCl-extractable concentrations and not with the DGT-measured concentrations. For Cu, bioaccumulation did not correlate with DGT-measured concentrations for any of the organisms and for Zn a positive correlation was found for one of the organisms (mussel).

From the above, it is clear that no generalizations can be made regarding the use of DGT as predictor of bioavailability in water and sediments. However, DGT certainly can be a useful research instrument in the study of metal biogeochemistry in sediments and in the interpretation of bioavailability data, as it allows detailed measurements of spatial and temporal changes in metal solubility (see Chapter 8).

9.4.4 DGT as Predictor of Bioavailability of Elements to Plants or Soil Organisms

9.4.4.1 General Considerations

In the context of bioavailability to plants, DGT measurements in soil have often been reported as the effective concentration c_E (e.g. Figure 9.1). This is the hypothetical soil solution concentration required to supply the measured DGT flux if there was no solid phase buffering. The ratio between c_{DGT} and c_E (R_{diff}) depends on deployment time, porosity, soil density and diffusion coefficient, and is usually around 0.1 for a 24-h deployment time [141, 142], i.e., c_E is usually about tenfold larger than c_{DGT}. As a result, c_{DGT} and c_E are generally very strongly correlated. In some datasets that cover soils with contrasting porosities, there may be more variation in the R_{diff} values and hence a less strong correlation between c_{DGT} and c_E. However, the few studies that assessed correlation of plant response with both c_{DGT} and c_E did not find any benefit in using c_E [15, 143]. Since interpretation of c_{DGT} is easier (the flux-averaged concentration of DGT-labile species at the soil:DGT interface) and does not require additional calculation of R_{diff}, we have opted to use c_{DGT} in the further discussion.

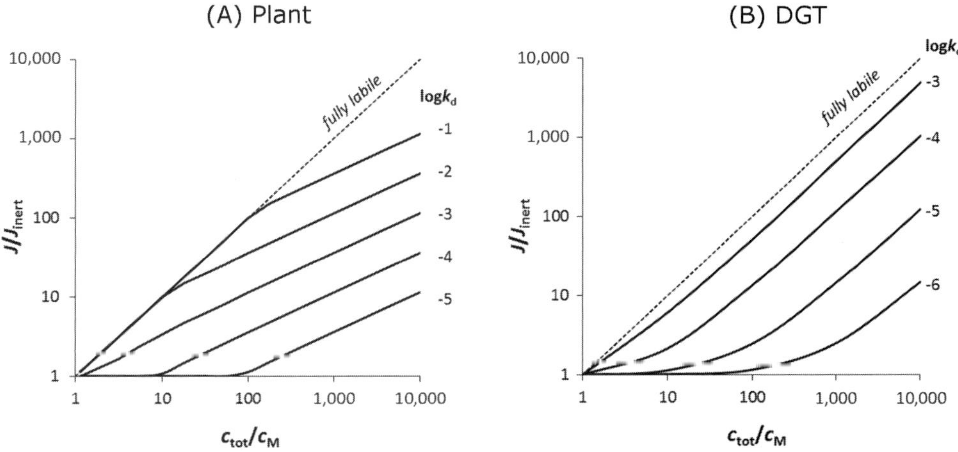

Figure 9.9   Effect of free metal ion buffering by complexes in solution on the diffusion flux in case of zero-sink diffusion for (A) plant uptake or (B) DGT uptake. Predicted increase in the flux relative to the flux at same free ion concentration but in absence of labile complexes (J_{inert}) as a function of ratio of total dissolved (= complex + free ion) to free ion concentration, for different dissociation rates of the complex (shown as log of the dissociation rate constant, k_d, in s^{-1}). In (A) a diffusion layer thickness of 0.4 mm is assumed. In (B) a diffusion layer thickness (Δg) of 0.8 mm and resin layer thickness of 0.4 mm are assumed. Adapted from ref. [21].

As discussed above, plant uptake is diffusion limited if the internalization flux (demand) is higher than the maximal diffusive flux (supply) of the species taken up, usually the free ion. Many factors affect these fluxes but, as a general rule, diffusion limitation is more likely at low solution concentrations (no saturation in uptake), for elements with potentially high internalization rates (e.g., fast water exchange in case of metal ions), in absence of competing ions, and for plant species with high demand (high transporter density). Diffusion limitations are unlikely to arise under toxic conditions, because of saturation in the uptake (cf. Figure 9.8).

There are clear parallels between DGT fluxes and diffusion-limited plant uptake. In both cases, a high-demand sink results in depletion of the free ion near the interface between DGT/plant and soil, and the thickness of the diffusion layers is of similar magnitude. While DGT fluxes are expected to correlate with plant uptake flux under diffusion-limited conditions, the fluxes are not necessarily equal, because of differences in geometry, soil water content, deployment time and physicochemical conditions [21]. Plant roots modify the soil surrounding the roots. Changes in the pH of the rhizosphere soil and root exudation may affect the solubility of mineral elements in rhizosphere [144, 145]. Furthermore, the presence of the Chelex layer in the DGT technique promotes the lability of the complex, as is illustrated by the model calculations in Figure 9.9 and discussed in detail in Chapter 5. It can be seen that for a same complex to free ion ratio and for a same dissociation rate of the complex, the complex contributes much more to the flux in the case of DGT. The complex diffuses into the resin layer, which greatly promotes its lability. A more rigorous analysis

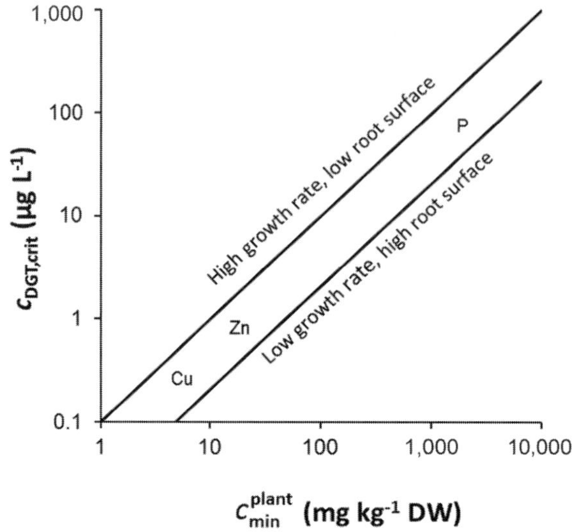

Figure 9.10   Theoretical relationships between critical DGT concentrations for deficiency ($c_{DGT,crit}$) and critical minimum concentration in the plant, c_{min}^{plant} (typical values for Cu, Zn and P indicated), for low growth rate and high root surface area or for high growth rate and low root surface area (per unit biomass). Adapted from ref. [21].

of this effect of the resin layer has been given by Mongin et al. [146], who also indicated that the thickness of the resin layer has a huge influence on the degree of lability. Indeed, Garmo et al. [147] observed that increasing the thickness of the resin layer increased the lability of lanthanide complexes.

Diffusion limitations are likely to occur under nutrient-deficient conditions if internalization is fast. Based on theoretical considerations, critical DGT concentrations have been estimated which correspond to the flux that would be just high enough to maintain the maximal plant growth (Figure 9.10). This critical concentration is proportional to the minimum internal concentration of the nutrient (c_{min}^{plant}) and the maximal growth rate, and inversely proportional to the root surface area per unit biomass [21]. Experimental evidence has indeed been found that DGT is a good predictor for P deficiency (see Section 9.4.4.3).

9.4.4.2 Trace Metals and Arsenic

There is extensive literature relating DGT measurements to availability of trace metals to plants [148]. Before discussing studies that assess relationships between c_{DGT} and plant uptake or concentrations across a range of soils, we first focus on studies that further illustrate the concepts of diffusion versus internalization limitation. As discussed above, DGT is considered to mimic the plant uptake under diffusion-limited conditions (diffusive transport to a high demand sink) but not under internalization-limited conditions (uptake determined by interactions with the biotic surface), although correlation between both can still be found in the latter case. Experiments in hydroponic solutions, which allow greater

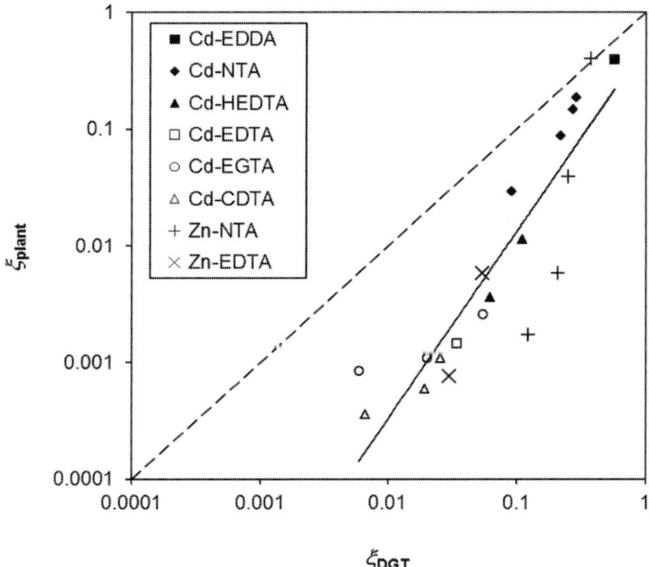

Figure 9.11   Relationship between the degree of lability ξ (Eq. 5.5) of Cd or Zn complexes with synthetic ligands as determined from plant uptake (ξ_{plant}) or from DGT measurements (ξ_{DGT}). Adapted from ref. [103] (uptake of Cd by spinach) and ref. [105] (uptake of Zn by wheat).

control of solution composition than soil-based experiment, have strongly suggested that plant uptake of Cd, Zn and Cu are limited by diffusion at low concentrations [103, 105, 112]. At constant and low free ion concentration, uptake of these metals was found to increase with increasing concentration or increasing dissociation rate of complexes with synthetic ligands. Furthermore, strong correlations with DGT-measured concentrations or fluxes were observed (see Figure 9.11), though DGT fluxes were generally higher than plant uptake fluxes, which can be explained by the presence of the Chelex layer promoting dissociation of the complexes during DGT deployment, i.e., DGT samples more of the labile complexes than plants (cf. Figure 9.9).

At high concentrations, plant uptake starts to saturate. As a result, uptake is more likely to be internalization limited, in which case the correlation between DGT and plant uptake flux may break down, as is illustrated in Figure 9.12, which shows uptake of Cd by spinach in nutrient solution at low or high free Cd^{2+} concentration in absence or presence of Cd-chloride complexes [9]. While Cd-chloride complexes contributed fully to the DGT flux both at low and high concentrations, they only contributed to the Cd uptake by spinach at the lower concentration, which was explained by saturation of the plant uptake (and hence internalization-limited uptake) at the higher concentration. Very similar results were observed in a soil experiment [9], in which a soil was amended with Cd at various concentrations and NaCl or $NaNO_3$. Chloride amendment, which resulted in increased Cd concentration in soil solution due to Cd–chloride complexation, increased Cd uptake by

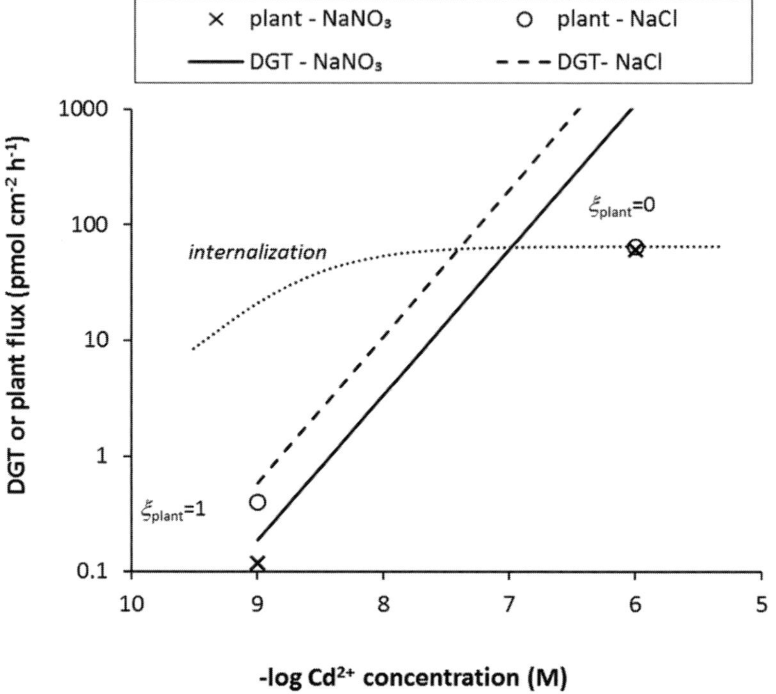

Figure 9.12   Uptake of Cd by spinach in nutrient solution (symbols) and the DGT-measured diffusion flux (lines) as function of free ion concentration for solutions with 40 mM $NaNO_3$ or NaCl. The free ion concentration was identical with both background salts but the total Cd concentration was threefold larger for the NaCl solution (68 percent of Cd complexed with chloride). The Cd–chloride complexes contributed fully to the DGT flux irrespective of the concentration, but only contributed to plant uptake at the low free ion concentration ($\xi_{plant} = 1$) and not at the high concentration ($\xi_{plant} = 0$). The dashed line shows the hypothesized internalization flux. Data from ref. [9]; graph adapted from ref. [21].

spinach at low Cd but not at high Cd concentrations. As a result, DGT fluxes and plant uptake fluxes increased proportionally at low but not at high Cd concentration.

In contrast with Cd and Zn, hydroponic experiments assessing the effect of contribution of complexes and stirring on Ni uptake have suggested that, even at low concentrations (in the linear part of the uptake curve), Ni uptake is on the border between internalization and diffusion control [109, 149], likely related to the slower water exchange kinetics of Ni (Table 9.4). This does not necessarily imply that DGT concentrations will not correlate to plant uptake. As discussed before, across a dataset, DGT-measured concentrations often correlate well with dissolved concentrations in the solution and for metals that are mainly present as free ion in soil solution (such as Ni) also with free ion concentrations. However, in case of internalization limitation, labile complexes will not contribute to plant uptake, unless they are taken up directly, and competition effects (which do not play a role in the DGT-measured concentrations) may affect uptake [109].

Soil experiments are much more complicated for interpretation and usually do not allow to distinguish unequivocally between diffusion- and internalization-limited uptake. Not only is it difficult to control solution composition as can be done in hydroponic experiments, but even the determination of solution speciation (e.g., free ion concentration) is challenging. Most studies relating DGT measurements in soils to metal availability have focused on the relationship between c_{DGT} and plant concentrations or uptake. Zhang and colleagues [3] were the first to correlate DGT measurements to plant uptake across a wide range of soils, and found that Cu concentrations in the plant correlated strongly to DGT-measured concentrations (Figure 9.1). In the last decade, several studies have assessed relationships between metal concentrations in plant and DGT-measured concentrations, with various but in general mostly positive results [148]. Good correlations between DGT measurements and concentrations of Cu, Ni, Pb and Zn in wheat roots were found in a pot trial with thirty soils covering a wide range of soil properties. In a pot trial with fourteen agricultural soils [8], good correlations between tissue (youngest leaf) concentrations in barley and DGT-measured concentrations were observed for Cu and Zn. In a pot trial with thirteen contaminated soils [5], only a weak correlation was found between concentrations in shoot of wheat and DGT-measured concentrations for Cu, while much better correlations were found for Cd and Zn. Black et al. [150] compared different soil metal bioavailability tests, including soil solution and DGT, in pot trials with long-term sewage sludge amended soils with moderate metal contamination. None of the tests explained more than 35 percent of the variation in shoot Cu, which was likely related to the strong physiological control on root-to-shoot translocation. For Cd, DGT performed better than other tests, whereas $Ca(NO_3)_2$ extraction was superior to the other tests for Ni and Zn. Several studies have also shown strong correlations between DGT measurements and metal uptake in field grown crops, with the DGT measurements either carried out in the field or in the laboratory [10, 11, 151].

There are several reasons why DGT concentration may or may not correlate with plant uptake. If uptake is internalization limited, as is expected to be the case at higher solution concentrations (e.g., under toxic conditions), DGT does not mimic the plant uptake process. Nevertheless, good correlations between DGT-measured concentrations and plant uptake may still be observed. As pointed out above, various metal indices are often strongly correlated within a dataset. This is certainly true when amending a single soil with metals, in which case strong correlations between total, dissolved, free and DGT-measured concentrations exist [152]. However, even in a dataset covering a wide range of soils, various metal indices often co-vary. These co-variations are often not reported, but explain why often nearly equally good relationships with plant uptake are obtained for different indices. For instance, Nolan et al. [5] found that uptake of Cd and Zn by wheat strongly correlated with the porewater concentration, free ion concentration and DGT concentration of these metals. Both Song et al. [4] and Tandy et al. [8] found that the correlation with Cu uptake by plants was almost equally strong for DGT-labile Cu and porewater concentration. Soriano-Disla et al. [6] reported similar correlation coefficients with Ni and Cu concentrations in plant for DGT-measured and porewater concentrations. Strong co-variations between DGT and porewater (total dissolved) concentrations are not surprising. Most species in soil solutions are DGT-labile (cf. Section 9.2.1.1), and the DGT measurements therefore reflect to a great extent the dissolved concentrations, especially when the dissolved concentrations span a wide range. For metals that are mostly present as free ion in solution (Cd, Zn, Ni at pH < 6.5), the DGT measurements are also expected to correlate strongly with free ion concentrations.

Some studies have specifically tested DGT as a predictor of toxicity, under which conditions the uptake is very likely not limited by diffusion. Nevertheless, the DGT technique may give good predictions of toxicity because it correlates well with soil solution metal concentrations, as argued above. For example, Zhao et al. [19] compared toxicity of Cu to tomato and barley plants in eighteen different soils. The differences in Cu toxicity thresholds (50 percent effect concentrations, EC50) between soils were large when expressed based on total Cu concentrations and even larger when expressed as Cu^{2+} ion activities. Soil solution Cu and DGT-measured concentrations improved the prediction of toxicity relative to total Cu and Cu^{2+} concentrations. For Zn toxicity to plants, DGT has been found to give a poorer prediction of toxicity than soil solution concentrations or concentrations in a 10^{-2} M or 10^{-3} M $CaCl_2$ extracts, but a better prediction than total concentrations or other indices of the 'quantity' (including adsorbed Zn) [20–22]. Along the same lines, toxicity of the molybdate anion in soil to growth of several plants was better explained by DGT than by total concentrations, but soil solution concentrations were somewhat better than DGT at explaining the toxicity [23].

While DGT-measured concentrations often correlate strongly with total dissolved concentrations, they may give superior predictions to total dissolved concentrations if total dissolved concentrations include inert colloidal species. This is illustrated in Figure 9.13, which shows Zn concentrations in shoot of wheat (grown in a pot experiment) as a function of soil solution or DGT-measured concentrations for laboratory-spiked or field-contaminated soils. In both cases, soil solution concentrations at the lowest Zn levels were much higher than the DGT-measured concentrations and showed poor relationship with plant concentrations, most likely due to the presence of inert colloidal Zn in the soil solution, which can be a major species at low Zn concentrations.

If uptake is diffusion limited, good correlations between DGT-measured concentrations and total plant uptake are to be expected, although this relationship may break down in cases where metal solubility is strongly modified by the presence of plant roots. For instance, Bravin et al. [7] found that DGT provided better predictions of Cu uptake by wheat than the FIAM or a terrestrial BLM, but DGT was a less good predictor than a rhizotest which measured fluxes within the rhizosphere and, therefore, accounted for the local pH (alkalinization) in the rhizosphere of acid soils. It is noteworthy that, under the conditions of this rhizotest, the rhizosphere alkalinization by the nitrate-fed plants yielded a near-neutral pH with little variation in rhizosphere pH across soils, despite a large variation in bulk soil pH [154]. The solid:liquid distribution coefficient (K_d) of Cu at neutral pH is expected to be similar to the solid:liquid distribution coefficient of (humified) organic matter [56], implying similar K_d values of Cu in the rhizosphere for soils with similar organic

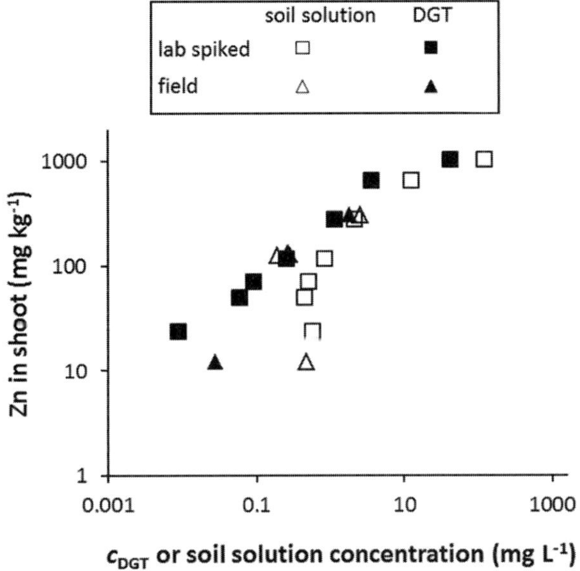

Figure 9.13   Zn concentrations in shoot of wheat as a function of DGT-measured or porewater concentration for a soil (Woburn, pH(CaCl$_2$) 6.4) amended in the lab with various ZnCl$_2$ concentrations up to a total Zn concentration of 1,900 mg Zn kg^{-1} or for soils from a field-contaminated transect (De Meern, pH(CaCl$_2$) 5.4–7.0, Zn up to 1,300 mg Zn kg^{-1}). Data from ref. [153].

matter content. In that case, total dissolved concentrations and potential diffusion fluxes in the rhizosphere would be related to total Cu concentrations, as was indeed experimentally observed in the study of Bravin et al. [7] and that explained why total Cu was a better predictor than free or dissolved Cu concentrations in the bulk soil, which reflected the more acid pH in the bulk soil. However, the stronger correlation of Cu uptake with total Cu than with dissolved or DGT-measured Cu is not generally applicable. While one other study [5] also found that total Cu in soil was a better predictor of Cu uptake by wheat than free or total dissolved Cu, several other studies have found that DGT was a better predictor than total, EDTA or resin-extractable Cu for Cu uptake by plants [3, 4, 8, 10, 155].

Another complication in the comparison between DGT and plant concentrations is that often relationships between DGT-measured concentrations and concentrations in above-ground tissue are assessed. Concentrations in the above-ground tissue do not only depend on the uptake but also on the translocation within the plant. As a result, while DGT concentrations may relate to concentrations in the total plant, the relationship with shoot or grain concentrations may be less pronounced, especially for metals for which the concentrations in above-ground tissue are strongly regulated (e.g., Cu) or that generally show limited translocation to the shoot (e.g., Pb, Cr). For instance, Soriano-Disla et al. [6] found that DGT-measured concentrations strongly correlated with concentrations of Cu ($r = 0.81$), Cr ($r = 0.80$) and Pb ($r = 0.78$) in wheat roots across a wide range of soils ($n = 30$), but no significant correlations with shoot concentrations, which were much lower than root concentrations, were observed. Similarly, Tian et al. [156] found weaker (though still highly significant) correlations with DGT-measured concentrations for shoot than for root concentrations of Cu, Pb and Zn.

Most studies relating bioavailability of trace elements to DGT measurements have focused on uptake of divalent cationic trace metals (Cd, Cu, Zn, Ni, Pb) by plants, using the standard DGT device with a Chelex resin layer, but there are a few studies that have compared bioavailability with DGT for other elements. Wang et al. [157] measured As concentrations in edible rape grown in forty-three different soils collected in China and assessed the relation with different soil tests, including different extraction methods, DGT (with ferrihydrite in the binding layer) and soil solution measurements. The DGT-measured and soil solution concentrations of As gave equally good predictions of the plant As concentrations and much better than for the extraction techniques (including acid digest). Williams et al. [143] related As concentration in flooded rice grain from the field to various As measurements in the soil for a laboratory experiment with thirty-nine soils. The DGT device used ferrihydrite as the binding layer that binds both the anionic arsenate (As(V), dominant form in aerobic soils) and the neutral arsenite (As(III), formed under anaerobic condition). It was found that As solubility in the anaerobic soils increased with increasing DOC concentrations. When assessing single correlations, As in the irrigation water gave better predictions of As in grain than any other soil property, including c_{DGT}, while a model including both c_{DGT} and DOC predicted As concentrations in grain most reliably. Increasing DOC increased As solubility in soil, likely because of the formation of As–DOM complexes (possibly bridged by Fe/Al), but decreased the availability to plant (grain As) at equal c_{DGT} for reasons yet unclear. A possible explanation might be that the DOM-associated As contributes more to the DGT flux than to the biouptake.

A couple of studies have used DGT in the study of uranium (U) availability. The speciation of uranium in soil solutions is complex, with U(IV) occurring as UO_2^{2+} or as complexes with carbonates or DOM. Duquene et al. [158] assessed the relationship between U uptake by ryegrass and DGT-measured concentrations in eighteen U-spiked soils in a greenhouse experiment (using a DGT device with a Chelex binding layer). Soil solution U concentrations and c_{DGT} were strongly correlated ($r = 0.85$), but both measurements only showed a weak correlation with plant uptake. An earlier study on U uptake by the same group used six soils that were either spiked and aged or derived from U contaminated sites [159]. Ryegrass U concentrations were equally well correlated to soil solution U as to c_{DGT} and better than selective soil extractions (0.11 M CH_3COOH or 0.4 M). Both studies concluded that the DGT method has little additional value in assessing uranium bioavailability compared to conventional tests.

Liu et al. [160] measured concentrations of CH_3Hg^+ in grain rice grown in contaminated paddy field and carried out DGT measurements using DGT probes inserted in the soil. Concentrations in grain rice showed a strong positive relationship with the DGT-measured concentrations, and it was suggested that DGT could be a useful monitoring tool to assess and mitigate risk of CH_3Hg^+-contaminated rice.

9.4.4.3 Phosphorus

As argued above (Section 9.4.4.1), DGT might serve as an indicator of nutrient deficiency and has been tested and applied to predict P deficiency in Australian soils. Menzies et al. [13] tested the yield response of tomato to addition of P fertilizers for twenty-four soils and found that the DGT-measured concentration allowed excellent separation between non-responsive and responsive soils. Soil solution P, and to lesser extent resin P, also provided a good separation, while this was not the case for Colwell P. The critical c_{DGT} was around 120 µg L^{-1}. In a pot trial with twenty-eight soils in which the response of wheat to liquid and granular fertilizers was compared [14], DGT also well predicted responsiveness to fertilizers, with a critical c_{DGT} around 40 µg L^{-1}. In a large study comparing the response of wheat to application of fertilizer P under field conditions (thirty-five response field trials across Australia) [15], it was again found that the DGT method predicted plant responsiveness to fertilizer P more accurately than Colwell and resin P. No significant relationship between relative yield and Colwell P was observed and only a weak relationship for resin P, while DGT explained 75 percent of the variation in response for both early dry matter and grain. The critical c_{DGT} was 255 µg L^{-1} for early dry matter and 66 µg L^{-1} for grain. These values fall within the range expected based on theoretical analysis (Figure 9.10) [21]. Six et al. [16] determined growth response of maize and upland rice to P fertilizer application in a pot trial with nine tropical soils (Figure 9.1). For maize, DGT and CaCl$_2$ extraction (both intensity measures) gave the best prediction of relative yield, with a critical c_{DGT} of 73 µg L^{-1}. In contrast, quantity-based indices (e.g., Olsen, Bray-1) gave a better prediction for upland rice and the critical c_{DGT} was variable between soils and much lower than for maize (around 8 µg L^{-1}). It was suggested that the lower value for rice may be due to the lower growth rate and hence lower P demand of rice.

Most studies relating DGT to P availability in soil analyzed the correlations with yield response, but it has been suggested that DGT may also give a good indication of P uptake under non-deficient conditions. Tandy et al. [8] found that plant tissue concentrations (youngest leaf of barley) determined in a pot trial with fourteen agricultural soils correlated strongly with DGT-measured concentrations ($r^2 = 0.72$), while only a weak correlation was observed with soil solution P ($r^2 = 0.43$) and no significant correlation with Olsen P.

Two recent hydroponic studies have also indicated that nanoparticulate P species, either associated with synthetic Al-oxide NP [114] or with natural soil colloids present in Andisols [12], may contribute to both plant uptake and DGT fluxes. While the degree of lability of these colloidal species was low (ξ_{plant} from 0.04 to 0.12), the plant uptake was up to eightfold increased in presence of colloids because of the abundance of the nanocolloidal species [12].

9.4.5 New Developments

DGT is a relatively new technique: the first demonstration of its potential as predictor of bioavailability was shown in 1999 [161] and its applications in this area are still expanding. A new type of DGT to measure potassium has been developed by two independent groups

[17, 18]. Good correlations between K concentrations in shoots of winter barley and DGT-measured concentrations were found in a pot trial with fourteen soils, but ammonium-acetate extractable K gave an even better correlation [17]. Preliminary experiments have also been carried out in developing and testing a DGT method to measure available S [162].

New developments do not only include the application of DGT for other elements, but also in other media. Pelfrêne et al. [163] used the DGT technique to determine lability of metals (Cd, Zn, Pb) in synthetic gastrointestinal fluids. Verheyen et al. [164] compared DGT fluxes to uptake fluxes by Caco-2 cells (a cell model for intestinal absorption) and found that the uptake fluxes at low concentrations increased in presence of labile complexes and correlated well with DGT fluxes at low Cd^{2+} concentrations, suggesting that the uptake was diffusion limited.

Recently, some studies have also used DGT in the study of manufactured nanomaterials. However, this is a challenging application, since the pore size of the diffusive layer plays a critical role and determining diffusion coefficients of nanoparticles is not straightforward [165]. It has been suggested that the DGT technique might be useful to measure available concentrations of manufactured nanomaterials, but further research is required [135, 165]. As described above, there has been recent progress in understanding the role of colloids (nanoparticles) on P uptake in plants by observing that there is contribution of P asso-ciated with natural or synthetic nanoparticles to DGT flux and to uptake fluxes in plant roots [12, 114].

9.5 Concluding Remarks

DGT measures the diffusive flux towards a zero sink, and is therefore considered to mimic, to some extent, uptake by biota under diffusion-limited conditions. Under such conditions, the DGT can be a practical and successful tool to diagnose and predict the bioavailability. These conditions are met for elements with low mobility in the environment combined with a relatively high internalization rate (e.g., P) in soils migrating to plant roots. Not surprisingly, DGT works relatively successful for diagnosing P deficiency in soil and DGT has found its way into commercial applications of soil testing in Australia. Diffusion controlled uptake in biota does, however, not warrant that DGT will be successful. Even though the diffusion layer thicknesses may be similar, complexes are more labile during DGT deployment because of the presence of the binding layer. In addition, biota can modify their surroundings (e.g., rhizosphere) which may affect the solubility of the element of interest. Furthermore, if DGT is used as predictor of uptake/concentrations within part of an organism (e.g., grain), translocation processes within the organism, which are not accounted for by DGT, may also contribute to the differences between soils. Conversely, if diffusion is not limiting the uptake, DGT does not mimic the processes, but may still correlate with uptake, albeit that fluxes will differ in absolute values. Indeed, many examples have been found where DGT correlated well with trace metal uptake/bioaccumulation at high (close to toxic) metal supply in soil, water or sediments. The DGT concentrations often correlate with dissolved and free ion concentrations across a dataset, and might have an advantage over these measurements because of DGT's ease of operation, lower detection limit, and exclusion of inert colloidal species.

Given the many processes that operate during biouptake, it is no surprise that no single technique can predict uptake/availability across all media and organisms [102]. However, DGT has often been found to correlate better with biouptake than other measurements. DGT may also be useful in assessing the role of nanoparticles (or colloids) carrying labile-sorbed elements on bioavailability, but this still needs further exploration. The same is true for diagnosis of Al toxicity, for which promising early results have been found. Complexes with Al^{3+} have slow association/dissociation kinetics, and DGT can be used to study these kinetics. In general, the DGT technique can be used as a convenient research tool to obtain additional information about physicochemical processes that play a role in bioavailability.

References

1. A. L. Nolan, E. Lombi and M. J. McLaughlin, Metal bioaccumulation and toxicity in soils – Why bother with speciation?, *Aust. J. Chem.* 56 (2003), 77–91.
2. W. Davison, P. S. Hooda, H. Zhang and A. C. Edwards, DGT measured fluxes as surrogates for uptake of metals by plants, *Adv. Environ. Res.* 3 (2000), 550–555.
3. H. Zhang, F.-J. Zhao, B. Sun, W. Davison and S. P. McGrath, A new method to measure effective soil solution concentration predicts copper availability to plants, *Environ. Sci. Technol.* 35 (2001), 2602–2607.
4. J. Song, F. J. Zhao, Y. M. Luo, S. P. McGrath and H. Zhang, Copper uptake by Elsholtzia splendens and Silene vulgaris and assessment of copper phytoavailability in contaminated soils, *Environ. Pollut.* 128 (2004), 307–315.
5. A. L. Nolan, H. Zhang and M. J. McLaughlin, Prediction of zinc, cadmium, lead and copper bioavailability to wheat in contaminated soils using chemical speciation, diffusive gradients in thin films, extraction, and isotope dilution techniques, *J. Environ. Qual.* 34 (2005), 496–507.
6. J. M. Soriano-Disla, T. W. Speir, I. Gómez et al., Evaluation of different extraction methods for the assessment of heavy metal bioavailability in various soils, *Water Air Soil Pollut.* 213 (2010), 471–483.
7. M. N. Bravin, A. M. Michaud, B. Larabi and P. Hinsinger, RHIZOtest: A plant-based biotest to account for rhizosphere processes when assessing copper bioavailability, *Environ. Pollut.* 158 (2010), 3330–3337.
8. S. Tandy, S. Mundus, J. Yngvesson et al., The use of DGT for prediction of plant available copper, zinc and phosphorus in agricultural soils, *Plant Soil* 346 (2011), 167–180.
9. C. Oporto, E. Smolders, F. Degryse, L. Verheyen and C. Vandecasteele, DGT-measured fluxes explain the chloride-enhanced cadmium uptake by plants at low but not at high Cd supply, *Plant Soil* 318 (2009), 127–135.
10. J. O. Agbenin and G. Welp, Bioavailability of copper, cadmium, zinc, and lead in tropical savanna soils assessed by diffusive gradient in thin films (DGT) and ion exchange resin membranes, *Environ. Monit. Assess.* 184 (2012), 2275–2284.
11. P. N. Williams, H. Zhang, W. Davison et al., Evaluation of in situ DGT measurements for predicting the concentration of Cd in Chinese field-cultivated rice: Impact of soil Cd: Zn ratios, *Environ. Sci. Technol.* 46 (2012), 8009–8016.

12. D. Montalvo, F. Degryse and M. J. McLaughlin, Natural colloidal P and its contribution to plant P uptake, *Environ. Sci. Technol.* 49 (2015), 3427–3434.
13. N. W. Menzies, B. Kusomo and P. W. Moody, Assessment of P availability in heavily fertilized soils using the diffusive gradient in thin films (DGT) technique, *Plant Soil* 269 (2005), 1–9.
14. T. M. McBeath, M. J. McLaughlin, R. D. Armstrong et al., Predicting the response of wheat (Triticum aestivum L.) to liquid and granular phosphorus fertilisers in Australian soils, *Aust. J. Soil Res.* 45 (2007), 448–458.
15. S. Mason, A. McNeill, M. J. McLaughlin and H. Zhang, Prediction of wheat response to an application of phosphorus under field conditions using diffusive gradients in thin-films (DGT) and extraction methods, *Plant Soil* 337 (2010), 243–258.
16. L. Six, E. Smolders and R. Merckx, The performance of DGT versus conventional soil phosphorus tests in tropical soils – maize and rice responses to P application, *Plant Soil* 366 (2013), 49–66.
17. S. Tandy, S. Mundus, H. Zhang et al., A new method for determination of potassium in soils using diffusive gradients in thin films (DGT), *Environ. Chem.* 9 (2012), 14–23.
18. Y. Zhang, S. Mason, A. McNeill and M. J. McLaughlin, Optimization of the diffusive gradients in thin films (DGT) method for simultaneous assay of potassium and plant-available phosphorus in soils, *Talanta* 113 (2013), 123–129.
19. F. J. Zhao, C. P. Rooney, H. Zhang and S. P. McGrath, Comparison of soil solution speciation and diffusive gradients in thin-films measurement as an indicator of copper bioavailability to plants, *Environ. Toxicol. Chem.* 25 (2006), 733–742.
20. O. Sonmez and G. M. Pierzynski, Assessment of zinc phytoavailability by diffusive gradients in thin films (DGT), *Environ. Toxicol. Chem.* 24 (2005), 934–941.
21. F. Degryse, E. Smolders, H. Zhang and W. Davison, Predicting availability of mineral elements to plants with the DGT technique: A review of experimental data and interpretation by modelling, *Environ. Chem.* 6 (2009), 198–218.
22. F. Hamels, J. Maleve, P. Sonnet, D. B. Kleja and E. Smolders, Phytotoxicity of trace metals in spiked and field-contaminated soils: linking soil-extractable metals with toxicity, *Environ. Toxicol. Chem.* 33 (2014), 2479–2487.
23. S. P. McGrath, C. Mico, F. J. Zhao et al., Predicting molybdenum toxicity to higher plants: Estimation of toxicity threshold values, *Environ. Pollut.* 158 (2010), 3085–3094.
24. O. Mojsilovic, R. G. McLaren and L. M. Condron, Modelling arsenic toxicity in wheat: Simultaneous application of diffusive gradients in thin films to arsenic and phosphorus in soil, *Environ. Pollut.* 159 (2011), 2996–3002.
25. A. Caillat, P. Ciffroy, M. Grote, S. Rigaud and J. M. Garnier, Bioavailability of copper in contaminated sediments assessed by a DGT approach and the uptake of copper by the aquatic plant Myriophyllum aquaticum, *Environ. Toxicol. Chem.* 33 (2014), 278–285.
26. J. Ren, J. Luo, H. Ma, X. Wang and L. Q. Ma, Bioavailability and oxidative stress of cadmium to Corbicula fluminea, *Env. Sci. Process. Impact.* 15 (2013), 860–869.
27. J. Roulier, M. Tusseau-Vuillemin, M. Coquery, O. Geffard and J. Garric, Measurement of dynamic mobilization of trace metals in sediments using DGT and comparison with bioaccumulation in Chironomus riparius: First results of an experimental study, *Chemosphere*, 70 (2008), 925–932.
28. H. Yin, Y. Cai, H. Duan, J. Gao and C. Fan, Use of DGT and conventional methods to predict sediment metal bioavailability to a field inhabitant freshwater snail (Bellamya aeruginosa) from Chinese eutrophic lakes, *J. Hazard. Mater.* 264 (2014), 184–194.

29. S. L. Simpson, H. L. Yverneau, A. Cremazy et al., DGT-induced copper flux predicts bioaccumulation and toxicity to bivalves in sediments with varying properties, *Environ. Sci. Technol.* 46 (2012), 9038–9046.

30. D. M. Costello, G. A. Burton, C. R. Hammerschmidt and W. K. Taulbee, Evaluating the performance of diffusive gradients in thin films for predicting Ni sediment toxicity, *Environ. Sci. Technol.* 46 (2012), 10239–10246.

31. V. I. Slaveykova, I. B. Karadjova, M. Karadjov and D. L. Tsalev, Trace metal speciation and bioavailability in surface waters of the Black Sea coastal area evaluated by HF-PLM and DGT, *Environ. Sci. Technol.* 43 (2009), 1798–1803.

32. P. Bradac, R. Behra and L. Sigg, Accumulation of cadmium in periphyton under various freshwater speciation conditions, *Environ. Sci. Technol.* 43 (2009), 7291–7296.

33. C. D. Luider, J. Crusius, R. C. Playle and P. J. Curtis, Influence of natural organic matter source on copper speciation as demonstrated by Cu binding to fish gills, by ion selective electrode, and by DGT gel sampler, *Environ. Sci. Technol.* 38 (2004), 2865–2872.

34. M. A. Jordan, P. R. Teasdale, R. J. Dunn and S. Y. Lee, Modelling copper uptake by Saccostrea glomerata with diffusive gradients in a thin film measurements, *Environ. Chem.* 5 (2008), 274–280.

35. D. Ferreira, P. Ciffroy, M.-H. Tusseau-Vuillemin, A. Bourgeault and J.-M. Garnier, DGT as surrogate of biomonitors for predicting the bioavailability of copper in freshwaters: An ex situ validation study, *Chemosphere* 91 (2013), 241–247.

36. O. Clarisse, G. R. Lotufo, H. Hintelmann and E. Best, Biomonitoring and assessment of monomethylmercury exposure in aqueous systems using the DGT technique, *Sci. Total Environ.* 416 (2012), 449–454.

37. A. Sakellari, S. Karavoltsos, D. Theodorou, M. Dassenakis and M. Scoullos, Bioaccumulation of metals (Cd, Cu, Zn) by the marine bivalves M. galloprovincialis, P. radiata, V. verrucosa and C. chione in Mediterranean coastal microenvironments: association with metal bioavailability, *Environ. Monit. Assess.* 185 (2013), 3383–3395.

38. A. J. Martin and R. Goldblatt, Speciation, behavior, and bioavailability of copper downstream of a mine-impacted lake, *Environ. Toxicol. Chem.* 26 (2007), 2594–2603.

39. O. Røyset, B. O. Rosseland, T. Kristensen et al., Diffusive gradients in thin films sampler predicts stress in brown trout (Salmo trutta L.) exposed to aluminum in acid fresh waters, *Environ. Sci. Technol.* 39 (2005), 1167–1174.

40. E. Navarro, F. Piccapietra, B. Wagner et al., Toxicity of silver nanoparticles to Chlamydomonas reinhardtii, *Environ. Sci. Technol.* 42 (2008), 8959–8964.

41. D. M. Templeton, F. Ariese, R. Cornelis et al., Guidelines for terms related to chemical speciation and fractionation of elements. Definitions, structural aspects, and methodological approaches (IUPAC recommendations 2000), *Pure Appl. Chem.* 72 (2000), 1453–1470.

42. K. J. Wilkinson and J. R. Lead. *Environmental colloids and particles: Behaviour, separation and characterisation* (Chichester: Wiley, 2007), 702pp.

43. C. Gustafsson and P. M. Gschwend, Aquatic colloids: Concepts, definitions, and current challenges, *Limnol. Oceanogr.* 42 (1997), 519–528.

44. E. Tipping, Humic ion-binding model VI: an improved description of the interactions of protons and metal ions with humic substances, *Aquat. Geochem.* 4 (1998), 3–48.

45. D. G. Kinniburgh, C. J. Milne, M. F. Benedetti et al., Metal ion binding by humic acid: Application of the NICA-Donnan model, *Environ. Sci. Technol.* 30 (1996), 1687–1698.

46. L. Weng, E. Temminghoff and W. Van Riemsdijk, Determination of the free ion concentration of trace metals in soil solution using a soil column Donnan membrane technique, *Eur. J. Soil Sci.* 52 (2001), 629–637.
47. E. J. Kalis, L. Weng, F. Dousma, E. J. Temminghoff and W. H. Van Riemsdijk, Measuring free metal ion concentrations in situ in natural waters using the Donnan membrane technique, *Environ. Sci. Technol.* 40 (2006), 955–961.
48. M. J. Ellwood and C. M. Van den Berg, Zinc speciation in the northeastern Atlantic Ocean, *Mar. Chem.* 68 (2000), 295–306.
49. T. P. Knepper, Synthetic chelating agents and compounds exhibiting complexing properties in the aquatic environment, *Trends Anal. Chem.* 22: (2003), 708–724.
50. W. W. Bedsworth and D. L. Sedlak, Sources and environmental fate of strongly complexed nickel in estuarine waters: The role of ethylenediaminetetraacetate, *Environ. Sci. Technol.* 33 (1999), 926–931.
51. S. Baken, F. Degryse, L. Verheyen, R. Merckx and E. Smolders, Metal complexation properties of freshwater dissolved organic matter are explained by its aromaticity and by anthropogenic ligands, *Environ. Sci. Technol.* 45 (2011), 2584–2590.
52. M. Pesavento, G. Alberti and R. Biesuz, Analytical methods for determination of free metal ion concentration, labile species fraction and metal complexation capacity of environmental waters: a review, *Anal. Chim. Acta* 631 (2009), 129–141.
53. N. Odzak, D. Kistler, H. Xue and L. Sigg, In situ trace metal speciation in a eutrophic lake using the technique of diffusion gradients in thin films (DGT), *Aquat. Sci.* 64 (2002), 292–299.
54. E. R. Unsworth, K. W. Warnken, H. Zhang et al., Model predictions of metal speciation in freshwaters compared to measurements by in situ techniques, *Environ. Sci. Technol.* 40 (2006), 1942–1949.
55. R. J. Dunn, P. R. Teasdale, J. Warnken and R. R. Schleich, Evaluation of the diffusive gradient in a thin film technique for monitoring trace metal concentrations in estuarine waters, *Environ. Sci. Technol.* 37 (2003), 2794–2800.
56. F. Degryse, E. Smolders and D. R. Parker, Partitioning of metals (Cd, Co, Cu, Ni, Pb, Zn) in soils: Concepts, methodologies, prediction and applications – A review, *Eur. J. Soil Sci.* 60 (2009), 590–612.
57. H. Ernstberger, W. Davison, H. Zhang, A. Tye and S. Young, Measurement and dynamic modeling of trace metal mobilization in soils using DGT and DIFS, *Environ. Sci. Technol.* 36 (2002), 349–354.
58. M. P. Harper, W. Davison and W. Tych, DIFS – A modelling and simulation tool for DGT induced trace metal remobilisation in sediments and soils, *Environ. Model. Softw.* 15 (2000), 55–66.
59. F. Degryse, E. Smolders, I. Oliver and H. Zhang, Relating soil solution Zn concentration to DGT measurements in contaminated soils, *Environ. Sci. Technol.* 27 (2003), 3958–3965.
60. M. Senila, E. A. Levei and L. R. Senila, Assessment of metals bioavailability to vegetables under field conditions using DGT, single extractions and multivariate statistics, *Chem. Cent. J.* 6 (2012), 119.
61. S. Scally, W. Davison and H. Zhang, Diffusion coefficients of metals and metal complexes in hydrogels used in diffusive gradients in thin films, *Anal. Chim. Acta* 558 (2006), 222–229.
62. G. Du Laing, J. Rinklebe, B. Vandecasteele, E. Meers and F. Tack, Trace metal behaviour in estuarine and riverine floodplain soils and sediments: a review, *Sci. Total Environ.* 407 (2009), 3972–3985.

63. L. Weng, E. J. Temminghoff and W. H. Van Riemsdijk, Contribution of individual sorbents to the control of heavy metal activity in sandy soil, *Environ. Sci. Technol.* 35 (2001), 4436–4443.

64. J. Buekers, F. Degryse, A. Maes and E. Smolders, Modelling the effects of ageing on Cd, Zn, Ni and Cu solubility in soils using an assemblage model, *Eur. J. Soil Sci.* 59 (2008), 1160–1170.

65. Y. Yin, C. A. Impellitteri, S.-J. You and H. E. Allen, The importance of organic matter distribution and extract soil: Solution ratio on the desorption of heavy metals from soils, *Sci. Total Environ.* 287 (2002), 107–119.

66. N. S. Bolan, D. C. Adriano and D. Curtin, Soil acidification and liming interactions with nutrient and heavy metal transformation and bioavailability, *Adv. Agron.* 78 (2003), 215–272.

67. O. Jacquat, A. Voegelin, A. Villard, M. A. Marcus and R. Kretzschmar, Formation of Zn-rich phyllosilicate, Zn-layered double hydroxide and hydrozincite in contaminated calcareous soils, *Geochim. Cosmochim. Acta* 72 (2008), 5037–5054.

68. W. Salomons, N. De Rooij, H. Kerdijk and J. Bril, Sediments as a source for contaminants?, *Hydrobiologia* 149 (1987), 13–30.

69. A. Widerlund and W. Davison, Size and density distribution of sulfide-producing microniches in lake sediments, *Environ. Sci. Technol.* 41 (2007), 8044–8049.

70. P. M. Haygarth, M. S. Warwick and W. A. House, Size distribution of colloidal molybdate reactive phosphorus in river waters and soil solution, *Water Res.* 31 (1997), 439–448.

71. M. Hens and R. Merckx, Functional characterization of colloidal phosphorus species in the soil solution of sandy soils, *Environ. Sci. Technol.* 35 (2001), 493–500.

72. C. V. Moorleghem, L. Six, F. Degryse, E. Smolders and R. Merckx, Effect of organic P forms and P present in inorganic colloids on the determination of dissolved P in environmental samples by the diffusive gradient in thin films technique, ion chromatography, and colorimetry, *Anal. Chem.* 83 (2011), 5317–5323.

73. K. Sakadevan and H. Bavor, Phosphate adsorption characteristics of soils, slags and zeolite to be used as substrates in constructed wetland systems, *Water Res.* 32 (1998), 393–399.

74. K. Börling, E. Otabbong and E. Barberis, Phosphorus sorption in relation to soil properties in some cultivated Swedish soils, *Nutr. Cycl. Agroecosys.* 59 (2001), 39–46.

75. J. Gerke, Humic (organic matter)-Al (Fe)-phosphate complexes: An underestimated phosphate form in soils and source of plant-available phosphate, *Soil Sci.* 175 (2010), 417–425.

76. P. E. Fixen, A. E. Ludwick and S. R. Olsen, Phosphorus and potassium fertilization of irrigiated alfalfa on calcareous soils. 2. Soil phosphorus solubility relationships, *Soil Sci. Soc. Am. J.* 47 (1983), 112–117.

77. E. Lombi, K. G. Scheckel, R. D. Armstrong et al., Speciation and distribution of phosphorus in a fertilized soil: A synchrotron-based investigation, *Soil Sci. Soc. Am. J.* 70 (2006), 2038–2048.

78. D. M. Nash, P. M. Haygarth, B. L. Turner et al., Using organic phosphorus to sustain pasture productivity: A perspective, *Geoderma* 221–222 (2014), 11–19.

79. M. M. Lasat, A. J. M. Baker and L. V. Kochkian, Physiological characterization of root Zn absorption and translocation to shoots in Zn hyperaccumulator and nonaccumulator species of Thlaspi, *Plant Physiol.* 112 (1996), 1715–1727.

80. F. M. M. Morel, R. J. M. Hudson and N. M. Price, Limitation of productivity by trace-metals in the sea, *Limnol. Oceanogr.* 36 (1991), 1742–1755.

81. P. G. Campbell and C. Fortin. Biotic ligand model, In *Encyclopedia of aquatic eco-toxicology*, eds. J. F. Férard and C. Blaise (Dordrecht: Springer, 2013), pp. 237–246.

82. F. M. M. Morel and J. Hering. *Principles and applications of aquatic chemistry* (New York: Wiley, 1993), 608pp.

83. P. G. C. Campbell. Interactions between trace metals and aquatic organisms: A critique of the free-ion activity model, In *Metal speciation and bioavailability in aquatic systems*, eds A. Tessier and D. R. Turner (Chichester: Wiley, 1995), pp. 45–102.

84. D. M. Di Toro, H. E. Allen, H. L. Bergman et al., Biotic ligand model of the acute toxicity of metals. 1. Technical basis, *Environ. Toxicol. Chem.* 20 (2001), 2383–2396.

85. H. Lineweaver and D. Burk, The determination of enzyme dissociation constants, *J. Am. Chem. Soc.* 56 (1934), 658–666.

86. K. J. Wilkinson and J. Buffle. Critical evaluation of physicochemical parameters and processes for modelling the biological uptake of trace metals in environmental (aquatic) systems, In *Physicochemical kinetics and transport at biointerfaces*, eds. J. Buffle and H. P. van Leeuwen (Chichester: Wiley, 2004), pp. 445–533.

87. I. A. Worms and K. J. Wilkinson, Ni uptake by a green alga. 2. Validation of equilibrium models for competition effects, *Environ. Sci. Technol.* 41 (2007), 4264–4270.

88. K. Lock, H. Van Eeckhout, K. A. C. De Schamphelaere, P. Criel and C. R. Janssen, Development of a biotic ligand model (BLM) predicting nickel toxicity to barley (Hordeum vulgare), *Chemosphere* 66 (2007), 1346–1352.

89. K. Lock, K. A. C. De Schamphelaere, S. Becaus et al., Development and validation of a terrestrial biotic ligand model predicting the effect of cobalt on root growth of barley (Hordeum vulgare), *Environ. Pollut.* 147 (2007), 626–633.

90. S. Thakali, H. Allen, D. Di Toro et al., A terrestrial biotic ligand model. 1. Development and application to Cu and Ni toxicities to barley root elongation in soils, *Environ. Sci. Technol.* 40 (2006), 7085–7093.

91. K. Lock, K. De Schamphelaere, S. Becaus et al., Development and validation of an acute biotic ligand model (BLM) predicting cobalt toxicity in soil to the potworm Enchytraeus albidus, *Soil Biol. Biochem.* 38 (2006), 1924–1932.

92. O. Errecalde, M. Seidl and P. G. Campbell, Influence of a low molecular weight metabolite (citrate) on the toxicity of cadmium and zinc to the unicellular green alga Selenastrum capricornutum: an exception to the free-ion model, *Water Res.* 32 (1998), 419–429.

93. C. Fortin and P. G. Campbell, Thiosulfate enhances silver uptake by a green alga: Role of anion transporters in metal uptake, *Environ. Sci. Technol.* 35 (2001), 2214–2218.

94. J. T. Phinney and K. W. Bruland, Uptake of lipophilic organic Cu, Cd, and Pb complexes in the coastal diatom Thalassiosira weissflogii, *Environ. Sci. Technol.* 28 (1994), 1781–1790.

95. C. S. Hassler and K. J. Wilkinson, Failure of the biotic ligand and free-ion activity models to explain zinc bioaccumulation by Chlorella kesslerii, *Environ. Toxicol. Chem.* 22 (2003), 620–626.

96. C. S. Hassler, V. I. Slaveykova and K. J. Wilkinson, Some fundamental (and often overlooked) considerations underlying the free ion activity and biotic ligand models, *Environ. Toxicol. Chem.* 23 (2004), 283–291.

97. C. M. Zhao and K. J. Wilkinson, Biotic ligand model does not predict the bioavailability of rare earth elements in the presence of organic ligands, *Environ. Sci. Technol.* 49 (2015), 2207–2214.

98. R. T. Checkai, L. L. Hendrickson, R. B. Corey and P. A. Helmke, A method for controlling the activities of free metal, hydrogen and phosphate ions in hydroponic

solutions using ion exchange and chelating resins, *Plant Soil* 99 (1987), 321–334.

99. P. F. Bell, R. L. Chaney and J. S. Angle, Free metal activity and total metal concentrations as indices of micronutrient availability to barley [Hordeum-vulgare (L.) 'Klages'], *Plant Soil* 130 (1991), 51–62.

100. M. McLaughlin, E. Smolders, R. Merckx and A. Maes. Plant uptake of Cd and Zn in chelator-buffered nutrient solution depends on ligand type, In *Plant nutrition – for sustainable food production and environment*, eds. T. Ando, K. Fujita, T. Mae et al. (Dordrecht: Kluwer Academic Publishers, 1997), pp. 113–118.

101. E. J. Berkelaar and B. A. Hale, Cadmium accumulation by durum wheat roots in ligand-buffered hydroponic culture: uptake of Cd-ligand complexes or enhanced diffusion?, *Can. J. Bot.*, 81 (2003), 755–763.

102. H. P. Van Leeuwen, Metal speciation dynamics and bioavailabillty. Inert and labile complexes, *Environ. Sci. Technol.* 33 (1999), 3743–3748.

103. F. Degryse, E. Smolders and R. Merckx, Labile Cd complexes increase Cd availability to plants, *Environ. Sci. Technol.* 40 (2006), 830–836.

104. F. Amery, F. Degryse, W. Degeling, E. Smolders and R. Merckx, The copper-mobilizing-potential of dissolved organic matter in soils varies 10-fold depending on soil incubation and extraction procedures, *Environ. Sci. Technol.* 41 (2007), 2277–2281.

105. P. Wang, D. Zhou, X. Luo and L. Li, Effects of Zn-complexes on zinc uptake by wheat (Triticum aestivum) roots: a comprehensive consideration of physical, chemical and biological processes on biouptake, *Plant Soil* 316 (2009), 177–192.

106. A. Gramlich, S. Tandy, E. Frossard, J. Eikenberg and R. Schulin, Availability of zinc and the ligands citrate and histidine to wheat: does uptake of entire complexes play a role?, *J. Agric. Food. Chem.* 61 (2013), 10409–10417.

107. R. J. M. Hudson, Which aqueous species control the rates of trace metal uptake by aquatic biota? Observations and predictions of non-equilibrium effects, *Sci. Total Environ.* 219 (1998), 95–115.

108. D. Winne, Unstirred layer, source of biased Michaelis constant in membrane transport, *Biochimica et Biophysica Acta (BBA)-Biomembranes*, 298 (1973), 27–31.

109. F. Degryse, A. Shahbazi, L. Verheyen and E. Smolders, Diffusion limitations in root uptake of cadmium and zinc, but not nickel, and resulting bias in the Michaelis constant, *Plant Physiol.* 160 (2012), 1097–1109.

110. J. Luo, H. Zhang, F. J. Zhao and W. Davison, Distinguishing diffusional and plant control of Cd and Ni uptake by hyperaccumulator and nonhyperaccumulator plants, *Environ. Sci. Technol.* 44 (2010), 6636–6641.

111. H. P. Van Leeuwen, R. M. Town, J. Buffle et al., Dynamic speciation analysis and bioavailability of metals in aquatic systems, *Environ. Sci. Technol.* 39 (2005), 8545–8556.

112. F. Degryse, E. Smolders and D. R. Parker, Metal complexes increase uptake of Zn and Cu by plants: implications for uptake and deficiency studies in chelator-buffered solutions, *Plant Soil* 289 (2006), 171–185.

113. J. Buffle, K. Startchev and J. Galceran, Computing steady-state metal flux at microorganism and bioanalogical sensor interfaces in multiligand systems. A reaction layer approximation and its comparison with the rigorous solution, *Phys. Chem. Chem. Phys.* 9 (2007), 2844–2855.

114. J. Santner, E. Smolders, W. W. Wenzel and F. Degryse, First observation of diffusion-limited plant root phosphorus uptake from nutrient solution, *Plant Cell Environ.* 35 (2012), 1558–1566.

115. S. Jansen, R. Blust and H. P. Van Leeuwen, Metal speciation dynamics and bioavailability: Zn(II) and Cd(II) uptake by mussel (Mytilus edulis) and carp (Cyprinus carpio), *Environ. Sci. Technol.* 36 (2002), 2164–2170.

116. R. J. M. Hudson and F. M. M. Morel, Trace-metal transport by marine microorganisms – implications of metal coordination kinetics, *Deep-Sea Res.* 40 (1993), 129–150.

117. P. Pypers, J. Delrue, J. Diels, E. Smolders and R. Merckx, Phosphorus intensity determines short-term P uptake by pigeon pea (Cajanus cajan L.) grown in soils with differing P buffering capacity, *Plant Soil* 284 (2006), 217–227.

118. M. Silberbush and S. Barber, Prediction of phosphorus and potassium uptake by soybeans with a mechanistic mathematical model, *Soil Sci. Soc. Am. J.* 47 (1983), 262–265.

119. N. Claassen, K. Syring and A. Jungk, Verification of a mathematical model by simulating potassium uptake from soil, *Plant Soil* 95 (1986), 209–220.

120. T. Sterckeman, J. Perriguey, M. Cael, C. Schwartz and J. L. Morel, Applying a mechanistic model to cadmium uptake by Zea mays and Thlaspi caerulescens: Consequences for the assessment of the soil quantity and capacity factors, *Plant Soil* 262 (2004), 289–302.

121. T. Adhikari and R. K. Rattan, Modelling zinc uptake by rice crop using a Barber-Cushman approach, *Plant Soil* 227 (2000), 235–242.

122. J.-M. Custos, C. Moyne, T. Treillon and T. Sterckeman, Contribution of Cd-EDTA complexes to cadmium uptake by maize: a modelling approach, *Plant Soil* 374 (2014), 497–512.

123. K. Oorts, U. Ghesquiere, K. Swinnen and E. Smolders, Soil properties affecting the toxicity of CuCl and NiCl for soil microbial processes in freshly spiked soils, *Environ. Toxicol. Chem.* 25 (2006), 836–844.

124. S. L. Simpson and G. E. Batley, Predicting metal toxicity in sediments: a critique of current approaches, *Integrated Environ. Assess. Manag.* 3 (2007), 18–31.

125. D. E. Howard and R. D. Evans, Acid-volatile sulfide (AVS) in a seasonally anoxic mesotrophic lake: Seasonal and spatial changes in sediment AVS, *Environ. Toxicol. Chem.* 12 (1993), 1051–1057.

126. G. A. Burton Jr, Sediment quality criteria in use around the world, *Limnology* 3 (2002), 65–76.

127. D. M. Di Toro, J. D. Mahony, D. J. Hansen et al., Acid volatile sulfide predicts the acute toxicity of cadmium and nickel in sediments, *Environ. Sci. Technol.* 26 (1992), 96–101.

128. G. T. Ankley, D. M. Di Toro, D. J. Hansen and W. J. Berry, Technical basis and proposal for deriving sediment quality criteria for metals, *Environ. Toxicol. Chem.* 15 (1996), 2056–2066.

129. D. M. Di Toro, J. A. McGrath, D. J. Hansen et al., Predicting sediment metal toxicity using a sediment biotic ligand model: methodology and initial application, *Environ. Toxicol. Chem.* 24 (2005), 2410–2427.

130. E. Smolders, J. Buekers, I. Oliver and M. J. McLaughlin, Soil properties affecting toxicity of zinc to soil microbial properties in laboratory-spiked and field-contaminated soils, *Environ. Toxicol. Chem.* 23 (2004), 2633–2640.

131. S. L. Simpson, Exposure-effect model for calculating copper effect concentrations in sediments with varying copper binding properties: A synthesis, *Environ. Sci. Technol.* 39 (2005), 7089–7096.

132. P. H. Barry and J. M. Diamond, Effects of unstirred layers on membrane phenomena, *Physiol. Rev.* 64 (1984), 763–872.

133. M. H. Tusseau-Vuillemin, R. Gilbin, E. Bakkaus and J. Garric, Performance of diffusion gradient in thin films to evaluate the toxic fraction of copper to Daphnia magna, *Environ. Toxicol. Chem.* 23 (2004), 2154–2161.

134. S. Meylan, R. Behra and L. Sigg, Influence of metal speciation in natural freshwater on bioaccumulation of copper and zinc in periphyton: a microcosm study, *Environ. Sci. Technol.* 38 (2004), 3104–3111.

135. W. Davison and H. Zhang, Progress in understanding the use of diffusive gradients in thin films (DGT) – Back to basics, *Environ. Chem.* 9 (2012), 1–13.

136. W. J. Peijnenburg, P. R. Teasdale, D. Reible et al., Passive sampling methods for contaminated sediments: State of the science for metals, *Integrated Environ. Assess. Manag.* 10 (2014), 179–196.

137. K. Sand-Jensen, N. P. Revsbech and B. B. Jørgensen, Microprofiles of oxygen in epiphyte communities on submerged macrophytes, *Mar. Biol.* 89 (1985), 55–62.

138. A. W. Larkum, E.-M. Koch and M. Kühl, Diffusive boundary layers and photosynthesis of the epilithic algal community of coral reefs, *Mar. Biol.* 142 (2003), 1073–1082.

139. H. F. Kaspar, Oxygen conditions on surfaces of coralline red algae, *Mar. Ecol. Prog. Ser.* 81 (1992), 97–100.

140. O. Clarisse and H. Hintelmann, Measurements of dissolved methylmercury in natural waters using diffusive gradients in thin film (DGT), *J. Environ. Monit.* 8 (2006), 1242–1247.

141. P. N. Williams, H. Zhang, W. Davison et al., Organic matter – Solid -phase interactions are critical for predicting arsenic release and plant uptake in Bangladesh paddy soils, *Environ. Sci. Technol.* 45 (2011), 6080–6087.

142. M. P. Harper, W. Davison and W. Tych, Estimation of pore water concentrations from DGT profiles: A modelling approach, *Aquat. Geochem.* 5 (1999), 337–355.

143. J. Luo, H. Cheng, J. Ren, W. Davison and H. Zhang, Mechanistic insights from DGT and soil solution measurements on the uptake of Ni and Cd by radish, *Environ. Sci. Technol.* 48 (2014), 7305–7313.

144. H. Marschner, V. Römheld, W. J. Horst and P. Martin, Root-induced changes in the rhizosphere: Importance for the mineral nutrition of plants, *Z. Pflanzenernähr. Bodenkd.* 149 (1986), 441–456.

145. D. L. Jones, Organic acids in the rhizosphere – A critical review, *Plant Soil* 205 (1998), 25–44.

146. S. Mongin, R. Uribe, J. Puy et al., Key role of the resin layer thickness in the lability of complexes measured by DGT, *Environ. Sci. Technol.* 45 (2011), 4869–4875.

147. O. A. Garmo, N. J. Lehto, H. Zhang et al., Dynamic aspects of DGT as demonstrated by experiments with lanthanide complexes of a multidendate ligand, *Environ. Sci. Technol.* 40 (2006), 4754–4760.

148. H. Zhang and W. Davison, Use of DGT for studies of chemical speciation and bioavailability, *Environ. Chem.* 12 (2015), 85–101.

149. F. Degryse and E. Smolders, Cadmium and nickel uptake by tomato and spinach seedlings: plant or transport control?, *Environ. Chem.* 9 (2011), 48–54.

150. A. Black, R. G. McLaren, S. M. Reichman, T. W. Speir and L. M. Condron, Evaluation of soil metal bioavailability estimates using two plant species (L. perenne and T. aestivum) grown in a range of agricultural soils treated with biosolids and metal salts, *Environ. Pollut.* 159 (2011), 1523–1535.

151. B. Nowack, S. Koehler and R. Schulin, Use of diffusive gradients in thin films (DGT) in undisturbed field soils, *Environ. Sci. Technol.* 38 (2004), 1133–1138.

    DGT and Bioavailability

152. Q. Sun, J. Chen, S. Ding, Y. Yao and Y. Chen, Comparison of diffusive gradients in thin film technique with traditional methods for evaluation of zinc bioavailability in soils, *Environ. Monit. Assess.* 186 (2014), 6553–6564.

153. E. Smolders, J. Buekers, N. Waegeneers, I. Oliver and M. McLaughlin, Effects of field and laboratory Zn contamination on soil microbial processes and plant growth. Final Report to the International Lead and Zinc Research Organisation (ILZRO). (Katholieke Universiteit Leuven and CSIRO, 2003), 67pp.

154. M. Bravin, A. L. Marti, M. Clairotte and P. Hinsinger, Rhizosphere alkalisation – A major driver of copper bioavailability over a broad pH range in an acidic, copper-contaminated soil, *Plant Soil* 318 (2009), 257–268.

155. E. E. V. Chapman, G. Dave and J. D. Murimboh, Bioavailability as a factor in risk assessment of metal-contaminated soil, *Water Air Soil Pollut.* 223 (2012), 2907–2922.

156. Y. Tian, X. Wang, J. Luo, H. Yu and H. Zhang, Evaluation of holistic approaches to predicting the concentrations of metals in field-cultivated rice, *Environ. Sci. Technol.* 42 (2008), 7649–7654.

157. J. J. Wang, L. Y. Bai, X. B. Zeng et al., Assessment of arsenic availability in soils using the diffusive gradients in thin films (DGT) technique – A comparison study of DGT and classic extraction methods, *Environ. Sci.-Process Impacts*, 16 (2014), 2355–2361.

158. L. Duquène, H. Vandenhove, F. Tack, M. Van Hees and J. Wannijn, Diffusive gradient in thin FILMS (DGT) compared with soil solution and labile uranium fraction for predicting uranium bioavailability to ryegrass, *J. Environ. Radioact.* 101 (2010), 140–147.

159. H. Vandenhove, K. Antunes, J. Wannijn, L. Duquene and M. Van Hees, Method of diffusive gradients in thin films (DGT) compared with other soil testing methods to predict uranium phytoavailability, *Sci. Total Environ.* 373 (2007), 542–555.

160. J. Liu, X. Feng, G. Qiu, C. W. Anderson and H. Yao, Prediction of methyl mercury uptake by rice plants (Oryza sativa L.) using the diffusive gradient in thin films technique, *Environ. Sci. Technol.* 46 (2012), 11013–11020.

161. P. S. Hooda, H. Zhang, W. Davison and A. C. Edwards, Measuring bioavailable trace metals by diffusive gradients in thin films (DGT): Soil moisture effects on its performance in soils, *Eur. J. Soil Sci.* 50 (1999), 285–294.

162. S. Mason, A. McNeill, Y. Zhang, M. J. McLaughlin and C. Guppy, Application of diffusive gradients in thin-films (DGT) to measure potassium and sulphur availability in agricultural soils, Sixteenth Australian Agronomy Conference (2012).

163. A. Pelfrène, C. Waterlot and F. Douay, Investigation of DGT as a metal speciation tool in artificial human gastrointestinal fluids, *Anal. Chim. Acta* 699 (2011), 177–186.

164. L. Verheyen, F. Degryse, T. Niewold and E. Smolders, Labile complexes facilitate cadmium uptake by Caco-2 cells, *Sci. Total Environ.* 426 (2012), 90–99.

165. H. M. Pouran, F. L. Martin and H. Zhang, Measurement of ZnO nanoparticles using diffusive gradients in thin films: Binding and diffusional characteristics, *Anal. Chem.* 86 (2014), 5906–5913.

10

Practicalities of Working with DGT

DIANNE F. JOLLEY, SEAN MASON, YUE GAO AND HAO ZHANG

10.1 Introduction

The aim of this chapter is to set out the basic procedures for the use of DGT, from assembly to deployment through to handling and analysis. Many research papers have reported procedures and provided details, with some reviews collating some of this knowledge [1–3], but there has been a lack of a comprehensive, coherent guide with practical hints and tips. Detailed guidance on particular procedures is already provided on the web site of DGT Research Limited at dgtresearch.com. Here we will set out the basics for the use of DGT in solutions, soils and sediments, with a view to updating, improving and extending some of the technical details already available. In essence, procedures for using DGT are very simple, but there are tricks of the trade and potential pitfalls usually associated with inattention to detail. DGT was first used for the measurement of trace metals and this remains a major application. The description of methodology is primarily suited to these target elements, but, where possible, modifications of procedures for other analytes are provided. As the variety of analytes that are within the scope of DGT (see Chapter 4) continues to expand, this chapter cannot cover all possibilities, but hopefully it can establish sound principles, which will be useful in responding to future challenges.

The name Diffusive Gradients in Thin-films refers to a principle rather than a particular device, as reflected in the original patent [4], which described several possible devices for deployment in solutions, soils and sediments and suggested there could be a range of analytes embracing both inorganic and organic substances. Early DGT devices were bespoke items, manufactured locally in research workshops [5, 6], but once the laboratory at Lancaster University and then DGT Research Limited started supplying manufactured DGT holders, these designs were used predominantly by most researchers and so they will be the focus of this chapter.

10.2 DGT Components and Their Assembly for Measurements in Solutions

10.2.1 Solution Devices

For deployments in solutions, the most commonly used DGT device is based on a simple plastic piston design (Figures 1.1, 6.2c). The piston and its sleeve fit together to leave a

1.2-mm gap between the flat end of the piston and the lip of the sleeve adjacent to the 20-mm diameter window. To assemble the DGT device, three layers are placed on the surface of the piston: a binding gel, a diffusive gel and a filter membrane. This solution moulding can be purchased (DGT Research Limited, UK, dgtresearch.com) preloaded with the layers appropriate for a wide range of analytes and situations. These assembled DGT devices, which are ready for deployment, have the advantage of being prepared in a clean laboratory, dedicated for this purpose, by experienced technical staff. It is also possible to purchase separately the DGT holders and different types of binding layers, diffusive gels and filter membranes. DGT devices can then be self-assembled. Some research laboratories fabricate the binding and diffusive layers themselves for research purposes. There is potential for errors, due to contamination and incorrect process control, being introduced with each step in the production, so self-fabrication and assembly will usually require practise to develop adequate expertise.

The most common configurations of DGT use a 0.4-mm thick Chelex binding layer for trace metals or an oxide (ferrihydrite, TiO_2, ZrO_2) binding layer for oxyanions, including phosphate, with a 0.8-mm APA (acrylamide cross-linked with an agarose derivative) diffusive gel and a 25-mm diameter, 0.45-μm, hydrophilic polyethersulphone membrane filter (Supor, Gelman Laboratories, USA). For a fuller list of possible analytes see Chapter 4. As the filter membrane is typically 0.14 mm thick, the total thickness of the layers is 1.34 mm, which allows for some compressibility within the 1.2 mm space to provide a good seal around the exposure window.

Other gel thicknesses can be used with appropriate use of the holder. If the total binding and diffusive layer thickness (including the membrane) is less than 1.34 mm, a layer of diffusive gel of an appropriate thickness should be placed between the binding layer and the piston surface to compensate for the difference. If the total layer thickness is greater, an annular spacer can be placed at the bottom of the sleeve where it contacts the base of the piston to ensure that the diffusive and binding layers within the device are not compressed excessively. An alternative piston design is marketed by DGT Research Limited for use in soils. It differs from the regular solution device in having a slightly tapered cap, which facilitates better contact with soil, and an 18-mm diameter window, which allows for a wider lip and therefore a better seal to prevent fine soil particles from entering the device. It can be used equally well for deployments in solution and is useful when measurements are made using different diffusive or binding layer thicknesses because it can be supplied with three different sizes of gap between piston and lip, capable of accommodating material diffusion gels of 0.4, 0.8 and 2.0 mm along with commonly used resin gels and filters. For diffusive gel thicknesses of 1.2 mm and 1.6 mm, the device with the largest gap (normally used to accommodate 2.0-mm diffusive gel) can be used with appropriate spacers between the piston surface and the binding gel layer.

10.2.2 Cleaning, Storage and Handling DGT Components

After fabrication and conditioning, gel components are quite robust, but it is very easy to contaminate them, especially the Chelex binding gel, which readily binds any trace metals it contacts, thus increasing the value of binding gel blanks. The same applies for other analytes. For example, caution must be exercised to prevent exposure to sources of P when handling oxide-based gels intended for phosphate analysis, as they will abstract P from any solution or surface they contact. Where contamination results in an inconsistent increase in analyte concentration in the binding gel blanks, method detection limits are likely to be increased.

Given the low concentrations at which trace elements often occur in natural waters, their quantification is analytically challenging and clean laboratory facilities and protocols are required to avoid or limit contamination problems. It is also essential that such clean procedures are followed when assembling DGT devices intended for measuring metals at trace levels. As a minimum, all operations should be performed in a laminar flow cabinet with filtered air. Clothing which protects the samples from human contamination is necessary, including laboratory coats that fasten to the neck, and laboratory head covers that contain all hair. The use of cosmetics should be avoided. Wherever possible gels should only be manipulated using cleaned plastic forceps, and gloves that are known to be free from trace metal contamination should always be worn. It is essential that gloves should not touch anything outside the clean bench area. All containers and surfaces that may contact the gels should be cleaned. Laboratories have developed different procedures, but one that has been successful for the plastic DGT holders is soaking overnight sequentially in (1) 2 per cent Decon90 (detergent); (2) 1.5 M HNO_3 (Anal-R); (3) 1.5 M HNO_3 (Aristar); and (4) 4 M HCl (Aristar) [7]. Between each overnight step, they should be completely rinsed with high purity water (HPW) which complies with ASTM (D1193–91) Type I requirements [7], for example Milli-Q (Millipore, USA) with a resistivity of 18.2 MΩ. Decon90 is a phosphate-free, surface-active detergent, cleaning agent supplied by Decon Laboratories Limited, Sussex, UK. Anal-R and Aristar denote purity classifications of BDH Laboratory Supplies, Dorset, UK, indicating analytical reagent grade and ultra-pure, respectively. It is essential that all residual acid is removed from the devices (rinsing until the rinse water has a pH <7). Usually filter membranes can be used as supplied from the manufacturer after soaking and washing with Milli-Q water, but if blanks from assembled and deployed holders (see later) are high, filters should be checked for contamination and if necessary cleaned by washing with high purity acid followed by copious rinsing. Only cleaned plastic ware should be used in a trace metal laboratory. Some laboratories without clean room facilities have found that high blanks can be reduced by washing assembled DGT units (20–30) in a large volume (e.g. 2 L) of 0.01–0.05 M HCl for at least 6 h followed by copious rinsing in HPW (S Mason, pers. comm.). Such an approach may work for some measurements, such as in soils, but is unlikely to be successful for low-level measurements in waters.

Chelex gels for trace metals and oxide gels for P should be stored in HPW. Diffusive gels should be stored in 10–100 mmol L^{-1} $NaNO_3$ or NaCl. To avoid contamination, these solutions should be prepared using HPW and high purity salts. All storage should be at room temperature in darkened conditions (e.g. a cupboard). Where the APA gel is used, they can be stored in these conditions for up to twelve months [8]. The restricted gel should only be stored for up to six months [8].

10.2.3 Diffusive Gel Selection and Preparation

The standard DGT gel stock solution used in synthesis of both diffusive and binding gels is prepared by combining 47.5 mL of HPW with 37.5 mL of 40 per cent acrylamide solution. The solution needs to be mixed thoroughly before adding 15 g of an agarose-derived cross-linker[1] (2 per cent, DGT Research Ltd, Lancaster, UK). The gel solution should be further mixed and stored at 4°C in the dark.

The most common diffusive gel types are agarose and APA-type polyacrylamide gels. APA gels are used if open-pored, pre-cast gels are needed as continuous sheets in DET, and DGT pistons and probes, while agarose gels are used for constrained DET probes (see Section 10.4.4).

Preparation of agarose for DET devices is described in Section 10.4.4. APA gel is typically prepared by polymerising 10 mL of gel stock solution by adding 70 µl 10 per cent ammonium persulphate (APS) and 20 µl N,N,N',N'-tetramethylethylenediamine (TEMED). The solution is well mixed and immediately cast between two glass plates separated by a plastic spacer with an appropriate thickness and allowed to set at a temperature of $42 \pm 2°C$ for at least 45 min (controlled by the temperature and the concentration of initiator and catalyst). It is important to avoid air bubbles in the gel, so gentle but effective mixing of the gel solution is recommended, and any foam formed during the mixing should be allowed to collapse before the solution is cast. However care should be taken not to wait too long, as the TEMED will initiate polymerisation of the gel solution. The set gels should then be carefully removed from the plates, hydrated in HPW >24 h, which promotes swelling, with at least four changes of water (until the rinse water pH <7, which indicates that the unreacted amides have been removed), then stored in NaCl or $NaNO_3$ solution at room temperature, with a shelf life of twelve months. A successful casting of 10 mL gel solution can produce approximately fifty diffusive gel discs (at 0.8 mm thickness). If gels fail to set, the 10 per cent APS solution is often the culprit, and we recommend that this is freshly prepared each day. On some occasions, small air bubbles may form in the diffusive gels, creating a pitted gel surface. This is most likely caused by high dissolved oxygen in the solutions, which can be avoided by using fresher reagents, or by degassing the solution beforehand, for example by placing the gel solution under suction for 5 min while keeping it in an ultrasonic bath.

Diffusive gels with smaller pore sizes (greater degree of cross-linking) are used to favour the measurement of inorganic metal species. This approach is viable because metal complexes formed with humic and fulvic acids are larger than simple inorganic complexes, and consequently they diffuse substantially more slowly through gels (Figure 3.5). A gel known as the restricted gel (RG or RES) was developed to allow diffusion of inorganic species, while restricting the access of organic species to the binding gel [9]. The restricted gel (RG) solution is prepared by combining acrylamide monomer (15 vol-%) cross-linked with bis(acrylamide) (0.8 vol-%) [9]. To prepare the gel, 70 µL of 10 per cent ammonium persulphate and 20 µL of TEMED are added to the gel solution (10 mL). The solution is

---

[1] DGT Research Limited only supplies this cross-linker for research purposes and requires a signed agreement that there is no actual or intended commercial use.

well mixed and immediately cast between two glass plates separated by a plastic spacer with an appropriate thickness and allowed to set at a temperature of 5°C for at least 60 min.

10.2.4 Binding Resin Gel Selection and Preparation

10.2.4.1 Selection of Binding Gels

Natural waters (including porewaters) exhibit a variety of properties that can affect the interaction between the binding layer and the analyte (e.g. major ion concentrations and composition, including H^+ ions, which are often present at several orders of magnitude higher concentrations than trace analytes). These can lead to competitive binding effects, reducing the effective capacity of the binding layer for the target analyte, and consequently restricting the effective deployment times. In Chapter 4 binding layers are comprehensively evaluated across a range of pH and ionic concentrations, while Chapter 5 models the parameters that affect binding.

The most common binding and diffusive layer configurations are similar, irrespective of whether DGTs are deployed in solutions, soils and sediments. The binding layer composition is selected based on the target analytes, including Chelex-100 for trace metals or an oxide (ferrihydrite, TiO_2, ZrO_2) for oxyanions. A comprehensive list of binding layers has been included in Chapter 4 (Table 4.1). To expand the range of analytes determined in a single DGT deployment, several researchers have developed mixed binding layers [10, 11]. Binding layer selection should consider factors such as (i) the selectivity of the binding agent for the analyte, (ii) kinetic effects, particularly the rate of binding, (iii) the binding capacity of the binding layer and (iv) the performance of the binding agent as specified in (i–iii) at the physico-chemical conditions of the deployment site (e.g. ionic strength, pH). These factors are discussed in Chapter 4, Section 4.2 and warrant consideration, as many of the binding layers listed in Table 4.1 are yet to be tested in saline waters. Preparation protocols for several of the common binding layers are presented below, and the appropriate elution solutions are described in Table 4.2.

10.2.4.2 Preparation of Binding Gels

Chelex-100 resin (Na form, 200–400 bead size, Bio-Rad Laboratories), which binds transition metal cations to its iminodiacetic acid groups, is the most widely used binding agent in DGT applications for trace metal analysis. The resin–gel is prepared by combining 4 g (wet weight) of resin in 10 mL of gel solution. The volumes of 10 per cent APS (50 μL) and TEMED (15 μL) are less than used for the diffusive gel, to prolong the setting process and allow the resin to settle by gravity to one side of the gel to achieve a uniform plane of closely packed beads [12]. A successful casting of 10 mL Chelex-based gel solution can produce approximately sixty gel discs.

The ferrihydrite binding layers were originally prepared using a ferrihydrite slurry, made by mixing freshly prepared ferrihydrite slurry into the gel solution prior to polymerisation [13]. This approach has been modified, to provide a higher capacity for P than the slurry gels, by precipitating ferrihydrite within a diffusive gel [14]. These precipitated ferrihydrite binding layers are prepared by immersing a 0.4 mm diffusive gel layer in 0.15 M $Fe(NO_3)_3$ for >2 h with regular gentle agitation. The layer is briefly rinsed with HPW then immersed for 1 h in 100 mL of 0.05 M 2-(N-morpholino)ethanesulphonic acid buffer (MES, $\geq$99 per cent, Sigma-Aldrich), pre-adjusted to pH 6.3 with 1 M NaOH. Gels should be rinsed thoroughly and stored in HPW at 4°C. They should be deployed within four weeks of synthesis [13, 14]. A successful casting of 10 mL ferrihydrite-based gel solution can produce between 60–70 gel discs.

Titanium dioxide binding resins are prepared from a commercially available agglomerated nanocrystalline titanium dioxide (anatase) powder (Metsorb™ Heavy Metal Removal Powder, Graver Technologies). This binding agent can be used for total inorganic arsenic, antimonate, molybdate, phosphate, selenate, vanadate, tungstate and dissolved aluminium species [15, 16]. The gels are prepared by combining 1 g of Metsorb™ with 10 mL of standard DGT gel stock solution, sonicating for 5 min, mixing well on a magnetic stirrer for 1 min, and then casting between two glass plates separated by an inert plastic spacer (0.25 mm). The gels are set in a preheated oven at 45°C for $\geq$30 min, removed from the plates and hydrated in HPW with at least three to four changes of water over 24 h. The binding agent settles on one side of the gel and this side should be placed facing up when assembling in DGT devices. A successful casting of 10 mL titanium dioxide-based gel solution can produce at least fifty gel discs.

Two types of Zr-oxide binding gel are available. For the slurry version, 25 g of amorphous zirconium hydroxide is required. This is prepared by dissolving $ZrOCl_2 \cdot 8H_2O$ in 600 mL of HPW and titrating with 25 per cent ammonia solution under vigorous stirring until the pH stabilizes at 7.0 ± 0.1. The precipitate slurry is rinsed ($\sim$10 times) with 1,000 mL HPW, centrifuged, and the solids dried (hair dryer) to a moisture content of 45–55 per cent (by mass). The solids are ground to achieve a small particle size and the amorphous zirconium hydroxide powder is stored at room temperature. The gel is prepared by combining 1.8 g amorphous zirconium hydroxide with 3.6 mL of gel solution (28.5 per cent acrylamide (w/v) and 1.5 per cent N,N'-methylene bisacrylamide (w/v)), and stirring well (20 min). Polymerisation is initiated using 100 μL APS (10%, w/v) and 4 μL TEMED, and the solution is mixed and immediately cast between glass plates separated by 0.4-mm plastic spacers. Gels are incubated at 10 ± 1°C for 30 min (allowing the zirconium hydroxide to settle to one side of the gel), then at 60 ± 1°C for 1 h. The gel sheet removed from the glass plates is soaked in HPW for $\geq$ 24 h (two to three water replacements) and stored in HPW [17, 18]. A successful casting of 3.6 mL Zr-oxide based gel solution can produce approximately twenty gel discs. For the precipitated version, which has finer particles [18], 3.22 g of $ZrOCl_2 \cdot 8H_2O$ are dissolved in HPW to 100 mL. A 0.4-mm thick APA diffusive gel is immersed for at least 2 h. On removal it is rapidly (1–2 s) rinsed with HPW and then immersed in 100 mL of 0.05 M MES (2-(N-morphilino)-ethane sulphonic acid) at pH 6.7. It is removed after 40 min, copiously washing in HPW for 24 h and then stored in 10 mM $NaNO_3$.

Several methods for preparing silver iodide binding gels have been reported [19–21], with the approach of precipitation within the gel [19, 21] having gained acceptance. A gel stock solution is prepared by dissolving 7.5 g N, N′-methylene bisacrylamide and 142.5 g acrylamide in HPW. Binding gels are polymerised by mixing 9.30 mL HPW, 9.30 mL gel stock solution, 2.4 mL 1 M $AgNO_3$ and 55 μL 10% APS. The mixed solutions are immediately pipetted between glass plates separated by a 0.5 mm plastic spacer, set in an oven at 44°C for 10 min and cooled to room temperature. Once removed from the plates, binding gels are immersed in 0.2 M KI solution and stored in the dark until opaque (typically several hours). During this step, the silver impregnated in the gel during the casting step reacts with the iodide in the bath to precipitate AgI evenly within the gel. Once opaque, binding gels are stored in HPW, which is changed >3 times in the first 24 h. Note that the AgI in the binding gel is prone to darkening upon prolonged exposure to light. The gel should be minimally exposed to light during and after setting and stored in a dark container [19, 20]. These binding gels are designed for sediment applications [21]; thus the number of gels obtained from this procedure will depend on the size of the deployment devices to be used.

Commercially available 3-mercaptopropyl-functionalised silica resin is selective for measuring As^{III} and if deployed in DGT alongside the Metsorb-DGT (for total inorganic arsenic), As^{III} can be obtained directly and As^V by difference [22]. This thiol resin has also been used for the measurements of dissolved mercury and methylmercury in natural waters [23–25]. For the preparation of mercapto-silica binding gels, bisacrylamide-cross-linked polyacrylamide is used in place of the standard polyacrylamide cross-linked with an agarose derivative, as it was found to provide a more homogeneous distribution of the mercapto-silica [22]. In brief, 1 g of dry mass of mercapto-silica is added per 10 mL of bisacrylamide-cross-linked polyacrylamide gel stock solution, and polymerised with 200 μL of ammonium persulphate and 8 μL of TEMED. The mixture is stirred well and cast between two perspex plates (the gels are more easily removed from perspex than glass) separated by a 0.5 mm spacer. This spacer was chosen because, as the bisacrylamide gels shrink slightly upon hydration, a 0.5 mm spacer provides a 0.4-mm thick binding gel. The mercapto-silica binding agent settles on one side of the gel, and this side is placed facing up, in contact with the diffusive gel, when assembled in DGT devices. Gels are fragile and require careful handling to avoid breakage [22]. A successful casting of 10 mL 3-mercaptopropyl-functionalised silica-based gel solution can produce approximately twenty gel discs.

A range of additional protocols for binding resins are available in the literature, with some successful mixed binding layers that provide powerful options for obtaining simultaneous multi-element information [10, 11] (see Chapter 4 and Table 4.1). The preparation of mixed binding layers often requires more consideration than simply combining the recipes for two single binding resins, so reference to the current literature is recommended.

10.2.5 Assembly and Storage

Pre-cut gel discs with a diameter of 25 mm can be obtained from the synthesised sheets. By a happy coincidence the plastic (Sterilin) tubes in which gel sheets are supplied have an opening diameter of 25 mm and can be used as a gel cutter. To cut the gel sheet, first remove it as a bulk sheet from the container and place it on an acid-cleaned plastic cutting board. With the aid of a drop of HPW it can be teased flat with forceps. Avoid lifting a single layer of gel as it tears easily. Press the gel cutter directly onto the gel. To obtain a clean cut, it may be necessary to press firmly and twist at the same time. It is important to inspect the cast gel before cutting to ensure that the gel is in good condition and that there is a consistent distribution of the binding resin, discarding damaged or 'patchy' uneven areas of the gel.

The most commonly used filter membrane is a 0.45 μm polyethersulphone (Supor, Gelman Laboratories, USA), which means most of the testing data is for this product, including determination of diffusion coefficients [9]. Although some reports indicate that the polyethersulphone filter is better able to resist field conditions [26], other membranes, including cellulose nitrate and acetate, have also been successfully used [5]. Provided that the DGT device is calibrated or diffusion coefficients are determined, pore size is not critical unless it is sufficiently small to impede diffusion of larger molecules, effectively introducing size fractionation (see Chapter 3).

Assembly of DGT devices is straightforward, but it is important that the gels and filter membranes are wet to prevent entrainment of air. In most binding layers, the binding agent is uniformly distributed within the gel, but for Chelex gels there is a layer of Chelex beads close to one surface of the gel. This is because the higher density of the Chelex causes the resin particles to settle to the base of the plate while the gel is polymerising. The Chelex gel must be placed on the base of the DGT device with this rougher side of the resin gel facing upwards towards the window (Figure 1.1, Chapter 1). Normally you can see the roughness when you put the gel on a flat surface. If it is not obvious you can touch a small area of the gel surface with the tip of a clean paper tissue, which will take up excess water from the surface. It will then be easy to see the roughness due to the resin particles. Once the resin gel is correctly placed on the plastic base, a diffusive gel disc can be put on top, followed by the filter membrane. After checking each layer is correctly aligned with the base, the plastic cap can be placed on top, ensuring it is horizontal. An even force must be used to press it down until the cap firmly sits on the base. The good contact between the lip of the cap and the filter membrane should be evident. If it is not, use a clean gloved-finger to test that the filter membrane cannot be moved.

Once prepared the DGT devices should not be allowed to dry out. This is best achieved by placing them in a small, air-tight, plastic container or bag with a few drops of 0.01 M NaNO$_3$ solution. Storage in solution is not advised as it increases the risk of contamination from the solution and container. Refrigeration (4°C) is advisable for storage of longer than a week, as it minimises potential biological growth or loss of moisture, but it is not essential. The main problem associated with storage is moisture loss. Without moisture loss devices work perfectly well when stored for the recommended shelf life of six months. If self-sealing plastic bags are used they should be checked weekly for moisture loss, as indicated by any slight gap between the lip of the cap and the membrane filter. If water loss occurs or the shelf life is exceeded the devices should be rehydrated, taking care to avoid contamination. This can be achieved by soaking the DGT devices in HPW for 5–6 h followed immediately by 24 h or overnight immersion in trace metal clean 0.01 M $NaNO_3$ solution. To avoid contamination, containers must be scrupulously acid cleaned and copiously rinsed with HPW. Volumes of HPW and $NaNO_3$ solution should be restricted to those necessary for immersion and the $NaNO_3$ solution must be trace metal clean. This can be achieved using the highest possible purity reagent and/or exposing 1 L of the solution to a few mg of Chelex beads and stirring overnight. Devices should be deployed immediately after revival.

10.2.6 Quality Control: Performance Tests and Method Detection Limit

10.2.6.1 Performance Tests

Each laboratory that uses DGT should verify its performance. This not only establishes that the DGT devices are performing well, but also verifies that the handling and analytical procedures used in the laboratory are appropriate. The simplest way to do this is to deploy replicate DGT devices in a solution with a known composition in the laboratory. It is essential that all containers and solutions are scrupulously clean. The procedure is described for Cd, which is less prone to contamination and does not adsorb to the diffusive layer or membrane filter. A mixed metal solution prepared from neutral salts could be used or a single metal solution of choice. Because Cu and Pb can adsorb to the diffusion layer materials, the deployment time at their concentration of 10 μg L^{-1} needs to be 24 h to ensure accuracy [27]. An alternative supporting electrolyte to $NaNO_3$ could also be used, including synthetic lakewater or seawater (see [28, 29] for preparation protocols).

Prepare a deployment solution of 10 μg L^{-1} of Cd in 10 mM $NaNO_3$ by diluting 20 mL of 1 M $NaNO_3$ solution to 2 L in a 3-L plastic container. Add an appropriate small aliquot of Cd standard solution. Check the pH is between 5 and 7. Place a magnetic follower in the solution and put the container onto a magnetic stirrer. Mount the DGT devices in a suitable holder that can be placed against the container wall, while maintaining the plane of the front filter membrane vertical. A simple plastic assembly will suffice, such as that shown in Figure 2.5, Chapter 2. Immerse in the solution, which should then be stirred well. Use an appropriate stirrer which does not heat up during use and whose speed is well regulated. Simple stirrers can be used, but the best are those that can maintain a constant stir rate that can be displayed. A good guide is that the solution should not cavitate, but should be on the verge of doing so. Take a small sample for analysis and note the time and temperature. After approximately 4 h, sample the solution again, measure the temperature and then retrieve the DGT devices, noting the time of retrieval. Rinse the devices with HPW and then dismantle, separating the cap from its base by twisting a screwdriver or similar flat blade inserted in the grove in the base. Place the Chelex gel in a clean sample tube, such as a 1.5-mL microcentrifuge vial, and add 1 mL of 1 M HNO_3, leaving for 24 h or at least overnight. The Cd concentration in this eluent should be about 20 μg L^{-1} assuming the test temperature was about 25°C. For analysis by ICP-MS or GF-AAS, a 10× dilution of the eluent is recommended.

10.2.6.2 Calculations

For this simple test, which uses a standard DGT device, with 0.8 mm thick diffusive gel and a membrane filter, deployed in a well-stirred solution, a simple calculation procedure can be used. Chapter 2 provides various calculation options, depending on the device parameters and deployment conditions and Chapter 1 provides the basic equations that can be used here. The accumulated mass, M, is calculated from the measured concentration of metal, c_e in the 1 M HNO_3 eluent, taking account of the volume of the binding layer, V^{bl}, 0.16 mL in this case, and the volume of the eluent solution, V_e, 1 mL in this case (equation 10.1).

$$M = c_e(V^{bl} + V_e)/f_e \tag{10.1}$$

The elution factor, f_e, compensates for the fact that only a proportion of metal is released on elution. For metals, a typical value is 0.8, but Chapter 4 provides values of elution factors obtained for a range of binding agents and analytes. When using equation 10.1, remember to express concentration in units of mass per millilitre and distance in centimetre, if the volumes are in millilitre. The concentration of metal measured by DGT, c_{DGT}, can be calculated using equation 10.2.

$$c_{DGT} = \frac{M \Delta g}{D^{mdl} A_p t} = \frac{M(\delta^g + \delta^f)}{D^{mdl} A_p t} \tag{10.2}$$

The combined thickness of the diffusive gel ($\delta^g = 0.08$ cm) and filter ($\delta^f = 0.014$ cm) is Δg (also known as δ^{mdl}), D is the diffusion coefficient of metal through this material diffusion layer and A_p is the physical area of the exposure window (3.14 cm^2). Values of D can be calculated for a given temperature using equation 2.25, Chapter 2 and the values at a given temperature provided in Table 3.2, Chapter 3, where details are also provided on approaches to measuring diffusion coefficients. Tables of parameters for simply calculating diffusion coefficients at different temperatures are available in Appendix 1 of this book and values of D^g at 1-degree intervals are listed on the website of DGT Research Ltd (dgtresearch .com). Diffusion coefficients in a standard polyethersulphone filter membrane are the same as in the standard APA diffusive gel. The units of c_{DGT} will be in mass per cubic centimetre (e.g. ng per millilitre) if Δg is in centimetre, t in seconds and D in cm^2 s^{-1}. An example calculation is provided in Chapter 1, Table 1.1.

10.2.6.3 Target Performance

The measured value of c_{DGT} should be within 10 per cent of the solution concentration measured directly in the samples of immersion solution collected during the experiment. If there is a greater difference, first check the calculation and then rerun the experiment, being especially aware of contamination issues and adequate stirring. The value of c_{DGT} for replicate discs from the experiment should also be very similar, with standard deviations of <10 per cent. Blank values will be useful in assessing possible contamination. Acceptable values for the mass (ng) measured on the blank discs are: Cd and Co, 0.02; Mn, Pb and U, 0.1; Cu, Ni and P, 1; Zn, 10.

10.2.6.4 Blanks and Method Detection Limits

Three different blanks can be considered. The Chelex gel blank is simply a gel disc after its preparation or supply. The eluent blank is the blank measured in the reagent used for elution. The DGT blank is obtained using a Chelex gel disc after it has been retrieved from a DGT device that was not deployed.

DGT blanks should be used to calculate method detection limits expressed as mass (three times the standard deviation of the blank) on a per experiment basis using all the DGT deployment conditions. The deployment time that would be necessary to reach the method detection limit is calculated for a hypothetical DGT piston, fitted with a diffusive gel and filter membrane that would be used, at the considered deployment temperature (equivalent procedures apply for sediment probes). Calculations should be performed using the appropriate sample dilution, elution efficiency and diffusion coefficient for each analyte.

10.3 DGT Components and their Assembly for Measurements in Soils

The first DGT deployments in soils used the same piston units that were routinely employed for measurements in solutions [30, 31]. Although they generally worked well, the measured accumulated mass was occasionally erroneously high. This was attributed to fine soil particles being forced through the seal between the membrane filter and lip of the cap when the device was pressed onto the soil. Consequently a new piston device was developed. It has a redesigned more springy lip, which has a larger surface area contacting the filter. The exposure window of the soil deployment unit has an area of 2.54 cm^2. The components used in the soil device are exactly the same as those used for solution devices and handling and assembly procedures are the same.

Most measurements in soils have been on samples hydrated in the laboratory, although standard piston holders have been used to deploy DGT in situ in saturated soils under field conditions [32] (see Section 10.5.2.4). Where information on spatial changes in analyte concentrations has been required, flat DGT probes have been used [33, 34]. Essentially they are the same as those used in sediments, as described in the next section.

10.4 DGT Components and their Assembly for Measurements in Sediments

The original idea for DGT came from sediment work, yet advancements in the application of DGT to sediments have been much slower than for aqueous deployments. This is because the physical, chemical and biological characteristics of sediments are heterogeneous, presenting a challenging matrix to sample and interpret. However, significant progress has been made in this field over the past decade owing to two-dimensional imagery of sediment profiles [33–36], three-dimensional models of sediments simulating localised DGT signals [37], and the distribution of solutes in microniches [38], as discussed in Chapters 5, 7 and 8.

10.4.1 Sediment Probes

There are a variety of sediment probe designs, many of them tailored for specific research goals. The two original planar designs produced by DGT Research Ltd are still the most commonly used devices: the sediment probe and the constrained probe. The plastic sediment probe can accommodate the DGT and DET gel systems, and the use of these probes in sediments is well established.

The sediment probe is commercially available through DGT Research Ltd, and thus easily obtained. As with the pistons, the planar probe consists of a two-piece assembly unit produced from pre-made moulds. The larger of the pieces is the base plate. The base plate is a single unit with a moulded handle on the top to facilitate probe handling. The sides of the base plate have small holes to enable the top plate to lock in place once the device is loaded. The bottom of the base plate has a gently bevelled edge to allow easier deployment and less physical disturbance when the device is inserted into the sediment. The body dimensions are 18 cm × 4 cm × 0.5 cm (Figure 8.4a), extending a further 5.5 cm for the handle (total length 24 cm). There is a cavity in the base plate in which the binding and diffusive gels are placed (16.2 × 2.8 cm). The front piece of the assembly unit has similar dimensions with an open window of 1.8 cm × 15 cm, and male connectors that secure it to the base plate once the probe is loaded.

The constrained sediment probe has the same external dimensions as the planar sediment probe and the same open window dimensions as the DGT probe (1.8 cm × 15 cm). The primary difference in design is that the plastic assembly has seventy-five slits (0.1 cm × 0.1 cm × 1.8 cm) embedded in the base plate, arranged in a single column, with 1 mm distance between each one (Figure 8.4a). The device is designed for the diffusive equilibration in thin-films (DET) technique, in which the recesses are filled with agarose gel (does not expand after polymerisation and is dimensionally stable within the slits during deployment). The agarose gel is used for capturing major anions, cations and metals with high concentrations (such as Fe and Mn) in porewaters. The cleaning and handling protocols discussed earlier for DGT pistons (Section 10.2.2) also apply to sediment probes.

10.4.2 Assembly and Storage

Using a Teflon (polytetrafluoroethylene, PTFE)-coated razor and acid-washed clear plastic ruler, the binding resin and diffusive gel can be prepared by cutting into large rectangles (16 cm × 2.8 cm) to fit within the cavity of the base plate. The 0.4-mm binding resin layer is placed on the base plate, overlain with a 0.8-mm APA diffusive gel and a 0.14-mm thick 0.45-μm pore size filter membrane. In order to avoid the introduction of nitrates into sediment deployment systems, polysulphone filter membranes (0.45-μm pore size) should be used instead of cellulose nitrate. Rolls or sheets of 0.45-μm filter membranes can be commercially purchased, then cut to size and washed before use (soaked in 10 per cent HNO_3 for approx. 24 h, rinsed multiple times with HPW to ensure pH is neutral, and stored in HPW until used).

During probe assembly, extreme care must be taken to ensure that no air bubbles are trapped within the gel layers and that each layer is correctly aligned with the base plate. The total thickness of the combined layers is approximately 1.34 mm for DGT, which allows for some compressibility within the 1.2 mm space to provide a good seal around the exposure window. The front plate should be aligned (male connectors with the backing plate holes) and firmly clipped into place with an even force. The secure fit can be checked from the back of the device, with the male connectors visibly in place. All preparations should be performed in a metal-free environment, such as a laminar flow hood. Prepared probes should be stored as described in Section 10.2.3, ideally by placing them in a small, airtight, plastic container or bag with a few drops of 0.01 M NaCl solution.

Slight variations in probe assembly pattern and handling have often been adopted, usually driven by the characteristics of the binding gel and the goal of the deployment program. For example, when using the combined silver iodide–iron oxide binding layer [39] the binding layer is placed onto a polysulphonate backing filter to aid removal of the binding layer from the probe assembly after deployment. The filter is also a useful support for the dimensional integrity of the binding phase during handling and drying [39].

Additional care must also be taken when preparing planar sediment probes for particular analyses, such as sulphides. The AgI in the binding gel is prone to darkening upon prolonged exposure to light, because of the photoreduction of silver halides to silver metal. This was mainly found to be a problem during setting of the gel. However, when exposure to light was minimised, during and after this step, and the binding gel stored in a dark container, it was stable. When assembling the DGT probe, the binding gel (AgI surface up against the diffusive gel) is overlaid with the diffusive gel and the 0.45-μm membrane filter. On retrieval, the diffusion layer is removed from the sampler immediately, while ensuring that exposure of the binding gel to sunlight is minimised [19, 20].

10.4.3 Assembly of the Constrained DET and DGT Sediment Probe

DET constrained sediment probes are usually filled with an agarose gel, as it does not expand on hydration [40]. An agarose solution is prepared by thoroughly dissolving 1.5 g of agarose in 100 mL of HPW, preheated to 80°C, while stirring continuously. The hot agarose gel solution is cast into the 0.1 cm $\times$ 0.1 cm $\times$ 1.8 cm wells on the base plate of the constrained sediment probes using a 5–10-mL pipette, while ensuring that there are no air bubbles in the wells. A Teflon-coated razor blade should be used immediately to scrape any excess gel from the surface of the plastic dividing the compartments, so avoiding any diffusive connection between wells. Cast gels should be set at room temperature. Once set, a 0.45-μm pore-size polysulphone membrane filter is placed over the agarose gel area and clamped in place by a retaining plate.

There is also a variant of a DGT constrained probe used for measuring U, Fe and Mn, in which a mixture of Spheron-Oxin® resin and agarose gel is used to fill the recesses [41]. The binding gel mixture is prepared by combining 0.5 g of Spheron-Oxin® 1000 (ion-exchange resin, 40–63 μm) in 5 mL of the hot agarose solution (described above), and immediately cast into the constrained probe wells. Once cooled to room temperature, it is covered with a 130-μm thick filter membrane of cellulose nitrate (Sartorius, pore size 3 μm) and/or polyethersulphonate (Pall Corporation, USA) and clamped in place by a retaining plate. These membranes were exclusively used because they were sufficiently inert to diffusing metal ions. Biotrace PVDF membrane filters (Pall Corporation, USA) should be avoided because they strongly adsorbed the metal ions [41].

10.5 Deployment

The preparation, deployment and retrieval of the DGT devices require consideration and care. Simple oversights can result in erroneous data, which subsequently leads to inappropriate interpretation of the site or sample. The following section highlights some of the considerations and challenges associated with DGT deployments in both field and laboratory settings, including device conditioning, deployment scenarios (laboratory and field), deployment duration, the role of temperature, time and oxygen, device retrieval steps and the preparation of gels for analysis. The quality control measures to ensure more reliable outcomes are also addressed.

10.5.1 Deployments in Natural Waters

10.5.1.1 General Considerations

For DGT analysis in natural waters the same protocol as described in Sections 10.2.2 and 10.2.3 should be followed. The chosen analyte(s) in natural water systems will determine the choice of binding layer. Two main binding layers have been used. Chelex binding layers have been accurately quantified to measure Cd, Cu, Fe, Mn, Ni, Zn Pb, Co, Al, Ga, Y and the rare earth elements [2, 12, 16, 42] (see Chapter 4). The other main binding layer, containing Fe oxide, has been evaluated for P, V, As, Mo, Se, Sb and W [10, 11], which have been measured in freshwater. Alternative binding agents for oxyanions are titanium dioxide available in proprietary form as Metsorb [15, 43, 44] and Zr-oxide [17, 18] (Chapter 6). These alternative binding layers for measurement of oxyanions offer advantages over the initially used Fe oxide binding layer for P [13] in particular, as they have greater capacities for P and are little affected by anions (Cl^-, HCO_3^-, CO_3^{2-}) competing with phosphate oxyanions in the binding step. As outlined in Chapter 4, binding agents have been developed and characterised for the measurement of specific analytes, such as thiols for Hg [24, 45, 46] and methyl-Hg [23, 24], TEVA- for Tc [47], MnO_2 for Ra [48] and ammonium molybdophosphate for Cs [49].

Normally DGT solution devices, with a 3.14 cm^2 exposure window, are used for deployments in natural water systems, but DGT devices manufactured for soil deployments ($A_p = 2.54$ cm^2) can also be used.

10.5.1.2 Choice of Diffusive Layers

One of the most important aspects of correctly using DGT technology for element assessment in natural waters is the choice of diffusive layer thickness and pore size. The recommended open pore APA diffusive layer will allow hydrated metal and anion complexes that are generally smaller than 5 nm to diffuse through to the binding layer (see Chapter 3) [50]. Complexes >5 nm in size will have minimal contribution towards the measured DGT value. Often organic ligands with diameters <5 nm are present in natural water systems. As diffusion coefficients for their metal complexes are significantly lower than for free metal ions, they will contribute proportionally less to the measured mass.

10.5.1.3 Thickness of Diffusive Layers and DBL

The diffusive boundary layer (DBL) effect has been well documented in previous chapters (2, 6) and will not be discussed in detail here. Briefly, the diffusion layer thickness is extended when DGTs are deployed in solutions that have inadequate flow at the DGT interface. The effect of the DBL on DGT calculations will be most pronounced when the material diffusion layer is thin. Natural water systems, such as lakes, may have low flow environments. In these cases it is recommended that the DBL is assessed with each deployment by using multiple DGT devices with different thicknesses of diffusion layers (see Chapter 2 for calculation details). Where there is a good flow of water, ensuring the DBL thickness is small and constant, the simple DGT equation 10.2 with the physical area provides a reasonably accurate estimate of concentration. In some environments the concentrations of elements of interest are very low and can challenge DGT detection limits. In these circumstances, thinner diffusive layer thicknesses can be a suitable option to enable greater element accumulation, but it is then important that the DBL effect is quantified.

10.5.1.4 Practicalities of In Situ Deployment

To allow for accurate assessment of analyte concentrations and fluxes in natural water systems, there may be a need to deploy multiple DGT devices at any one time. One of the main hurdles is finding an appropriate holder or system that facilitates stable DGT devices in moving water, while not interfering with the natural flow of water at the DGT interface. Designs for deploying multiple DGT devices have been illustrated in Chapter 6. Other approaches have included simply dangling single DGT devices on a line [51], or mounting them inside tubes submerged on stream beds [7]. Mounting several devices on flat plastic plates has its dangers, as a large plate is likely to impede flow and increase the DBL thickness [52]. An alternative is to mount on plastic cubes [53] or to simply tie six DGT devices together with fishing line so that they form a cuboid cluster with all devices freely exposed [7]. Depending on the purpose of the work and the type of water system, DGT devices are sometimes deployed for a combination of different times and at different depths. It is recommended that at least two, preferably three, DGT replicates are used at each location and time. The maximum length of deployment time where accumulation proceeds according to equation 10.2 will be element specific, due to the different characteristics of binding layers, and it will also depend on the concentration of analyte and possible interfering ions present. Temperature and the DGT deployment time are essential for the calculation of c_{DGT}. If temperature is likely to vary substantially during deployment, a temperature logger can be used at each site of DGT deployment (see Chapter 6). After the required deployment time has elapsed DGT devices should be retrieved and rinsed immediately with HPW. At this stage it is important to inspect the DGT interface for any evidence of biofouling (see Chapter 6), which can potentially affect concentration gradients through the diffusion layer.

After rinsing, the DGT device can be either stored or the elution step can be performed at the site of deployment. If DGT samplers are to be returned to a laboratory it is essential that protocols for DGT storage and avoiding contamination are strictly followed (Section 10.2.4). Either on site or back in the laboratory, the front cap of the DGT device is removed as described in Section 10.2.6.1, which outlines the procedure for analysis, with calculation according to Section 10.2.6.2, or Chapter 2 when the DBL has to be considered.

10.5.2 Deployments in Soils

Most studies where DGT has been used in soils have used Chelex or Fe oxide binding layers and therefore they will be the focus of the discussion on deployment techniques. The use of DGT in soil environments is increasing and within Australia it is offered as a commercial service for the measurement of P. With the progression to a commercial service the method had to be streamlined and consistency ensured between laboratories. The following sections will report various progressions of methodology that have worked and can be utilised in a research environment.

10.5.2.1 Soil Preparation

Analysis of soil samples by DGT has been assessed against other soil test measures performed in the laboratory after collection of samples from the field. For appropriate comparative performance, soil samples collected from the field must undergo similar treatments. Samples are usually routinely dried (40°C or air dried) and sieved to <2 mm. For glasshouse studies the soil used has usually been through this process and requires little further preparation. The amount of soil used is not an important factor for standard reported DGT deployment times (~24 h), but it is important that there is an adequate depth of soil adjacent to the DGT sampling window. Predicted DGT depletion zones within a 24 h period usually only extend to a few millimetre, but it could be substantially more if there is little resupply of the analyte from the solid phase (Chapter 7). It is recommended that depths of 1 cm or greater are used. Common amounts of soil used per DGT device have ranged from 20–200 g. Containers for each soil sample must be free from any contamination. Size requirements of the container will be a balance between the amount of soil required to create 1-cm depth and enough hand space to allow for easy deployment and retrieval.

In general, disposable plastic cups with a diameter of 7 cm and a depth of 2–3 cm are appropriate for 30–50 g of soil.

10.5.2.2 Soil Moisture

Previously an accurate determination of the soil's Maximum Water Holding Capacity (MWHC) was recommended to enable the soil to be wetted to this moisture content prior to DGT soil deployment [54]. For selected research studies involving a small number of soil samples (<10) this measurement is recommended. For a commercial service the additional time and cost associated with a MWHC measurement would make the DGT non-profitable and impractical. In this situation an approximation of a soil MWHC has been used successfully. Water is applied to soil samples until a glistening effect is viewed at the soil surface. It is assumed that the soil sample has then reached maximum MWHC. The glistening effect is a rapid method that has some merit, as Hooda et al. [54] showed there were minimal differences in DGT measured analytes between soil moisture levels at field capacity and MWHC. Required amounts of water to reach MWHC can also be approximated by clay content [55].

Methods of increasing soil moisture to levels that are appropriate for DGT deployment have varied from study to study. Initial literature [55] suggested that soil samples should be incubated at approximately 50 per cent maximum water holding capacity (MWHC) for two to three days prior to deployment. Water contents of the soil are then increased to MWHC the day prior to DGT deployment. This allows for the analyte of interest to equilibrate between soil solution pools and sorption sites on the soil surface, while minimising redox changes. In a commercial environment, comparable DGT results for P were obtained between six laboratories, by increasing soil samples from a dry state to MWHC the day prior to DGT deployment using the glistening end point. For non-wetting soils the soil moisture might require adjustment the morning of the next day. For elements other than P, the time required for equilibration might be longer, especially if their response times are long (see Chapter 7), and therefore required incubation times might be greater than one day. Unpublished work at the University of Adelaide that assessed the effect of incubation times on DGT P measurements found no real differences between increasing moisture on the day of deployment to increasing it one, two or five days prior to deployment (Mason pers. comm.). During incubation moist soil samples should be covered, allowing for some air space. The idea is to avoid reducing conditions, while limiting drying and potential rewetting effects, which could affect DGT measurements of some analytes.

10.5.2.3 Practicalities of Laboratory Deployment

It is recommended that laboratory deployments of DGT are performed in a controlled temperature environment or one where the change in temperature will be no more than a few degrees. On the day of DGT deployment it is important to have the DGT devices at room temperature, so they should be removed from the fridge (See Section 10.2.3) several hours before deployment. For accurate analysis there must be 100 per cent contact of the exposure window with the soil. Most common DGT devices are initially placed window facing down on top of the soil and good contact ensured by gently moving the DGT device in a circular motion. Some studies have alternatively placed relatively small amounts of soil on top on the DGT sampling window [56]. To ensure soil contact with the window using this method (and sometimes for face-down deployment) a small amount of soil is first smeared on the DGT filter membrane. The DGT devices can be checked on retrieval to verify that all the sampling area has been stained by the soil sample. The start time of deployment must be recorded along with the room temperature. Deployment times can vary, but have been typically around 24 h. A deployment time of 24 h should be sufficient to provide quantification of elemental resupply dynamics of P, but minimal deployment times could be as short as 6 h [13]. Sample containers must be covered (with air space) to control the soil moisture conditions throughout the deployment period. For metals, such as Cu and Pb, that can be affected by interactions with the diffusive gel and humic substances, longer deployment times of several days are recommended (see Chapters 3 and 5). Possible changes in redox conditions associated with longer deployments can be assessed by measuring redox-sensitive cations (Fe, Mn) or anions (NO_3^-, SO_4^{2-}) in soil solution or by DGT, as these are more informative than measurements of redox potential [57].

When the recommended deployment time has elapsed, DGT devices are removed from the soil and any adhering soil rinsed off with HPW from a wash bottle, taking care to avoid the water flow encouraging soil particles to enter under the seal between filter and cap. Finer texture soils tend to 'adhere' to the DGT devices, so that pulling them straight out often brings most of the sample with them, which makes the rinsing procedure very difficult. A useful trick to reduce the amount of attached soil is to squirt water into the soil sample surrounding the DGT device. Eventually the DGT device should be free from the soil sample. When cleaned, the DGT device front cap is removed and elution and analysis procedures described in Section 10.5.3 are followed.

10.5.2.4 Deployment In Situ in the Field

One of the attractive features of DGT is its potential for in situ deployment in the field, but unfortunately the reliability of simple piston-type measurements is not as good as those performed in the laboratory [58]. Potential problems that field conditions present to DGT measurements are associated with the lack of control, particularly with respect to moisture content, as soil moisture conditions can vary spatially and temporally. There is no standard wetting procedure, temperature can fluctuate, and, as soils are normally uncovered, they are subject to evaporation. There are special situations for which in situ deployment of DGT piston devices in soils is appropriate, such as water-logged soils, and rice paddies, where acceptable reproducibility has been achieved [32]. Deployments in a large-scale lysimeter showed there is considerable scope to use standard planar DGT devices in situ to examine the often complex distribution of solutes in soils [59] in a similar way to the work on sediments (see Section 10.5.3). An important recent development is the acquisition of high resolution, two-dimensional images of elemental distributions within the rhizosphere, as described in detail in Chapter 8.

10.5.2.5 DGT Calculations

For comparative studies with other soil tests DGT results have been presented as either c_{DGT} or c_E (Chapter 7 and Section 10.2.4), with the latter requiring use of the DIFS model. In most studies where DGT measurements are compared to plant uptake data it is acceptable to use c_{DGT}, which approximates to the mean DGT interfacial concentration, because, unless soils have markedly different porosities, c_{DGT} and c_E are highly correlated (Chapters 7 and 9).

10.5.2.6 Quality Control

It is recommended that all the quality-control measures reported in Section 10.2.6 for solutions are followed, which includes use of DGT blanks (3) and the calculation of the method detection limit for each run (mean $+ 3 \times$ standard deviation of the blanks). Further controls are recommended to ensure robustness of DGT measured values for soils. For an unknown batch of soil samples, duplicates should be introduced every fifteen to twenty samples, depending on batch size. It is also advisable that each laboratory has a control soil for which DGT values are known through previous analysis.

10.5.3 Deployments in Sediments

10.5.3.1 Preparing Probes for Deployment

The sediment probes need to be conditioned prior to deployment. This should be performed using solutions of 0.01 M NaCl for freshwater deployments and 0.12 M NaCl for seawaters. Nitrogen is a limiting nutrient in marine systems, and should be avoided as it may encourage fouling of the filter membrane. In addition, nitrate is a redox sensitive compound, actively used as an electron acceptor for microbial respiration, and may affect redox gradients in sediments.

Ideally conditioning solutions should be cleaned for ≥ 24 h prior to use by placing a sheet of the relevant binding gel (e.g. Chelex-100 for probes containing Chelex) in the conditioning solution. Probes should then be immersed for 24 h (or at least overnight) in the conditioning solution, through which high purity argon or nitrogen gas is bubbled to deoxygenate the probes. Deoxygenating the probes reduces the disruption of sediment chemistry, e.g. oxidative precipitation of Fe(II) to Fe(III) and dissolution of FeS, that would have resulted from the introduction of oxygen into suboxic and anoxic sediments [60]. Conditioning also ensures that the diffusive gel of the probes has a similar ionic strength to deployment conditions. To keep the gels moist, conditioned and oxygen-free, assemblies should be transported to the sites in a container filled with deoxygenated conditioning solution, and deployed immediately, as oxygen will rapidly return upon exposure to air. While the exact effect of some oxygen in a probe has not been systematically evaluated, the relatively small reservoir within the probe will be quickly consumed and anoxia will prevail throughout most of the deployment.

It is also very useful to place a mark on the side of the front plate of the DGT probe, adjacent to the sample window, using a black permanent marker prior to probe conditioning. This line can be used to indicate the sediment-water interface and hence the depth that the probe is inserted into the sediment (≥ 2 cm below the top of the sampling window). It is much easier to mark a dry clean probe before conditioning than a dirty wet probe post-deployment. Equipment preparation for the collection of relevant physical and chemical parameters also warrants consideration (for example, to ensure electrodes are clean and calibrated for measurements of temperature, pH, dissolved oxygen, and conductivity/salinity of both sediments and the overlying water).

10.5.3.2 Deployment Scenarios

Sediments are stratified substrates, formed by the deposition of particulate material from the overlying water body. This typically leads to varied spatial strata both with depth and horizontally. In productive sediments, microbe-driven redox gradients further complicate the system as they simultaneously generate steep concentration gradients of the involved species throughout the sediment profile. Thus, achieving a representative assessment of these regions is challenging. Deployments are commonly made in situ or in sediment core systems, typically performed within the laboratory. These approaches reflect the purpose of the deployment, for example as a component of a site assessment, a fundamental research study on a specific substrate or validation of a new analytical resin.

In cases where the DGT planar probe deployments will be in situ, there are several factors to consider: (i) the depth of the overlying water, (ii) the characteristics of the receiving substrate and (iii) the nature of the surrounding environment. Water depth is primarily a safety consideration. In shallow water, when it is possible to wade or snorkel into the system, care must be taken not to stand on the sediments at or immediately adjacent to the deployment site, as this can compact the sediments, altering their porosity and forcing porewaters to alter their original path. The sediment stratum may also be disturbed. For deeper water, SCUBA can be used, or it may be necessary to fix an extension device to the handle of the probe. This requires some skill as it is important to deploy the device as vertical as possible during the insertion. Attaching an underwater camera to a pole can be very helpful as it allows you to inspect the site both before and after probe deployment. DGT probes have also been deployed in deep ocean sediments using autonomous landers, but this requires very specialised equipment [61].

The physical characteristics of sediments vary between eco-regions and can pose some interesting challenges for in situ deployments. The most important point is to ensure that there is close contact between the face of the probe and the sediment. The ideal texture is a relatively homogeneous silt, sand or mixture of the two. These substrates are easy to work with and allow the probe to be gently pushed into the sediment to the desired depth, when the sediment–water interface mark on the probe is reached. Deployments are more difficult in sediments containing larger particle sizes, shells and other debris, as sharp edges may tear the DGT membrane and gels during deployment and retrieval. Various deployment strategies have been used, including pushing a planar device into the sediment (clear plastic or stainless steel ruler) then sliding the probe into the sediment behind the device (sampling window facing towards the device), and then carefully removing the device to allow the sediment to collapse against the DGT probe. Some specifically designed deployment devices have been developed, including the DET deployment system for coarse riverbed sediments [62] This approach utilises a stainless steel housing that is pushed or hammered into the riverbed. A second steel plate containing the DET gels (with windows facing towards the plate) is inserted into the housing. Upon probe deployment the protective plate is removed from the housing, exposing the DET windows and collapsing the gravel sediment against the probes.

The physical nature of the sampling site should also be considered. There is always a risk that the deployed probes cannot be recovered. In other cases, placing a marker (with or without a label describing the work) will draw attention to the area and devices, which can lead to the loss of equipment. The use of global positioning satellite (GPS) coordinates combined with location notes can enable less visible deployment of the probes and the ability to return to the exact deployment location. However, it is still worth using as much discretion as possible. We are fortunate that society is becoming more aware of our environment. However, the general public will engage in activities to keep our environment clean, including actively removing any visible 'intrusive' materials from waterways. Curious local residents have also been reported to show an interest in deployed probes, as we discovered when we deployed probes adjacent to an octopus' den, but some things cannot be predicted. The tides and weather are easily predicted and warrant consideration, as many deployed probes have fallen victim to unexpected high rainfall events.

Laboratory-based studies avoid many of the issues associated with challenging field environments, particularly working in deep waters and loss of deployed probes. These are referred to as sediment cores, mesocosms or box experiments. The technique of sediment collection will depend on both the project goals and the ease of access to the sediments. Intact sediment blocks will retain more of the properties associated with the collection site and will result in significantly less disturbance of the physical and chemical gradients within the sediment. However, in deep waters, collection of intact cores may be difficult, and grab samples may have to be used. For both collection approaches it is important to allow the sediments to re-establish equilibrium under circulating clean overlying water in containers that protect the sides of the sediment from light penetration. Equilibration times range from several days to several months, depending on the aims and available time within the project. Re-establishing redox stratification in productive sediments under warmer temperatures will be far more rapid than in sediments with lower nutrients at cooler temperatures. However, longer equilibration times will most likely provide the most environmentally realistic results, particularly when using sediments collected by grab sampling. Laboratory-based deployments do have their limitations, as laboratories rarely have the equipment necessary to mimic the natural environment, such as tide simulations, overlying water flow rates, light cycles, ambient temperature fluctuations and groundwater movement. Therefore, it is worth noting these limitations when interpreting the data from these deployments.

10.5.3.3 Duration of DGT Deployment

There is no single answer to the question 'how long should I deploy the sediment DGT probe?' There are many factors that influence the time needed to reach 'steady-state' uptake rates, as discussed in detail in earlier chapters. As a general rule, the deployment time in sediments should be at least two to three days for probes with 0.8-mm thick gels. In sediments that are highly sulphidic, it is recommended to limit DGT-ferrihydrite gel deployments to two days to avoid the formation of iron-sulphides on the binding gel, which may occur due to iron being a constituent of the binding resin and dissolved Fe(II) and hydrogen sulfide being present within the anoxic region of sediments. The DET technique relies on solutes equilibration with the water of the hydrogels, which are then removed for analysis. The minimum time required to reach equilibrium is determined by the diffusion coefficient of the solutes in the gel [63], but two to three days will generally suffice.

It is important to note the exact time of probe retrieval following deployment. Retrieved probes should be thoroughly rinsed with copious amounts of high purity water and stored separately in clean plastic bags with 1 mL of HPW at 4°C until processed for analysis. Care should be taken to protect the surface of the probe during transport and storage, to avoid contamination and damage.

10.5.3.4 Treatment of Gel Prior to Analysis

The ability of DET and DGT to measure one- and two-dimensional profiles in the millimetre and submillimetre ranges has provided previously unavailable insight into solute dynamics in soils and sediments. This is achieved by either slicing of the binding gel for fine resolution analysis, or by direct gel analysis techniques, the basis of which are discussed in Chapter 8, Section 8.2.2.

After retrieval, DGT binding layers are typically cut at the position corresponding to the sediment-water interface and then further cut to the desired resolution, typically performed by hand or using a mechanical system. In both cases PTFE-coated razor blades are used to effect a well-defined separation between segments with minimal danger of contamination. Binding gels can be sliced to give vertical one-dimensional resolution (e.g. 1 mm) or two-dimensional resolution (e.g. 3 mm × 3 mm) [64]. For higher spatial resolution measurements, the element in the resin gel can be analysed directly (see Section 10.6.2). The sliced sections of binding gel are subsequently eluted in the appropriate eluant for at least 24 h (see Table 4.2 for elution solutions and elution efficiencies for commonly used binding gels). An aliquot of this eluant is typically diluted for subsequent instrumental analysis, and the gel slice is removed, gently patted dry with a tissue and its mass recorded. DET gel slivers are typically excised from the recess using an acid-washed plastic toothpick (or similar), placed in a 2.5-mL Eppendorf tube and re-equilibrated in ≥ 0.1 M HNO_3 for 24 h with gentle agitation; the gel is removed prior to instrumental analysis.

10.5.3.5 Analysis of Eluting Solution

Analytical strengths of the DGT technique are the inbuilt pre-concentration provided by the binding gel combined with the selectivity, which means that analytes that cause analytical matrix interferences during analyses can often be eliminated (such as salts from marine samples). These characteristics have allowed the range of element quantification techniques to be expanded to include widely available and affordable detection systems such as ion chromatography (IC) and graphite furnace atomic absorption spectrometry (GF-AAS). The most common instrumental analyses for DGT and DET gels are performed with simultaneous multi-element inductively coupled plasma (ICP) techniques. Following the elution of the analytes from the binding gel, the eluant will need dilution with high purity water to reduce the acidity (or sodium ion concentration for NaOH-based eluants). For example, to avoid interferences during ICP-MS analyses, ferrihydrite and Chelex gel slices eluted in 1 M HNO_3 require a 1:10 dilution with HPW, and Metsorb™ gels eluted in 1.0 M NaOH requires a 1:20 final dilution [12, 14, 44].

The different binding agents, eluants and analyte compositions will determine the matrix of the final eluting solution. It is therefore appropriate to incorporate quality-control samples routinely with each batch of analyses, including blank solutions, certified reference materials and drift correction standards. When evaluating new binding gels and sample matrices, spiked recoveries should also be considered.

10.5.3.6 Direct Analysis of the Gel

The range of two-dimensional approaches that enable high-resolution analyses of planar probes is discussed in Chapter 8. For example, Computer-Imaging Densitometry (CID) is an affordable, rapid and accurate technique. Using a conventional flat-bed scanner as the detector system and appropriate data-processing software, colour densitometry was an effective tool for two-dimensional DGT imaging of AgI gels, yielding recoveries of 95 per cent and precisions close to 5 per cent for DGTs in spiked solutions of sulphide [19–21]. Laser ablation-inductively coupled plasma-mass spectrometry (LA-ICP-MS) is a popular approach as it offers two-dimensional high resolution coupled with simultaneous multi-element analyses [35, 36]. The technical capability, advantages, correction considerations for signal variability and optimisation requirements of LA-ICP-MS are extensively discussed in Section 8.2.4.

Together with planar optodes, a complementary, fluorescence-based imaging technique, DET and DGT are currently the state of the art in chemical imaging of solute distributions in aquatic and terrestrial biogeochemical hotspots (Chapter 8).

10.5.3.7 Calculations

The availability of metals in porewaters is a complex process, directly associated with diagenetic cycling within the sediments. Porewater concentrations of solutes, at apparent chemical equilibrium between the solid phase and adjacent porewaters, are controlled by porewater and sediment chemistry such as redox conditions, pH, mineralisation and complexation of metals by dissolved inorganic and organic ligands. The sediment chemistry of discrete areas may also be drastically altered due to the presence of benthic organisms. Creation of burrows can create oxic environments within anoxic regions of the sediments, often resulting in remobilisation of metals that had been buried there. These sediment complexities are explored in Chapter 8.

When a DGT probe is inserted into soil or sediment it comes into direct contact with porewaters. The probe perturbs the porewater system by the introduction of a localised sink and solutes in the porewaters adjacent to the probe diffuse through the diffusive gel to the binding layer, establishing a steady state linear concentration gradient in the diffusive layer. Therefore, the amount of solute that diffuses into a DGT probe during its deployment is determined by (1) the diffusive gradient in the diffusion layer, (2) the diffusive flux of solute from the porewater, (3) the ability of the solid phase to resupply solute if the concentration in the porewater becomes depleted and (4) the duration of deployment. The processes are treated quantitatively in Chapter 7.

Chapter 2 illustrates that the DGT technique has its theoretical basis in derivations of Fick's First Law of diffusion, which relates mass uptake by a device to the length of exposure, concentration gradient and area exposed. DGT directly measures flux, F, the rate of supply of labile ions that diffuse from bulk solution to a binding layer in a given time. By analysing the total amount of solute bound by the binding layer, the flux from the sediment ($pg\ cm^{-2}\ s^{-1}$) can be measured directly by DGT using equation 10.3, where M (pg) is the mass of diffused ions accumulated in the binding layer over a known deployment period of time (t, seconds) through a well-defined sampling area (A_p, cm^2).

$$F = M/A_p t \qquad (10.3)$$

However, manuscript reviewers frequently request that the DGT flux data derived from sediment deployments be converted to a time averaged concentration of that solute at the probe-sediment interface, as discussed in Chapter 7. The units of concentration are familiar across the discipline of science, and enable an easy comparison with solute data determined using alternative analytical means (such as grab sample analyses using ICP-MS). The conversion of the flux data to concentrations is possible using the standard equations provided in Chapter 2 (and equation 10.2). This concentration approximates to the mean concentration in the porewater at the surface of the DGT device, as discussed more fully in Chapter 7. The absence of a DBL that exists for solution deployments simplifies the approach. Irrespective of the calculation used, it is important to stipulate the deployment duration.

10.5.3.8 Quality Control

Throughout this chapter there has been repeated mention of quality-control considerations. All analytical techniques are susceptible to contamination in both the laboratory and the field, and vigilance is required to minimise this. When using DGT and DET, clean room techniques will reduce contamination during the preparation of gels, and the assembly and disassembly of all pistons and probes. Reporting gel blank concentrations provides confidence in the DGT and DET procedures, and these gel blanks should be used to calculate method detection limits (three times the standard deviation of the blank) on a per experiment basis using all the DGT deployment conditions. Routine use of analytical blanks will also identify any contamination in the extraction or diluent solutions, or laboratory equipment (e.g. glass or plastic ware). In addition to this, replicate deployments of DGT and DET devices are recommended. In well-mixed homogenous waters and soils replicate deployments would ideally have standard deviations of <10 per cent. However, such low variability is rarely observed in sediments, and reporting of the replicate data provides important information on the inherent variability that occurs in natural systems.

References

1. W. Davison, G. Fones, M. Harper, P. Teasdale and H. Zhang, Dialysis, DET and DGT: In situ diffusional techniques for studying water, sediments and soils, In *In situ monitoring of aquatic systems: Chemical analysis and speciation*, eds. J. Buffle and G. Horvai, IUPAC Series Analytical and Physical Chemistry of Environmental Systems, Vol. 6 (Chichester: Wiley, 2000), pp. 495–569.
2. K. W. Warnken, H. Zhang and W. Davison, In-situ monitoring and dynamic speciation measurements in solution using DGT, In *Passive sampling techniques in environmental monitoring*, eds. R. Greenwood, G. Mills and B. Vrana, Comprehensive Analytical Chemistry, Vol. 48 (Amsterdam: Elsevier, 2007), pp. 251–278.
3. W. Davison, H. Zhang and K. W. Warnken, Theory and applications of DGT measurements in soils and sediments, In *Passive sampling techniques in environmental monitoring*, eds. R. Greenwood, G. Mills and B. Vrana, Comprehensive Analytical Chemistry, Vol. 48 (Amsterdam: Elsevier, 2007), pp. 353–378.
4. W. Davison, Device for concentrating trace components in a liquid, GB2295229 (A) (1996).
5. W. Davison and H. Zhang, In-situ speciation measurements of trace components in natural-waters using thin-film gels, *Nature* 367 (1994), 546–548.
6. H. Zhang, W. Davison, S. Miller and W. Tych, In situ high resolution measurements of fluxes of Ni, Cu, Fe and Mn and concentrations of Zn and Cd in porewaters by DGT, *Geochim. Cosmochim. Acta* 59 (1995), 4181–4192.
7. K. W. Warnken, A. J. Lawlor, S. Lofts et al., In situ speciation measurements of trace metals in headwater streams, *Environ. Sci. Technol.* 43 (2009), 7230–7236.
8. H. Cheng, Investigation of diffusion and binding properties for extending applications of the DGT technique, PhD Thesis, University of Lancaster, UK (2014).
9. S. Scally, W. Davison and H. Zhang, Diffusion coefficients of metals and metal complexes in hydrogels used in diffusive gradients in thin films, *Anal. Chim. Acta* 558 (2006), 222–229.
10. S. Mason, R. Hamon, A. Nolan, H. Zhang and W. Davison, Performance of a mixed binding layer for measuring anions and cations in a single assay using the diffusive gradients in thin films technique, *Anal. Chem.* 77 (2005), 6339–6346.
11. J. G. Panther, W. W. Bennett, D. T. Welsh and P. R. Teasdale, Simultaneous measurement of trace metal and oxyanion concentrations in water using diffusive gradients in thin films with a Chelex-Metsorb mixed binding layer, *Anal. Chem.* 86 (2014), 427–434.
12. H. Zhang and W. Davison, Performance characteristics of diffusion gradients in thin films for the in situ measurement of trace metals in aqueous solution, *Anal. Chem.* 67 (1995), 3391–3400.
13. H. Zhang, W. Davison, R. Gadi and T. Kobayashi, In situ measurement of dissolved phosphorus in natural waters using DGT, *Anal. Chim. Acta* 370 (1998), 29–38.

14. J. Luo, H. Zhang, J. Santner and W. Davison, Performance characteristics of diffusive gradients in thin films equipped with a binding gel layer containing precipitated ferrihydrite for measuring arsenic(V), selenium(VI), vanadium(V), and antimony(V), *Anal. Chem.* 82 (2010), 8903–8909.

15. W. W. Bennett, P. R. Teasdale, J. G. Panther et al., New diffusive gradients in thin films technique for measuring inorganic arsenic and selenium(IV) using a titanium dioxide based adsorbent, *Anal. Chem.* 82 (2010), 7401–7407.

16. J. G. Panther, W. W. Bennett, P. R. Teasdale, D. T. Welsh and H. J. Zhao, DGT measurement of dissolved aluminum species in waters: comparing Chelex-100 and titanium dioxide-based adsorbents, *Environ. Sci. Technol.* 46 (2012), 2267–2275.

17. S. Ding, D. Xu, Q. Sun, H. Yin and C. Zhang, Measurement of dissolved reactive phosphorus using the diffusive gradients in thin films technique with a high-capacity binding phase, *Environ. Sci. Technol.* 44 (2010), 8169–8174.

18. D. X. Guan, P. N. Williams, J. Luo et al., Novel precipitated zirconia-based DGT technique for high-resolution imaging of oxyanions in waters and sediments, *Environ. Sci. Technol.* 49 (2015), 3653–3661.

19. Robertson, P. R. Teasdale and D. T. Welsh, A novel gel-based technique for the high resolution, two-dimensional determination of iron (II) and sulfide in sediment, *Limnol. Oceanogr. Methods* 6 (2008), 502–512.

20. P. R. Teasdale, S. Hayward and W. Davison, In situ, high-resolution measurement of dissolved sulfide using diffusive gradients in thin films with computer-imaging densitometry, *Anal. Chem.* 71 (1999), 2186–2191.

21. A. Widerlund and W. Davison, Size and density distribution of sulfide-producing microniches in lake sediments, *Environ. Sci. Technol.* 41 (2007), 8044–8049.

22. W. W. Bennett, P. R. Teasdale, J. G. Panther, D. T. Welsh and D. F. Jolley, Speciation of dissolved inorganic arsenic by diffusive gradients in thin films: Selective binding of As III by 3-mercaptopropyl-functionalized silica gel, *Anal. Chem.* 83 (2011), 8293–8299.

23. O. Clarisse and H. Hintelmann, Measurements of dissolved methylmercury in natural waters using diffusive gradients in thin film (DGT), *J. Environ. Monit.* 8 (2011), 1242–1247.

24. Y. Gao, E. De Canck, M. Leermakers, W. Baeyens and P. Van Der Voort, Synthesized mercaptopropyl nanoporous resins in DGT probes for determining dissolved mercury concentrations, *Talanta* 81 (2011), 262–267.

25. Y. Gao, S. De Craemer and W. Baeyens, A novel method for the determination of dissolved methylmercury concentrations using diffusive gradients in thin films technique, *Talanta* 120 (2014), 470–474.

26. H. Brandl and K. W. Hanselmann, Evaluation and application of dialysis porewater samplers for microbiological studies at sediment-water interfaces, *Aquatic Sci.* 53 (1991), 55–73.

27. O. A. Garmo, W. Davison and H. Zhang, Effects of binding of metals to the hydrogel and filter membrane on the accuracy of the diffusive gradients in thin films technique, *Anal. Chem.* 80 (2008), 9220–9225.

28. E. J. Smith, W. Davison and J. Hamilton-Taylor, Methods for preparing synthetic freshwaters, *Water Res.* 36 (2002), 1286–1296.

29. D. R. Kester, I. W. Duedall, D. N. Conners and R. M. Pytkowicz, Preparation of artificial seawater, *Limnol. Oceanogr.* 12 (1967), 176–179.

30. H. Zhang, W. Davison, B. Knight and S. McGrath, In situ measurement of solution concentrations and fluxes of trace metals in soils using DGT, *Environ. Sci. Technol.* 32 (1998), 704–710.

31. H. Zhang, F. J. Zhao, B. Sun, W. Davison and S. P. McGrath, A new method to measure effective soil solution concentration predicts copper availability to plants, *Environ. Sci. Technol.* 35 (2001), 2602–2607.
32. P. N. Williams, H. Zhang, W. Davison et al., Evaluation of in situ DGT measurements for predicting the concentration of Cd in Chinese field-cultivated rice: Impact of soil Cd:Zn ratios, *Environ. Sci. Technol.* 46 (2012), 8009–8016.
33. J. Luo, H. Zhang, B. Sun et al., Localised mobilisation of metals, as measured by diffusive gradients in thin-films, in soil historically treated with sewage sludge, *Chemosphere* 90 (2013), 464–470.
34. J. Santner, H. Zhang, D. Leitner et al., High-resolution chemical imaging of labile phosphorus in the rhizosphere of Brassica napus L. cultivars, *Environ. Exp. Bot.* 77 (2012), 219–26.
35. M. Motelica-Heino, C. Naylor, H. Zhang and W. Davison, Simultaneous release of metals and sulfide in lacustrine sediment, *Environ. Sci. Technol.* 37 (2003), 4374–4381.
36. Y. Gao, S. Van De Velde, P. Williams, W. Baeyens and H. Zhang, Two dimensional images of dissolved sulphide and metals in anoxic sediments by a novel DGT probe and optical scanning techniques, *Trends Anal. Chem.* 66 (2015), 63–71.
37. L. Sochaczewski, W. Davison, H. Zhang and W. Tych, Understanding small-scale features in DGT measurements in sediments, *Environ. Chem.* 6 (2009), 477–485.
38. L. Sochaczewski, A. Stockdale, W. Davison, W. Tych and H. Zhang, A three-dimensional reactive transport model for sediments, incorporating microniches, *Environ. Chem.* 5 (2008), 218–225.
39. A. Stockdale, W. Davison and H. Zhang, High-resolution two-dimensional quantitative analysis of phosphorus, vanadium and arsenic, and qualitative analysis of sulphide, in a freshwater sediment, *Environ. Chem.* 5 (2008), 143–149.
40. H. Docekalova, O. Clarisse, S. Salomon and M. Wartel, Use of constrained DET probe for a high-resolution determination of metals and anions distribution in the sediment pore water, *Talanta* 57 (2002), 145–155.
41. B. Docekal and M. Gregusova, Segmented sediment probe for diffusive gradient in thin films technique, *Analyst* 137 (2012), 502–507.
42. K. Andersson, R. Dahlqvist, D. Turner et al., Colloidal rare earth elements in a boreal river: changing sources and distributions during the spring flood, *Geochim. Cosmochim. Acta* 70 (2006), 3261–3274.
43. J. G. Panther, R. R. Stewart, P. R. Teasdale et al., Titanium dioxide-based DGT for measuring dissolved As(V), V(V), Sb(V), Mo(VI) and W(VI) in water, *Talanta* 105 (2013), 80–86.
44. H. L. Price, P. R. Teasdale and D. F. Jolley, An evaluation of ferrihydrite and Metsorb^{TM}-DGT techniques for measuring oxyanion species (As, Se, V, P): Effective capacity, competition and diffusion coefficients, *Anal. Chim. Acta* 803 (2013), 56–65.
45. H. Docekalova, and P. Divis, Application of diffusive gradient in thin films technique (DGT) to measurement of mercury in aquatic systems, *Talanta* 65 (2005), 1174–1178.
46. P. Divis, M. Leermakers, H. Docekalova and Y. Gao, Mercury depth profiles in river and marine sediments measured by the diffusive gradients in thin films technique with two different specific resins, *Anal. Bioanal. Chem.* 382 (2005), 1715–1719.
47. M. A. French, H. Zhang, J. M. Pates, S. E. Bryan and R. C. Wilson, Development and performance of the diffusive gradients in thin-films technique for the measurement of technetium-99 in seawater, *Anal. Chem.* 77 (2005), 135–139.
48. M. Leermakers, Y. Gao, J. Navez et al., Radium analysis by sector field ICP-MS in combination with the diffusive gradients in thin films (DGT) technique, *J. Anal. Atom. Spectrom.* 24 (2009), 1115–1117.

49. C. Murdock, M. Kelly, L.-Y. Chang, W. Davison and H. Zhang, DGT as an *in situ* tool for measuring radiocesium in natural waters, *Environ. Sci. Technol.* 35 (2001), 4530–4535.
50. W. Davison and H. Zhang, Progress in understanding the use of diffusive gradients in thin-films – Back to basics, *Environ. Chem.* 9 (2012), 1–13.
51. H. Zhang and W. Davison, Direct in situ measurements of labile inorganic and organically bound metal species in synthetic solutions and natural waters using diffusive gradients in thin films, *Anal. Chem.* 72 (2000), 4447–4457.
52. G. S. C. Turner, G. A. Mills, M. J. Bowes et al., Evaluation of DGT as a long term water quality monitoring tool in natural waters; uranium as a case study, *Environ. Sci. Process. Impacts* 16 (2014), 393–403.
53. L. Sigg, F. Black, J. Buffle et al., Comparison of analytical techniques for dynamic trace metal speciation in natural freshwaters, *Environ. Sci. Technol.* 40 (2006), 1934–1941.
54. P. S. Hooda, H. Zhang, W. Davison and A. C. Edwards, Measuring bioavailable trace metals by diffusive gradients in thin films (DGT): Soil moisture effects on its performance in soils, *European J. Soil Sci.* 50 (1999), 1–10.
55. S. Mason, A. McNeill, M. J. McLaughlin and H. Zhang, Prediction of wheat response to an application of phosphorus under field conditions using diffusive gradients in thin-films (DGT) and extraction methods, *Plant Soil* 337 (2010), 243–258.
56. I. Muhammad, M. Puschenreiter and W. W. Wenzel, Cadmium and Zn availability as affected by pH manipulation and its assessment by soil extraction, DGT and indicator plants, *Soil Tot. Env.* 416 (2012), 490–500.
57. H. Ernstberger, H. Zhang, A. Tye, S. Young and W. Davison, Desorption kinetics of Cd, Zn, and Ni measured in soils by DGT, *Environ. Sci. Technol.* 39 (2005), 1591–1597.
58. B. Nowack, S. Koehler and R. Schulin, Use of diffusive gradients in thin films (DGT) in undisturbed field soils, *Environ. Sci. Technol.* 38 (2004), 1133–1138.
59. J. Luo, H. Zhang, W. Davison et al., Localised mobilisation of metals, as measured by diffusive gradients in thin-films, in soil historically treated with sewage sludge, *Chemosphere*, 90 (2013), 464–470.
60. W. Davison, H. Zhang and G. W. Grime, Performance characteristics of gel probes used for measuring the chemistry of pore waters, *Environ. Sci. Technol.* 28 (1994), 1623–1632.
61. G. R. Fones, W. Davison, O. Holby et al., High-resolution metal gradients measured by in situ DGT/DET deployment in black sea sediments using an autonomous benthic lander, *Limnol. Oceanogr.* 46 (2001), 982–988.
62. S. Ullah, H. Zhang, A. L. Heathwaite et al., In situ measurement of redox sensitive solutes at high spatial resolution in a riverbed using diffusive equilibrium in thin films (DET), *Ecol. Eng.* 49 (2012), 18–26.
63. M. P. Harper, W. Davison and W. Tych, Temporal, spatial, and resolution constraints for in situ sampling devices using diffusional equilibration: dialysis and DET, *Environ. Sci. Technol.* 31 (1997), 3110–3119.
64. S. Tankéré-Muller, H. Zhang, W. Davison et al., Fine scale remobilisation of Fe, Mn, Co, Ni, Cu and Cd in contaminated marine sediment, *Mar. Chem.* 106 (2007), 192–207.

Appendix

Table for Calculating Diffusion Coefficients of Commonly Measured Analytes in the APA Gel and Water at any Temperature Between 0°C and 35°C

The temperature dependence of the diffusion coefficient, D, of each analyte is expressed as a second-order polynomial (quadratic) of the form $D = AT^2 + BT + C$ where T is the temperature in °C. The number obtained from the equation should be multiplied by 10^{-6} to obtain a value of D in cm^2 s^{-1}. The coefficients of the equation were obtained by fitting to values of D calculated using equation 2.25, Chapter 2. For each analyte there are either two or three equations represented by the subscripts of the coefficients. Subscript 3 provides values of D in water based on the value at 25°C given in the indicated references. Subscripts 1 and 2 refer to the APA diffusive gel. For subscript 1 the values of D^g at 25°C were those given in column 2 ($\times$ 10^{-6} cm^2 s^{-1}). For superscript (a), the values of D^g were calculated assuming the value of D in the APA gel at 25°C is 85 per cent of the value in water given in reference [1]. The calculated values for different temperatures agree with the tabulated values provided by DGT Research Ltd. For other analytes values of D^g were critically selected using the data provided in Table 3.2. For subscript 2 the values of D in water at 25°C were those given in ref. [2]. Differences between the two approaches are negligible or at most <5 per cent.

	D^g	A_1	B_1	C_1	A_2	B_2	C_2	A_3	B_3	C_3	Ref
[a]Ag	14.1	0.00275	0.2421	6.3396	0.00273	0.2404	6.2946	0.00282	0.2485	6.5058	[2]
[a]Al	4.75	0.00093	0.0815	2.1342	0.00090	0.0789	2.0668	0.00105	0.0928	2.4307	[2]
As(V)	5.5	0.00107	0.0944	2.4711				0.00176	0.1553	4.0661	[2]
[a]Cd	6.09	0.00119	0.1045	2.7362	0.00119	0.1048	2.7452	0.00140	0.1234	3.2304	[2]
[a]Co	5.94	0.00116	0.1019	2.6688	0.00121	0.1067	2.7946	0.00143	0.1256	3.2888	[2]
[a]Cr(III)	5.05	0.00098	0.0867	2.2689	0.00098	0.0867	2.2689	0.00116	0.1021	2.6733	[2]
Cr(VI)	8.5	0.00166	0.1459	3.8190				0.00218	0.1922	5.0321	[1]
[a]Cu	6.23	0.00121	0.1069	2.7991	0.00118	0.1042	2.7272	0.00139	0.1225	3.2080	[2]
[a]Fe	6.11	0.00119	0.1048	2.7452	0.00119	0.1048	2.7452	0.00140	0.1234	3.2304	[2]
[a]Mn	5.85	0.00114	0.1004	2.6284	0.00118	0.1038	2.7182	0.00139	0.1222	3.1990	[2]
Mo(VI)	6.0	0.00117	0.1030	2.6960				0.00193	0.1700	4.4525	[1]
[a]Ni	5.77	0.00112	0.0990	2.5924	0.00117	0.1028	2.6913	0.00137	0.121	3.1675	[2]
phosphate	6.05	0.00118	0.1038	2.7182							
[a]Pb	8.03	0.00156	0.1378	3.6078	0.00156	0.1378	3.6078	0.00184	0.1621	4.2458	[2]
Se(VI)	7.2	0.00140	0.1235	3.2349				0.00184	0.1623	2.2503	[1]
Sb(V)	5.5	0.00107	0.0944	2.4711				0.00161	0.1416	3.7067	[1]
$UO_2{}^+$	5.6	0.00109	0.0961	2.5161				0.00148	0.1304	3.4146	[3]
V(V)	6.7	0.00131	0.1150	3.0103							
W(VI)	5.5	0.00107	0.0944	2.4711				0.00180	0.15837	4.1470	[1]
[a]Zn	6.08	0.00118	0.1043	2.7317	0.00116	0.1026	2.6868	0.00137	0.1206	3.1585	[1]

References

1. Y. Li and S. Gregory, Diffusion of ions in seawater and in deep-sea sediments, *Geochim. Cosmochim. Acta* 38 (1974) 703–714.
2. J. Buffle, Z. Zhang and K. Startchev, Metal flux and dynamic speciation at (bio) interfaces. Part 1: Critical evaluation and compilation of physico-chemical parameters for complexes with simple ligands and fulvic/humic substances, *Environ. Sci. Technol.* 41 (2007), 7609–7620.
3. C. Liv, J. Shang and Z. M. Zhachara, Multi-species diffusion models: a study of uranyl species diffusion, *Wat. Resources Res.* 47 (2011), 1–16.

Index